Achieving carbon-negative bioenergy systems from plant materials

It is widely recognised that agriculture is a significant contributor to global warming and climate change. Agriculture needs to reduce its environmental impact and adapt to current climate change whilst still feeding a growing population, i.e. become more 'climate-smart'. Burleigh Dodds Science Publishing is playing its part in achieving this by bringing together key research on making the production of the world's most important crops and livestock products more sustainable.

Based on extensive research, our publications specifically target the challenge of climate-smart agriculture. In this way we are using 'smart publishing' to help achieve climate-smart agriculture.

Burleigh Dodds Science Publishing is an independent and innovative publisher delivering high quality customer-focused agricultural science content in both print and online formats for the academic and research communities. Our aim is to build a foundation of knowledge on which researchers can build to meet the challenge of climate-smart agriculture.

For more information about Burleigh Dodds Science Publishing simply call us on +44 (0) 1223 839365, email info@bdspublishing.com or alternatively please visit our website at www.bdspublishing.com.

Related titles:

Assessing the environmental impact of agriculture
Print (ISBN 978-1-78676-228-3); Online (ISBN 978-1-78676-228-3, 978-1-78676-231-3)

Climate change and agriculture
Print (ISBN 978-1-78676-320-4); Online (ISBN 978-1-78676-322-8, 978-1-78676-323-5)

Chapters are available individually from our online bookshop:
https://shop.bdspublishing.com

climate•SMART•publishing
IN AGRICULTURAL SCIENCE

BURLEIGH DODDS SERIES IN AGRICULTURAL SCIENCE
NUMBER 64

Achieving carbon-negative bioenergy systems from plant materials

Edited by Dr Chris Saffron, Michigan State University, USA

burleigh dodds
SCIENCE PUBLISHING

Published by Burleigh Dodds Science Publishing Limited
82 High Street, Sawston, Cambridge CB22 3HJ, UK
www.bdspublishing.com

Burleigh Dodds Science Publishing, 1518 Walnut Street, Suite 900, Philadelphia, PA 19102-3406, USA

First published 2020 by Burleigh Dodds Science Publishing Limited

Library of Congress Control Number: 2019956997

British Library Cataloguing in Publication Data
A catalogue record for this book is available from the British Library

ISBN 978-1-78676-252-8 (Print)
ISBN 978-1-78676-255-9 (PDF)
ISBN 978-1-78676-254-2 (ePub)
ISSN 2059-6936 (print)
ISSN 2059-6944 (online)

DOI: 10.19103/AS.2019.0027

Typeset by Deanta Global Publishing Services, Dublin, Ireland

Contents

Series list

Title	Series number
Achieving sustainable cultivation of maize - Vol 1 From improved varieties to local applications Edited by: Dr Dave Watson, CGIAR Maize Research Program Manager, CIMMYT, Mexico	001
Achieving sustainable cultivation of maize - Vol 2 Cultivation techniques, pest and disease control Edited by: Dr Dave Watson, CGIAR Maize Research Program Manager, CIMMYT, Mexico	002
Achieving sustainable cultivation of rice - Vol 1 Breeding for higher yield and quality Edited by: Prof. Takuji Sasaki, Tokyo University of Agriculture, Japan	003
Achieving sustainable cultivation of rice - Vol 2 Cultivation, pest and disease management Edited by: Prof. Takuji Sasaki, Tokyo University of Agriculture, Japan	004
Achieving sustainable cultivation of wheat - Vol 1 Breeding, quality traits, pests and diseases Edited by: Prof. Peter Langridge, The University of Adelaide, Australia	005
Achieving sustainable cultivation of wheat - Vol 2 Cultivation techniques Edited by: Prof. Peter Langridge, The University of Adelaide, Australia	006
Achieving sustainable cultivation of tomatoes Edited by: Dr Autar Mattoo, USDA-ARS, USA & Prof. Avtar Handa, Purdue University, USA	007
Achieving sustainable production of milk - Vol 1 Milk composition, genetics and breeding Edited by: Dr Nico van Belzen, International Dairy Federation (IDF), Belgium	008
Achieving sustainable production of milk - Vol 2 Safety, quality and sustainability Edited by: Dr Nico van Belzen, International Dairy Federation (IDF), Belgium	009
Achieving sustainable production of milk - Vol 3 Dairy herd management and welfare Edited by: Prof. John Webster, University of Bristol, UK	010
Ensuring safety and quality in the production of beef - Vol 1 Safety Edited by: Prof. Gary Acuff, Texas A&M University, USA & Prof. James Dickson, Iowa State University, USA	011
Ensuring safety and quality in the production of beef - Vol 2 Quality Edited by: Prof. Michael Dikeman, Kansas State University, USA	012
Achieving sustainable production of poultry meat - Vol 1 Safety, quality and sustainability Edited by: Prof. Steven C. Ricke, University of Arkansas, USA	013
Achieving sustainable production of poultry meat - Vol 2 Breeding and nutrition Edited by: Prof. Todd Applegate, University of Georgia, USA	014

Acknowledgements

We wish to acknowledge the following for their help in reviewing particular chapters:

- Chapter 1: Prof. Juan Carles Bruno Argilaguet, Universitat Rovira i Virgili, Spain
- Chapter 2: Prof. Robert Brown, Iowa State University, USA
- Chapter 3: Dr Stephen P. Gent, South Dakota State University, USA
- Chapter 5: Prof. H. Raheman, Indian Institute of Technology, India
- Chapter 6: Dr Peter McVetty, University of Manitoba, Canada
- Chapter 7: Dr Emily Heaton, Iowa State University, USA,
- Chapter 8: Dr Michael Casler, USDA-ARS, USA
- Chapter 9: Dr Tim Volk, State University of New York, USA
- Chapter 10: Prof. Graeme Walker, Abertay University, UK

Introduction

There is a need to develop next-generation bioenergy systems that exhibit net carbon capture. Biomass removes carbon dioxide from the atmosphere during growth and, if converted to biofuel, has the potential to be 'carbon negative', especially if combined with carbon capture and storage. Accurately accounting for below-ground soil carbon, owing to root exudation and decay in the plant's rhizosphere, is also important for achieving net carbon capture. To achieve ambitious targets for global reductions in greenhouse gas emissions, biomass crops should generate high yield from minimal input energy while minimising environmental impacts that could make crop production less sustainable. This collection reviews advances in producing next-generation biofuels from non-food plant materials. Part 1 discusses key technologies to achieve this goal such as biomass gasification, fast pyrolysis and torrefaction. Chapters 1–3 address advances in technology, applications and commercial development. Part 2 assess advances in the production of biofuels from non-food crops: jatropha, oilseeds (such as canola and rapeseed), *Miscanthus*, switchgrass, willow and seaweed.

Part 1 Technologies

The first chapter in the book reviews developments in thermal gasification. This very flexible technique allows many different biomass feedstocks to be converted into a wide variety of products such as heat, electricity, chemicals and transport fuels as well as high value ash and biochar co-products. The chapter reviews current research to optimize process performance, fuel and product flexibility, including types of gasifier and gasification operating variables. It assesses developments in biomass gasification for heat and power generation as well as biofuel production. The chapter concludes by assessing developments in flexible biorefineries with integrated systems capable of polygeneration of different core and co-products.

Chapter 2 discusses the use of fast pyrolysis for biofuel production. Pyrolysis is a thermal depolymerization and fragmentation process carried out at moderate temperatures in the absence of oxygen and with a wide range of reaction residence times. It can convert a wide range of solid materials, ranging from woody biomass and algae to waste plastics, into complex mixtures of low molecular weight organic molecules. The chapter begins by examining the chemical reaction kinetics of fast pyrolysis as well as fast pyrolysis process technologies, including the transition from hydroprocessing to catalytic upgrading. It then describes pyrolysis co-products and their uses. The chapter also reviews the use of both technoeconomic assessment (TEA) and life-cycle assessment (LCA) to assess process efficiency and sustainability. Finally, the chapter looks at the developments in commercialization of biomass fast pyrolysis.

The final chapter in Part 1, Chapter 3, discusses how biofuels can be produced using torrefaction. Significant investments have been made around the world to accommodate the use of dried and pelleted biomass materials in power plants. Substituting biomass for coal in coal-based boiler systems without degrading boiler capacity requires pretreatment to make its properties similar to the coal that it is displacing. The pretreatment technologies currently under active development include: torrefaction, steam explosion, and hydrothermal carbonization (wet torrefaction). This chapter reviews the factors associated with biomass pretreatment and the use of the advanced biomass-based fuels in power plants. It also provides a case study which describes the use of torrefied wood pellets at the Portland General Electric Boardman Facility in Oregon in the US.

Part 2 Materials

Part 2 of the book begins with a review of how biodiesel can be produced from renewable sources. There is increasing interest in the production of biodiesel from non-edible oil sources such as waste or recycled oil and animal fats. Transesterification is the most used method to convert triglycerides (TG) from different types of lipids. After reviewing the range of available feedstocks, Chapter 4 uses thermodynamic analysis to investigate the main three transesterification reaction systems: triolein and supercritical methanol (T&SCM), triolein and supercritical ethanol (T&SCE), and triolein and supercritical isopropanol (T&SCI). Using a case study, it identifies ways of optimizing supercritical transesterification technology for biodiesel production, as well as ways of improving the quality of the biodiesel manufactured from the process.

The subject of Chapter 5 is the production of biodiesel from oilseeds, in particular *Jatropha curcas*. The importance of edible oils for food applications makes non-edible oils a key feedstock for biodiesel production, including the seeds of *J. curcas*. The chapter begins by reviewing the characteristics of *Jatropha curcas* L. and how it can be used as a biodiesel feedstock through the use of transesterification. It then discusses the factors affecting the efficiency of enzymatic transesterification using immobilised lipase and provides a section on the life cycle assessment (LCA) of *Jatropha curcas* for biodiesel production. The chapter concludes by providing an economic analysis of biodiesel production using this feedstock.

Chapter 6 also reviews the production of biodiesel from oilseeds but instead focuses on canola and rapeseed. Currently, rapeseed oil is the dominant feedstock for biodiesel production in Europe, while canola oil is also extensively used for biodiesel production in the Pacific Northwest of North America. The chapter describes current technologies and constraints of biodiesel production and examines life cycle assessment of canola/rapeseed biodiesel production. Finally, the chapter looks ahead to future trends of research in this area and provides guidance on further reading.

Chapter 7 discusses the sustainable use of *Miscanthus* for biofuel. The chapter focuses on research demonstrating that Miscanthus embodies a range of attributes (e.g. C4 photosynthesis combined with cold tolerance, high energy output/input ratios, efficient nutrient recycling and high yield from minimal agronomic input) that make it an ideal sustainable biomass crop for biofuels. Sections on both nutrient use efficiency and water use/ water use efficiency are included. The chapter also discusses carbon flux and life cycle assessments for different end uses to improve the efficiency and sustainability of the process.

The next chapter addresses the sustainable use of switchgrass for biofuel. Switchgrass, a productive, perennial warm-season grass, native to much of North America, has significant potential as a sustainable biofuel crop. Chapter 8 begins by discussing switchgrass and environmental sustainability metrics. It then reviews the challenges that switchgrass faces in becoming a primary feedstock resource and provides a section on the production and agronomics of switchgrass in the context of sustainability. The chapter also reviews the economic, policy and social issues associated with the development of a sustainable biofuel industry. The chapter also includes a case study on burning switchgrass as a biofuel to illustrate biofuel production in practice.

Chapter 9 reviews the sustainable production of willow for biofuel use. Intensively managed plantations of willow (Salix spp.) are gaining interest worldwide, mainly due to their efficient and sustainable land use along with an increasing demand for biofuel resources. Willows have traditionally been used as biofuel in the form of wood chips for direct combustion in heat and power plants, but there is an increasing interest in using willow biomass as raw material for other biofuel supply chains. The chapter provides an overview of the challenges and key issues in the sustainable production and use of willow as feedstock for biofuel production in northern-temperate regions, including the issues of feedstock quality for biofuel use, feedstock productivity, cropping security, biodiversity, nutrient uptake and use, carbon accumulation and sequestration, and environmental impact. Finally, the chapter looks ahead to future research trends in this area and provides useful sources for further reading.

The final chapter of the book addresses the sustainable use of seaweed for biofuel. Chapter 10 describes the biological advantages and disadvantages of seaweed in terms of its use as an energy crop and examines global production trends The chapter reviews methods of seaweed cultivation, together with harvesting and supply chain logistics. It provides a systematic review of conversion routes to producing biofuels and other value-added products from seaweed. Finally, the chapter includes detailed case studies of potential industrial-scale processes for three different conversion routes and looks ahead to future research trends in this area.

Part 1

Technologies

Chapter 1

Biomass gasification for bioenergy

Maria Puig-Arnavat, Technical University of Denmark, Denmark; Tobias Pape Thomsen, Roskilde University, Denmark; and Zsuzsa Sárossy, Rasmus Østergaard Gadsbøll, Lasse Røngaard Clausen and Jesper Ahrenfeldt, Technical University of Denmark, Denmark

1 Introduction

Gasification is a partial thermal oxidation of carbon-rich materials yielding a non-condensable gas product (CO_2, CO, H_2, H_2O and other gaseous hydrocarbons) and smaller quantities of by-products such as char, ash and condensable fractions including water, tars and oils. Biomass gasification platforms are most commonly designed for conversion of wood in the form of wood pellets, wood chips or waste wood. However, it is possible to extend the potential range of organic material fractions converted in thermal gasification to cover various organic secondary resources. This includes agricultural, municipal and industrial by-products and residues such as cereal straw, fibre residues, sludge and organic waste fractions.

Biomass-based energy systems have a unique potential for genuine carbon-negative energy production. There are two main routes that may be pursued individually or in a complementary combination: (i) combination with carbon capture and storage (CCS) and (ii) processes with co-production of carbon-rich char/ash fractions as well as other by-products to replace current, fossil-based products in the market. CCS has a huge potential, but the technology

http://dx.doi.org/10.19103/AS.2019.0027.04

has to mature and the economic feasibility needs to improve for it to penetrate market barriers (Álamo et al., 2018). The technical potential for carbon sequestration via bioenergy processes with carbon-rich residuals is smaller than via CCS, but it is cheaper, robust and readily applicable. The potential to produce carbon-negative energy in this way, arise from a combination of high net energy efficiency and a potential for carbon sequestration via the ash/char fraction as illustrated in Fig. 1. Emissions from peripheral unit processes and upstream inputs influence the net potential of this approach. The level of direct carbon negativity will be given by the amount of long-term sequestered carbon minus all emissions related to biomass fuel procurement and pre-treatment, production and use of utilities (water, chemicals, fuels) as well as the production and maintenance of the plant and peripheral systems (Thomsen et al., 2011).

Additional contribution from downstream CCS added to this system will have a substantial potential to increase the level of carbon negativity. Even CCS processes with low coverage and/or efficiency may provide a sizable net contribution when combined with efficient low-carbon biomass-based energy

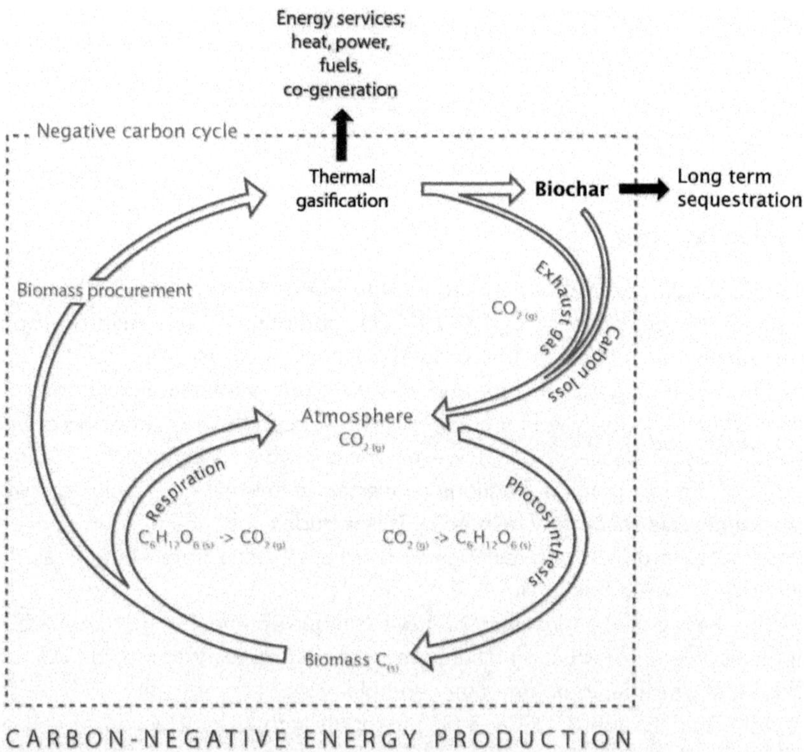

Figure 1 Schematics of the ideal carbon-negative biomass-based CHP system. Source: adapted with permission from Thomsen et al. (2011).

systems. Systems with high net energy efficiency will have the highest abatement potential. In thermal systems, these are most often cases with utilization of the inevitable heat production, that is biomass-based heat systems or biomass-based co-generation systems. Thermal gasification of biomass is unique in this regard, since the platform excels in overall efficiency and flexibility.

Gasification is no new invention. First steps in the development of gasification started in the late 1600s. Hundred years later, in 1792, coal producer gas was reported to be used for lighting and later on for cooking and heating. Almost simultaneously, the water-gas-shift reaction was described and within a few decades, the first commercially used gasifier was built. Since the early 1900s, application of producer gas rapidly spread into power production, combined heat and power (CHP) production and operation of vehicle motors as well as into gasification of wood and other biomasses (Behera and Varma, 2019).

Despite the advantages that biomass gasification offers, it has not yet been able to consolidate its role and become a fully mature technology. Thus, a majority of biomass gasification projects remain at the pilot or demonstration scale. This is due to inherent challenges of biomass as a fuel and insufficient resources allocation to address operational issues related to the following (Asadullah, 2014a; Heyne et al., 2013; Sansaniwal et al., 2017; Zhang et al., 2013a):

- Supply chain management (transportation and storage due to low-energy density);
- Handling, pre-treatment and fuel feeding;
- Optimizing the energy efficiency to high moisture content feedstock;
- Addressing variations in biomass fuel properties;
- Unrealistic fuel flexibility expectations;
- Unrealistic product property expectations;
- Upscaling of specific gasifier reactors;
- Problems with gas cleaning, conditioning and impurities such as sulphur compounds (e.g. H_2S, COS), hydrogen chloride, alkali and ammonia;
- Tar-related issues, for example condensation at lower temperatures which can lead to fouling and plugging of the plant pipelines, filters, catalyst units or engines; and
- Ash-related problems such as sintering, agglomeration, deposition, erosion and corrosion.

There is a firm belief in academia and industry that all mentioned technical problems can be overcome and that most of the solutions already exist. However, the foremost barrier for biomass gasification is associated with the economic risk in a rapidly changing energy market. Thus, policy measures ensuring long-term economic viability of the projects should be implemented.

The aim of this chapter was to provide an insight into the versatility and benefits of biomass gasification processes and its products in relation to production of carbon-negative bioenergy from plant-based materials. General key issues related to designing a gasification plant are reviewed and discussed, and several new concepts and solutions in the form of process integration schemes, polygeneration strategies and biochar uses are described.

2 Fundamentals of biomass gasification

Biomass gasification involves a sequence of several stages occurring at different temperatures: drying (100–200°C), pyrolysis (200–700°C), partial combustion (>800°C) and reduction or gasification (700–900°C). These stages, depending on the specific gasifier design, often overlap.

- Drying: Evaporation of the moisture contained in the feedstock. The amount of heat required in this stage is proportional to the feedstock moisture content. The drying often takes place as a preliminary step in the pyrolysis stage and the required heat for both sets of processes is usually supplied from the oxidation stage.
- Pyrolysis: Thermal decomposition of biomass in the absence of oxygen/air. The cracking of chemical bonds takes place, resulting in the formation of molecules with a lower molecular weight. Thus, different fractions are obtained: solid char, liquid/condensed tars and a gaseous fraction. The condensable fraction of the pyrolysis gas will usually mix with the evaporated water from the drying stage as well as released cell bound water. Product weight and energy ratios will depend on the nature of the biomass, the reactor design and operating conditions of the process. Pyrolysis is a net endothermic stage and the heat required comes from the oxidation stage of the process.
- Oxidation: Usually, gasification involves a stage with partial oxidation to provide heat for the drying, pyrolysis and char gasification. During the partial oxidation, a part of the organic material in the fuel is oxidized and heat is released from the net exothermic reactions in the process. A partial oxidation is conducted by supplying, for example, air or pre-heated air to the gas/tar or to the char in a sub-stoichiometric ratio.
- Reduction or gasification: During the reduction/gasification step, the carbon matrix of the pyrolysis char is decomposed by reaction with gasification agents CO_2 and/or H_2O at elevated temperatures under a reducing atmosphere leading to formation of combustible gases. The gas from pyrolysis and char gasification can be mixed subsequently or

prior to the char gasification stage. The reduction step is endothermic, and the temperature at which it takes place has a fundamental role in determining the composition of the final gas and its characteristics (LHV, tar presence).

The quality of the gas produced in biomass gasification varies according to the gasifying agent used, the feedstock, bed material, operational conditions (temperature, pressure, air-to-fuel ratio) and gasification technology.

2.1 Biomass resources

Biomass is a broad concept that includes 'the biodegradable fraction of products, waste and residues of biological origin from agriculture (including vegetal and animal substances), forestry and related industries, including fisheries and the aquaculture as well as the biodegradable fraction of industrial and municipal wastes' (2009/28/EC European Directive). Even though gasification platforms are often designed for conversion of conventional biomass types, primarily wood, the global increase in demand for bioenergy has facilitated growing efforts to extend the potential range of organic material fractions converted in thermal processes. As a result, there is a growing focus on thermal utilization of various secondary organic resources including: (Thomsen et al., 2016):

- Agricultural by-products and residues: For example, crop residues (stalk, leaf, cob etc.), manure fibres, muck and bedding and fibres from biogas slurry;
- Municipal by-products and residues: For example, source segregated organic waste, used textiles, grass and cuttings from road and park maintenance, beach cleaning waste and sewage sludge; and
- Industrial by-products and residues: For example, residues from breweries, food packaging, food retail, food preparation or residues from production of non-food products based on the partial conversion of animal and vegetable raw materials.

Optimized treatment of secondary resources has great potentials from an economic, environmental and resource political point of view (Thomsen et al., 2016). However, the diversity of biomass resources makes it essential to characterize them thoroughly, as their physical-chemical properties will have a direct impact on the design of the gasification plant as well as on the technical and economic feasibility of the process. This characterization should be carried out before choosing the gasification technology to avoid subsequent problems. The biomass properties that have the highest impact on gasification technology are:

- *Biomass particle size* (Molino et al., 2018; Sansaniwal et al., 2017; Sikarwar et al., 2016): Reducing the biomass particle size increases the pre-treatment cost of the feedstock, but it increases the surface area and decreases the diffusion resistance. This results in improved heat and mass transfer, increased reaction rates, fuel conversion, carbon conversion efficiency and decreased char and tar yields. Conventional gasifiers can take particle sizes in the range of 0.15–51mm. Particles of 0.15 mm are required for entrained flow gasifiers while fixed-bed reactors, with longer residence times, can tolerate particles up to 51 mm. Usually, bubbling fluidized-bed (BFB) reactors can take particles of up to 6 mm.
- *Moisture content* (Molino et al., 2016, 2018; Ruiz et al., 2013; Sansaniwal et al., 2017; Sikarwar et al., 2016): High moisture contents increase the energy required for water evaporation and steam gasification reactions that, in turn, may lower the gasifier operating temperature especially in the first stages. Reducing the moisture content may in some cases increase the energy efficiency, especially if low-temperature heat cannot be utilized. In addition, a dryer fuel will increase syngas quality, syngas HHV and decrease the conversion emissions. Fluctuations in biomass moisture content lead to variations in bed temperature, and this causes changes in the composition of the syngas generated.

  The required moisture content of the biomass depends on the gasifier design; updraft fixed-bed gasifiers can operate with moisture contents up to 60% (w/w), while downdraft fixed gasifiers can take maximum moisture contents of 25% (w/w). However, values around 10–20% (w/w) are generally recommended for conventional gasification technologies and will keep bed temperatures moderately stable.
- *Ash content* (Molino et al., 2018; Sansaniwal et al., 2017; Sikarwar et al., 2016): The ash content in biomass plays a significant role in thermal biomass utilization, especially if it contains alkali metals such as potassium or halides such as chlorine. Slagging, fouling and corrosion are important - and potentially critical, issues that may be caused by the inorganic species present in the biomass. It is not possible to give any general biomass criteria related to ash content as the potential problems relate to a combination of process design, process temperature profile, ash quantity and ash composition. However, generally, biomass with ash content lower than 2% (w/w) can be used as feedstock material in all types of gasifiers even though there may still be issues related to corrosion and long-term deposit build-up.
- *Bulk density:* Bulk density will have a strong influence on the transportation costs, handling and feeding of the biomass in the gasification plant. Biomass with a low bulk density might cause severe feeding problems to the gasification units because of getting stuck in the feeding line. These

problems can be overcome by densifying the bulky biomass; however, this will result in higher energy consumption and cost.

Organic resources that are initially found to be unsuited for thermal gasification may still have a fuel potential through fuel design or thermal pre-treatment. Torrefaction is a well-known process used to pre-treat biomass resources for various purposes and may render organic material more coal like. Many other problematic issues may be overcome by fuel design, for example through proper mixing. Recent studies on these issues have indicated many potential benefits related to moisture content, ash melting-related issues and ash/char by-product quality (Skoglund et al., 2016; Thomsen et al., 2016, 2017a,b,c).

2.2 Types of gasifiers

The design of a biomass gasifier depends ideally on the fuel characteristics (size, shape, moisture content, ash content) and the marketability of the end products. The principal design concepts, even though there are others available, can be classified in fixed beds, fluidized beds and entrained flow reactors (Table 1).

Fixed-bed reactors are often considered the best choice for small-scale plants of up to 10 MW. Within fixed-bed reactors, it is possible to distinguish updraft and downdraft configurations. In updraft gasifiers, the biomass is supplied from the top and the gasifying agent from the bottom (counter-current). For these gasifiers, the sequence for biomass is drying → pyrolysis → reduction → combustion. Instead, in downdraft gasifiers, the biomass and gasifying agent are supplied from the top (co-current). Thus, the sequence is drying → pyrolysis → combustion → reduction. The main difference is that in downdraft gasifiers the decomposition products from drying and pyrolysis go through thermal cracking in the oxidation zone, thus produce less tar and a higher-quality gas than updraft gasifiers, where the pyrolysis gas goes directly into the producer gas. Fixed-bed gasifiers are simple in construction and operate at low gas velocity with high carbon conversion and long residence time.

Fluidized-bed reactors, including BFB and circulating fluidized-bed (CFB) reactors, are based on the principle of fluidization in which both the fuel and the hot bed material (inert inorganic material and/or catalyst) is made to behave as a fluid by mixing the high-density bed and fuel material with a low-density fluidization agent, for example, air, steam or other gas. The semi-suspended conditions are kept by means of the gasifying agent added at the appropriate velocity. These gasifiers provide an excellent gas-solid mixing, a uniform temperature and solid/gas concentration in the entire bed. It is not possible to distinguish the different zones in single-reactor fluidized-bed systems (drying,

Table 1 Comparison of different types of gasifiers (Basu, 2018; E4tech, 2009; Molino et al., 2016, 2018; Ruiz et al., 2013)

Gasifier	Downdraft	Updraft	Bubbling fluidized bed	Circulating fluidized bed	Entrained flow bed
Technology	Simple, reliable and proven. Simple reactor with relatively low investment cost		Plants with higher investment and maintenance costs. Good ability to scale-up		Complex. High investment and maintenance costs
Fuel specifications	<51 mm	<51 mm	<6 mm	<6 mm	<0.15 mm
Tar (g/Nm³)	0.015–3.0	30–150	3.7–61.9	4–20	0.01–4
Ash and particles in syngas	Low	High	High	High	Low
Gasification temperature	900–1050°C	950–1150°C	800–1000°C		1300–1500°C
Syngas exit temperature	700°C	150–400°C	800–1000°C		>1260°C
Capacity	<5 MWe	<10 MWe	1–20 MWe	2–100 MWe	5–100 MWe
Carbon conversion efficiency	High	High	High. Loss of carbon in ash	High	High
Process flexibility	Very limited. Any change in the process variables needs a new design		Flexible to loads less than design		Very limited. Size and energy content of the fuel must be in a narrow range
Hot gas efficiency	85–90%	90–95%	89%	89%	80%

pyrolysis, reduction and combustion) due to the intense gas-solid mixing. Fluidized-bed gasifiers usually operate at relatively low temperatures (700–900°C) to prevent ash sintering and agglomeration and are the preferred option for large-scale plants due to superior scalability characteristics compared to fixed-bed systems.

Entrained flow reactors operate at high temperatures (1400°C) and high pressure (20–70 bar) by injecting powdered fuel into a high-speed stream of the gasifying medium and into the reactor. Thus, the gas stream entrains the fuel particles and directs the flow direction and speed through the reactor system.

2.3 Gasification operating variables

The operating parameters such as temperature, pressure, gasifying agent and equivalence ratio (ER) will have a very important role in determining the gasification performance, syngas yield and gas composition.

2.3.1 Temperature

Temperature is a crucial parameter in biomass gasification which, to a large extent, controls the gas composition, reaction rate, tar concentration, ash-related problems and so forth. Low temperatures generally result in higher tar content and low CO and H_2 content in the product gas. High-temperature gasification, on the other hand, leads to a desired high yield of CO and H_2 while reducing the tar content. However, high gasification temperatures present two major potential drawbacks: ash-melting problems, especially for high ash-containing biomass, and rigorous reactor specifications.

2.3.2 Pressure

Biomass gasification can be operated at atmospheric pressure or at higher pressures usually depending on the downstream application of the product gas. If the product gas is used in turbines or in Fischer–Tropsch synthesis, then pressurized gasification can be convenient even though it is more technologically complex. A reduction in the tar yield and in light hydrocarbons can be achieved by operation at higher pressures (Asadullah, 2014a; Molino et al., 2018).

2.3.3 Gasifying agent

Air, steam, pure oxygen, carbon dioxide or mixtures of them can be used as gasifying agents. The selection will depend on the gas quality requirements for the downstream applications and the economics of the overall process. Air is the cheapest and most cost-effective gasifying agent and has been extensively applied for heat and power generation purposes. It results in a highly nitrogen diluted and low-quality gas with low H_2 content (8–14 vol. %) and a heating value in the range of 4–7 MJ/Nm^3 (Matas Güell et al., 2012). Pure oxygen is the most expensive gasifying agent because it requires an air-separation unit, but it leads to a syngas with a heating value up to 28 MJ/m^3, higher concentrations of CO, H_2 and low concentration of tars (Asadullah, 2014a; Molino et al., 2018). The cost of steam is somewhere between air and oxygen and leads to a product gas with a heating value around 10–18 MJ/Nm^3 and with a higher concentration of H_2 as a result of the water-gas-shift reaction, but it reduces the thermal efficiency of the gasification. Steam-only gasification requires an external energy supply due to the endothermic reactions involved in the process. CO_2 gasification results in a product gas rich in CO because of the reaction of CO_2 and carbon; however, this reaction is slow and external heat supply is required.

2.3.4 Equivalence ratio (ER)

The ER can be defined as the ratio between air-to-fuel ratio of the gasification process and the air-to-fuel ratio for complete stoichiometric combustion. ER is a relevant parameter because it defines the degree of combustion and, thus, the temperature and distribution of gas, tar and char in the gasifier. ER optimal values are in the range of 0.2–0.3 for fixed-bed and fluidized-bed gasifiers, while entrained flow gasifiers usually require a 20% higher ER (Molino et al., 2018). ER lower than 0.2 results in an incomplete gasification, while ER higher than 0.4 approach the combustion. Decreasing the ER results in higher concentrations of H_2 and CO and higher calorific value of the product gas. Increasing the ER has the opposite effect and increases the CO_2 concentration but it reduces the tar yield. Thus, it is necessary to optimize the process depending on the downstream applications.

2.4 Synthesis gas requirements for further applications

The gas produced from biomass gasification is rarely used directly in downstream applications. Usually the gas needs to be cooled and cleaned to ensure smooth and efficient operation. The gasifier design, feedstock type, gasifying agent and other operating parameters will influence the gas composition and the quantity and type of impurities. The gas quality is a key driver in the technical and economic feasibility of different application schemes. In Table 2, a range of generalized quality measures accepted by engine, gas turbine and fuel-cell applications are provided.

Table 2 Acceptable range of impurities for different applications (Hasler and Nussbaumer, 1999; Sansaniwal et al., 2017; Singh et al., 2014)

	IC engine	Gas turbine	Fuel cell
Particles (mg/Nm^3)	<50	<30	-
Particulate matter (μm)	<100	<5	<1
Tar (mg/Nm^3)	<10	<5	<1
Alkali metals (mg/Nm)	-	<0.24	-

In fuel synthesis from gasification producer gas, the gas quality requirements may be even higher than for direct production of electricity and heat. The catalysts used in Fischer–Tropsch synthesis are sensitive to even very small amounts of impurities. In commercial operation, catalysts are replaced or regenerated after a certain operational period. Thus, the definition of the gas cleaning is therefore based on economic considerations: investment in gas cleaning versus accepting decreasing production due to poisoning of the catalyst. Requirements for feed gas quality of different downstream applications can be found elsewhere (Boerrigter et al., 2004; Molino et al., 2018; Tijmensen et al., 2002).

2.5 Gas cleaning processes

As previously mentioned, cleaning and conditioning of the product gas may be required to remove contaminants and potential catalyst poisons as well as to achieve the qualitative composition required by, for example, biofuel production processes. There are several papers available in the literature that have recently addressed and analysed the different possible cleaning processes (Abdoulmoumine et al., 2015; Asadullah, 2014b; Woolcock and Brown, 2013). In this section, a brief summary of commonly applied physical, thermal and catalytic gas cleaning technologies is presented.

2.5.1 Physical gas cleaning

Physical gas cleaning is one of the simplest cleaning methods and includes filtration and wet scrubbing to remove tar and particulate matter from the gas through gas/solid or gas/liquid interactions. Filtration can be conducted at high or low temperatures. High-temperature filtration requires temperature tolerable materials such as ceramics, fibre glass, composites or sand. Low-temperature filtration processes can use charcoal, cotton fibres as filtration media. Filtration techniques often face the problem of clogging and deposition of gas particles causing a pressure drop across the filtration. Back-flushing the filtration media with, for example, nitrogen may revitalize the performance of the filter.

Wet scrubbing is usually done at low or ambient temperature using water, oil or another solvent but faces the problem of producing high amounts of contaminated liquid that needs to be treated.

2.5.2 Thermal gas cleaning

Thermal decomposition consists of degrading (cracking) heavy aromatic tar compounds into smaller, lighter and non-condensable gas species such as methane, hydrogen and carbon dioxide at temperatures around 1000°C. The most problematic aspects of high-temperature cracking relate to the required temperature resistance of the setup that needs a complex and highly controllable heating system and intensive cooling system for the gas afterwards as well as possible ash-melting-related problems (Asadullah, 2014a).

2.5.3 Catalytic gas cleaning

It is possible that physical and thermal gas cleaning steps are not enough to fulfil stringent requirements of some downstream applications. Thus, the utilization of effective catalysts is often considered a method to decrease the concentration of tar and ammonia in the product gas. Catalytic decomposition usually occurs at much lower temperatures (600–800°C) than thermal cracking (>1000°C). Catalysts for tar removal have been tested and used directly in the gasifier reactor as well as in secondary catalytic reactors. Sometimes, additional oxidant (O_2 or H_2O) is injected into the secondary catalytic bed. Tars can be cracked or hydrocracked into light hydrocarbons or converted into CO and H_2 through steam and dry reforming reactions. The spectrum of studied catalytic materials covers acidic solids for cracking, basic solids for 'reforming' reactions and metal oxides with both acidic and metal sites for cracking and hydrocracking reactions (Torres et al., 2007). Desirable properties of a good catalyst for any of these reactions are (Sutton et al., 2001): (i) capability to efficiently convert tars into useful gas products in an environment containing high concentrations of H_2, CO, CO_2 and H_2O at temperatures of 600–800°C at ambient pressure (or even higher temperatures for pressurized reactors); (ii) resistance to deactivation and poisoning by species in the gas, such as sulphur-containing compounds; and (iii) resistance to attrition. It is also desirable to carry out the tar conversion process in such a way as to keep the deposition of carbonaceous materials on the catalyst surface under control.

In biomass gasification, most of the nitrogen ends up as ammonia (NH_3) and N_2 with lower concentrations of hydrogen cyanide (HCN), isocyanic acid (HNCO) and nitrogen oxides (NOx). Nitrogen-containing species are undesirable because they can poison catalysts or serve as precursors for NOx in downstream burners, gas engines or gas turbines. Most of the literature on the removal of ammonia refers to the use of calcined dolomite, nickel and iron catalysts, the same catalysts used for tar decomposition reactions.

3 Biomass gasification for heat and power generation

Biomass gasification for production of power or CHP is currently the most widely applied use of the technology. In recent studies on the status of biomass gasification in EU, it was found that almost 900 biomass gasifiers are currently in operation. In Germany alone, almost 400 biomass gasifiers have been identified. The majority of the systems are small-scale wood gasifiers build for CHP production in integrated gas engines with power production capacities averaging around 80 kW power per unit. The dominating manufacturer is Burkhardt who produces CHP gasifiers with approximately 30% electrical efficiency operated on wood pellets (Böcker-Riese and Bräkow, 2017; Hrbek, 2016).

However, large-scale facilities for biomass gasification and production of heat and power also exist, even though number and total capacity is limited. In a report on large-scale biomass applications from IEA task 40 from 2013, it was found that the ten largest biomass and waste gasifiers in Europe constituted approximately 600 MW thermal capacity in total, ranging from 160 to 20 MW thermal capacity. During 2012, the ten plants converted approximately 9 PJ of biomass out of a total capacity potential of approximately 18 PJ biomass. Most of the recent large-scale plants are built by the Finnish company Valmet, and in addition to the listed plants in Europe at least three other Valmet gasifiers are in operation in Indonesia and China. Additional biomass gasifiers are found in, for example, Cambodia (+35), Japan and the United States, but the aggregated capacity of these plants is limited (Shackley et al., 2012; Vakkilainen et al., 2013; Valmet, 2015).

As previously described, there are many different types of biomass gasifiers, and the application of the different systems is plentiful and innovative. Today, complete systems for production of heat and power as well as add-on systems for retrofitting the fossil fuel-based infrastructure exist.

The application of the biomass gasifier CHP system poses different possibilities and limitations to the design and operation. The following two systems can be used to illustrate two very different examples of this, that is as in Fig. 2: high efficiency, fully automatic small-scale wood gasification with integrated gas conversion or as in Fig. 3: retrofitting of existing coal-fired power plants with low-temperature gasification of low-value biogenic residues and wastes. These two opposing examples of biomass gasification for CHP production are described in more detail with case studies in the following.

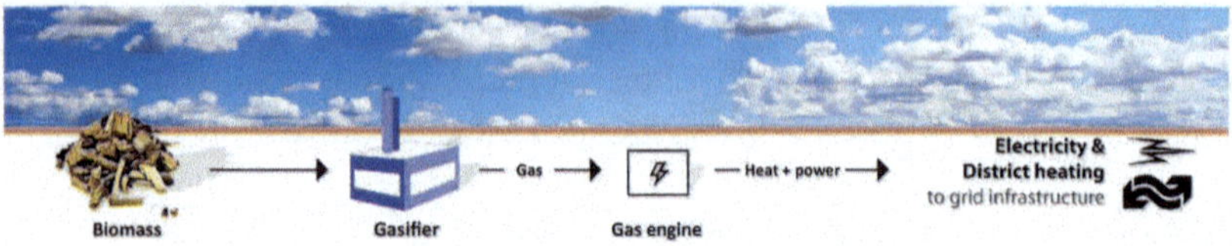

Figure 2 Illustration of a complete CHP setup with thermal gasification of biomass and conversion of the producer gas in a gas engine, full cell or gas turbine.

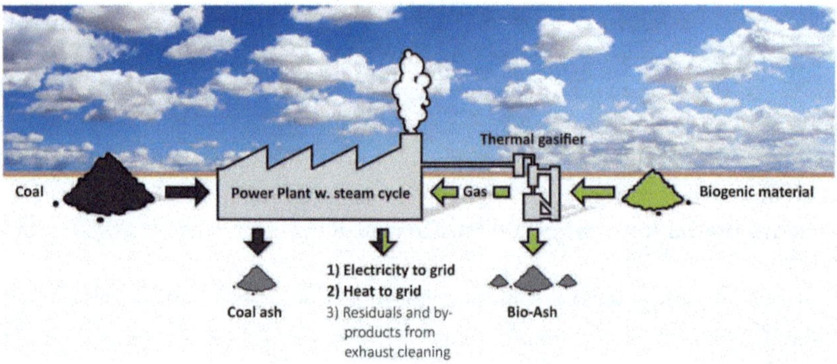

Figure 3 Illustration of a gasification system for retrofitting existing fossil fuel-powered steam cycles with medium- to large-scale thermal gasification of biomass.

3.1 Case studies

Three cases with very different biomass gasifiers in CHP systems are presented to illustrate the variety in design and application. Many more cases can be studied by addressing previously published literature on this issue (Ahrenfeldt et al., 2013a; Behera and Varma, 2019; Brandin et al., 2011; Sikarwar et al., 2016). The three cases are as follows:

3.1.1 Case study 1: Product flexibility: The TwoStage Downdraft Gasifier

The Biomass Gasification Group, at the Technical University of Denmark, has been working with staged gasification in fixed beds since 1989. This has led to the development of the TwoStage downdraft concept originally designed for gasification of wet wood chips in plants of up to 2 MW thermal capacity. The TwoStage process encompasses pyrolysis and gasification in two separate reactors with an intermediate high-temperature tar conversion zone. This staged design allows for efficient temperature control, and combined with catalytic tar cracking in the char bed of the gasification reactor, the intermediate thermal tar cracking leads to extremely low tar concentrations in the produced gas (Ahrenfeldt et al., 2006, 2013a; Henriksen et al., 2006). An illustration of the TwoStage process is provided in Fig. 4.

The TwoStage downdraft gasifier has been built in several different sizes from 50 kW to 1.5 MW thermal capacity. Different designs have been investigated, and currently alternative reactor designs are considered to allow for scaling up the plant (Ahrenfeldt et al., 2013b; Gadsbøll, 2018). The system is designed for automatic, unmanned operation and has achieved more than 4000 h of operation (more than 3500 with engine operation). The producer gas from this unit is very clean and the high quality allows for multiple uses of the gas in CHP systems with Solid Oxide fuel cells as well as gas engines or for production of liquid or gaseous fuels as described later (Gadsbøll et al., 2017).

In recent years, the fuel flexibility of the TwoStage downdraft design has been tested and additional relevant fuel candidates identified. Current focus of the technology is therefore expanded to include smaller energy island systems in rural areas of developing countries in Africa and Asia.

3.1.2 Case study 2: Operational stability: The Harboøre updraft gasifier

Celebrating 25 years of operation in 2018, the oldest biomass gasifier in operation today is the Harboøre 5.2 MW (thermal capacity) fixed-bed updraft gasifier. The plant went into operation in 1993, and in 2010, it had already

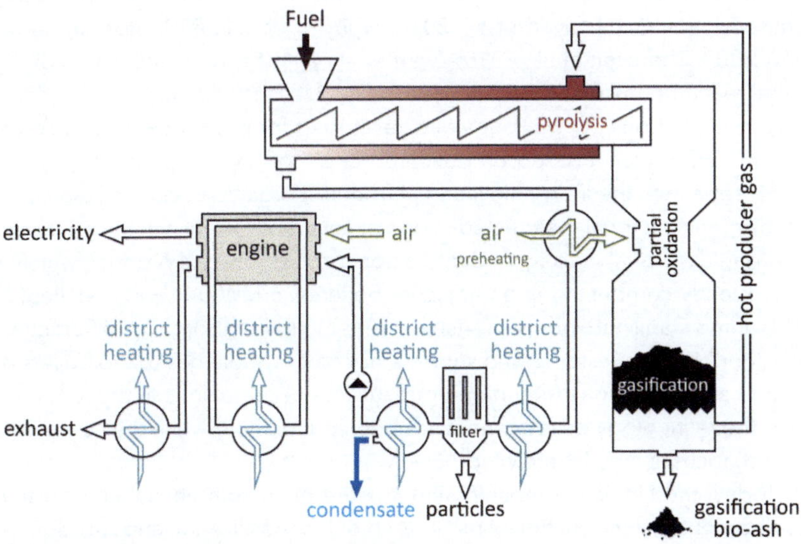

Figure 4 Simplified illustration of a TwoStage downdraft gasifier. Source: adapted with permission from Ahrenfeldt et al. (2013a).

operated for more than 110,000 h (Brandin et al., 2011; Heeb, 2010). In year 2000, the plant was upgraded with two Jenbacher gas engines for CHP production, and 10 years later, the engines had operated for more than 70,000 h each (Brandin et al., 2011; FORCE technology, n.d.).

The CHP plant in Harboøre is an air-blown updraft gasifier converting wood chips into producer gas, which is converted into heat and power in two gas engines after conditioning. The primary conditioning step is gas cooling and subsequent removal of water and tars in an electrostatic precipitation filter. The removed tar is stored and used for peak load situations in the heat demand by combusting the tar in adjacent boilers. The plant is designed for unattended operation, and holds a record in most hours of unattended operation of a biomass gasification plant. Unmanned operation and low downtime is possible due to extremely stable operation, which is an essential parameter for commercial CHP production (Jesper Ahrenfeldt et al., 2013a; Brandin et al., 2011; FORCE technology, n.d.).

3.1.3 Case study 3: Fuel flexibility: The Low Temperature Circulating Fluidized Bed (LT-CFB) gasifier

The low-temperature circulating fluidized-bed (LT-CFB) gasifier is a process developed by Danish Fluid Bed Technology A/S, the Biomass Gasification Group at the Technical University of Denmark and DONG Energy A/S (now Ørsted A/S). The technology has been under development for almost 20 years and is today commercialized under the registered alias Pyroneer. The LT-CFB process and operation has been described several times in previously published articles (Narayan et al., 2016; Nguyen et al., 2013; Stoholm et al., 2002, 2007; Thomsen et al., 2015; Zwart et al., 2010) and an illustration of the central reactor construction is provided in Fig. 9.

The LT-CFB gasifier was originally developed for pre-processing of cereal straw in CHP systems based on dust-fired coal boilers. In popular terms, the LT-CFB separates the inorganic material from the organic fraction of the straw. The inorganic material is extracted in one or several different ash fractions, and the organic material is converted into a hot combustible gas product, which is subsequently combusted in an adjacent boiler. A 6 MW unit was installed, in 2014, at the Danish power plant Asnæsværket in Kalundborg to convert straw and other biomass residues and supply a hot combustible gas to the boilers of unit 2 to substitute coal. After numerous successful campaigns in the following years, the technology development has been put on hold while the company's (Ørsted) focus is on offshore wind power.

Today, the Finnish company Valmet is the main commercial actor within development and production of plants for biomass gasification and subsequent gas combustion in adjacent boilers. Valmet has gathered profound experience and expertise within CFB systems and retrofitting as well as gas combustion

Table 3 Plant data estimations of three different types of biomass gasification plants for combined heat and power production. Using data from real plants or present state-of-the-art estimates (Ahrenfeldt et al., 2013a; Energistyrelsen and Energinet.dk, 2012; FORCE technology, n.d.)

Parameter	Unit	Staged, downdraft + IC gas engine	Updraft fixed bed + IC gas engine	Pre-gasifier for boilers
Maximum thermal capacity per unit	MW thermal input	1.5	5	100
Efficiency, fuel to hot gas	% of fuel energy	-	-	95
Net efficiency, fuel to heat	% of fuel energy	55	55	Depend on boiler
Net efficiency, fuel to power	% of fuel energy	35–40	28-32	Depend on boiler
Technical lifetime	Years	15-20	20	25
Construction time	Years	1	1.5	2
Investment costs	Million € per MW_{TH}	3.4-3.7	3.7	0.3-0.5
Operation and maintenance costs	€ per MWh_{TH}	18 + 69000 € per MW per year	19 + 185000 € per MW per year	2.4

TH = Thermal input.

in new boilers. Recent tests have shown than retrofitted coal boilers may run exclusively on biomass producer gas as well as with a combination of producer gas and coal dust (Valmet, 2015, n.d.). Table 3 provides a set of CHP data of the three different cases of biomass gasification plants presented previously.

4 Fuel synthesis from biomass gasification synthesis gas

4.1 The Fischer-Tropsch process for hydrocarbon production

The Fischer–Tropsch (FT) process is a set of chemical reactions that convert carbon monoxide and hydrogen (synthesis gas) into liquid hydrocarbons. The reaction was invented and developed by Franz Fischer and Hans Tropsch in 1925 in Germany, although the fuel production first started in 1935 at the Ruhrchemie Company. Industrial interest in the reaction fluctuated due to the changes in availability and price of oil. As the demands for environmentally friendly fuels are increasing, interest in FT synthesis is growing as well, and the technology gains attention to produce liquid fuels from natural gas, coal and biomass (Fechete and Vedrine, 2017; Krylova, 2014; Samavati, 2018). Bio-derived synthetic hydrocarbons from synthesis gas offers a cleaner and more sustainable solution compared to fossil alternatives, with significantly reduced greenhouse gas emissions (GHGs) and net climate impact. The hydrocarbons produced by the FT reactions may be used as liquid fuels such as gasoline, diesel and jet fuel or as building block chemicals, such as lower olefins. Today, technological platforms exist that use FT synthesis in conversion of various carbon-rich materials into liquid fuels including coal to liquid (CTL), biomass to liquid (BTL), gas to liquid (GTL) and waste to liquid (WTL; Krylova, 2014; Maitlis, 2013; Williams et al., 2011; Zhang et al., 2013b). Large-scale BTL plants are very scarce, hence product yields and process efficiencies are only estimated. Based on modelling studies, efficiencies in terms of chemical energy and carbon recovered in the produced hydrocarbons are in the range of 30–50% and 25–45%, respectively.

4.1.1 Reaction mechanism, environment and catalysts

The FT reactions, synthesis of hydrocarbons from CO and H_2 is a complex catalytic process, including a great number of consecutive and parallel reactions. The reaction in a general form is showed in Eq. 1. FT synthesis is performed at normal or elevated pressures (typically 15–40·10^5 Pa) with catalysts based on the Group VIII transition metals, mainly Co and Fe and to some extent also Ni and Ru. Two main regimes are used: low-temperature Fischer–Tropsch (200–250°C) results in long-chain molecules, and high-temperature Fischer–Tropsch (320–375°C) gives shorter-chain molecules (Maitlis, 2013).

$$CO + 2H_2 = [-CH_2-] + H_2O \tag{1}$$

4.1.2 Products via the Fischer-Tropsch reaction

The Fischer-Tropsch synthesis can produce a wide range of long-chain hydrocarbon products, such as gasoline, synthetic naphtha, synthetic middle oil distillates (diesel and kerosene), lubricating oils and synthetic waxes (Fechete and Vedrine, 2017; Krylova, 2014). The produced hydrocarbons are typically a mix of straight-chain alkanes or paraffins. The reaction results in hydrocarbons with no oxygen content, leading to fine compatibility with the current engine and fuelling infrastructure (Bergthorson and Thomson, 2015). The lack of aromatic molecules in the FT fuels may lead to issues such as poor lubricity and seal-swell problems. The reduced aromatic content, however, is also an advantage, since aromatic content is associated with poorer combustion efficiency (increased CO and unburned hydrocarbon emissions) and increased soot emissions. Biomass-derived FT fuels have essentially zero sulphur content, reduced amounts of other contaminants and extremely low amounts of air pollutants leave the plants (Liu et al., 2011). The FT fuel products may be modified by isomerization processing to produce branched hydrocarbons for gasoline and jet-fuel applications, or to produce aromatics and naphthas, which could lead to renewable biojet fuels. A standard was made (ASTM 7566) that allows blends of up to 50% of synthetic hydrocarbon alternative jet fuels to be used when mixed with conventional commercial jet fuel, but at present only from specific feedstocks and processes (Bergthorson and Thomson, 2015).

When biomass is utilized via gasification, the produced synthesis gas can be converted to hydrocarbons with the FT process. The resulting liquid hydrocarbon fuels (mainly diesel and gasoline compounds), however, have a low octane number because of high concentrations of linear alkanes, hence they need upgrading for better properties (Galadima and Muraza, 2015).

4.1.3 Use of CO_2 in a Fischer–Tropsh reaction mechanism

Great efforts are being invested to develop new and efficient technologies that will contribute to capture and convert CO_2 into fuel materials and decrease the GHG levels as well. Transformation of CO_2 into fuels and chemicals with renewable H_2 via heterogeneous catalysis is a promising way to reach that goal; however, there are still great challenges before these processes can be commercialized and applied at a large scale. The catalytic reduction of CO_2 by H_2 can lead to the formation of three types of products: CO via a reverse water-gas-shift (RWGS) reaction, methanol via selective hydrogenation and hydrocarbons via combining CO_2 reduction with FT reactions. The CO intermediate produced by RWGS reactions offers a flexible solution, since CO can be used in both methanol synthesis and downstream FT for fuels and chemicals. Direct hydrogenation of CO_2 can result in hydrocarbon formation, including both alkanes and olefins. Another route may be direct FT reactions from CO_2 and H_2 (CO_2-FT) by performing RWGS followed by FT in one reactor. The CO_2-FT process is very attractive because this way alkanes and olefins can be directly converted from CO_2 and H_2, but finding appropriate catalysts is very challenging (Gao et al., 2018; Porosoff et al., 2016).

4.1.4 Economic feasibility of Fischer–Tropsch using syngas from biomass gasification

The FT process has been developed for decades; however, large-scale applications are scarce, especially when the syngas is based on biomass conversion. Large-scale, long-time investments of such plant developments are very risky due to the high capital costs, high operational and maintenance costs and even environmental concerns, so plant constructions have been often announced and then shut down or halted (Baliban et al., 2013; Maitlis, 2013; Venvik and Yang, 2017). There are different approaches for cheaper and more efficient plant installations, for example the so-called microstructured or microchannel reactors (also referred to as microreactors) showing enhanced heat and mass transfer in combination with highly active, selective and stable catalysts (Venvik and Yang, 2017). It has also been showed that the production of FT fuels is more profitable when electricity is generated as a co-product. The new plant designs should aim for producing ultraclean synthetic fuels from coal and lignocellulosic biomass combined with CO_2 capture and storage (Liu et al., 2011).

4.2 Methanol synthesis

Methanol (CH_3OH) has been a widely utilized substance in the chemical industry for many decades. It is very versatile and used for a variety of industrial products including production of olefins, formaldehyde and acetic acid, which are important industrial intermediates. The global consumption of methanol in 2017 was approximately 92 Mt/year (Dalena et al., 2018). It is a colourless, tasteless and poisonous substance that is liquid at ambient conditions and is easily stored. In recent years, there has been an increasing interest in methanol as a fuel, as it might act as an efficient renewable energy carrier (provided that it is produced from sustainable sources). The interest is especially centred around its abilities as a transport fuel for internal combustion engines. It can either be blended with gasoline or be used as a primary fuel. Converting the vehicle fleet to methanol will only require minimal modifications to existing technology. Other methanol applications include fuel-cell operation and hydrogen carriers.

Methanol synthesis has been commercial for about 90 years, initially with plants based on steam gasification of coal. Today, the primary source is natural gas that accounts for 90% of the production (Dalena et al., 2018). This process features steam reforming of natural gas up to the GW_{th}-scale and the energy efficiency can be up to 69% (Danish Technological Institute, 2011). Methanol from biomass gasification is still in the developmental phase and, while there are technical solutions to the majority of challenges, there is still not a clear economic incentive to broadly commercialize the technology. There are, however, several pilot as well as early-commercial demo plants in operation (Molino et al., 2018). In relation to the natural gas-based efficiency, studies on methanol via biomass gasification have modelled efficiencies on the range of 55–59% (Danish Technological Institute, 2011; Puerari et al., 2014; Røngaard Clausen, 2011) – the difference in efficiency to natural gas reforming is namely because of the gasification step.

The key reactions in methanol synthesis are:

$$CO + 2H_2 \rightarrow CH_3OH\ (-91\ kJ/mol) \tag{2}$$

$$CO_2 + 3H_2 \rightarrow CH_3OH + H_2O \left(-50\ \text{kJ/mol}\right) \tag{3}$$

A significant amount of heat is released during the synthesis. The reactors are either isothermal (pressure-controlled boiling water reactors) or adiabatic (several reactors with intermediate cooling). The catalysts are based on Zn, Cu, Al and/or Cr components and offer very high selectivities >99.9%, which ensures high yields and low amounts of by-products. Following synthesis and partial recycling for increased conversion rates, the product is purified via distillation and cooling in three columns: (1) separation in a gas-liquid separator; (2) separation of dissolved gases in a topping column; and (3) separation of water and other liquid by-products. The key by-product is ethanol, which is removed along with the produced water (Røngaard Clausen, 2011).

Methanol can be further refined to, for example, di-methyl ether (DME) or gasoline. These conversion processes are typically connected, as the DME is the first step towards gasoline. While DME (CH_3OCH_3) can be directly synthesized from H_2 and CO, it is traditionally synthesized via methanol dehydration at similar reactor and process conditions:

$$2CH_3OH \rightarrow CH_3OCH_3 + H_2O\ (-23\ \text{kJ/mol}) \tag{4}$$

Further processing for gasoline can then feature further dehydration and the co-production of hydrocarbons (e.g. C_2-C_5 gases, paraffins, aromats) and water (Basu, 2013).

The overall process from product gas to methanol can be broken down into subprocesses as shown in Fig. 5. Tar represents the most significant challenge for gasification systems. Besides tar and inorganic components (that are discussed in the next section), the synthesis plant should be coupled to a gasifier that can ensure optimal conversion from biomass to methanol: high cold gas efficiency, no/negligible levels of methane and inert nitrogen and a high hydrogen content.

Particles and sulphur are the main issues in focus during the initial cleaning of the product gas, but also other inorganic compounds are of interest as the downstream equipment is extremely sensitive to impurities. Desulphurizing the gas is crucial and a well-known process because natural gas typically carries

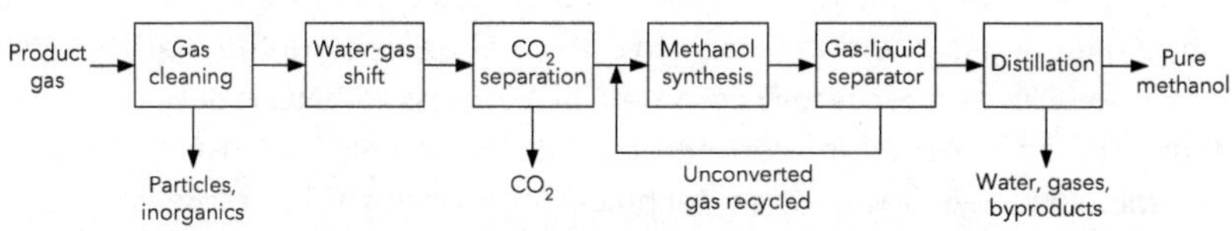

Figure 5 Typical flow diagram for methanol synthesis from product gas. The compression step can be at gasifier (fuel inlet) or prior to gas cleaning.

significant amounts of sulphur. The most used gas cleaning method is the commercial Rectisol process (Gatti et al., 2014; Sun and Smith, 2013). It is a convenient method as it utilizes pressurized cooled methanol (around −40°C) as a solvent for a range of impurities including CO_2, H_2S, COS, HCl and NH_3. The primary downsides of Rectisol are that it is mostly applicable to large-scale plants and that it requires a refrigeration plant consuming electricity. Other commercial gas cleaning processes for sulphur and/or CO_2 include Selexol and Purisol (Kapetaki et al., 2015; Vitasari et al., 2011).

In order to ensure a high yield of methanol in the synthesis reactor, the gas composition typically needs to be adjusted for H_2, CO and CO_2 content. One way to define the optimal relationship between these components is given on a molar basis in Eq. 5. Here, the value of *M* should be at least 2 and preferably higher. The catalytic water-gas-shift reactor is applied for this very purpose at 200–500°C. This reactor might be placed ahead of the primary gas cleaning if a sulphur-tolerant catalyst is chosen and/or if the cleaning step should also remove the produced CO_2.

$$M = \frac{H_2 - CO_2}{CO + CO_2} \quad (5)$$

The synthesis is done in one or more catalytic reactors at 240–400°C and 50–300 bar (below 100 bar is typical) depending on the process (reactor, gas composition, pressure). An analysis of the process conditions is shown with the CO conversion rate in Fig. 6 for methanol and DME. The highest yield of methanol at equilibrium is at lower temperatures, but a higher value is often chosen to ensure a high reaction rate. High pressures are favourable as both equilibrium and the reactions are increased.

4.3 Synthesis of synthetic natural gas (SNG)

Synthetic natural gas (SNG or biomethane, CH_4) has gathered, like methanol, significant interest in the framework of the energy system transition, as it already has large infrastructure in place. SNG is nearly identical to traditional natural gas, as it possesses identical properties, but has a CO_2-neutral footprint as it originates from biomass. SNG production is therefore expected to be relatively feasible, as the existing infrastructure can be largely kept. This also means that the requirements for quality/purity of the SNG are closely tied to those of the existing natural gas grid that acts as the physical gateway to the market. While the coal-to-SNG process is fully commercial, biomass-to-SNG is at pilot or early commercial demonstration (Rönsch et al., 2016), mainly because of a lack of economic incentive more than technical challenges. The biomass-to-SNG efficiency is relatively high and models show values in the range of 55–65% (Rönsch et al., 2016; van der Meijden et al., 2010).

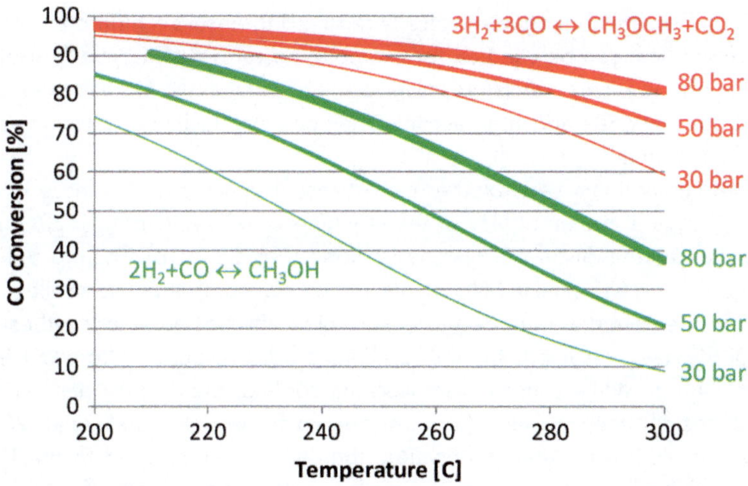

Figure 6 Equilibrium CO conversion as a function of the reactor temperature and the reactor pressure. Used with permission from Røngaard Clausen (2011). Methanol and DME (CH_3OCH_3) curves are calculated via optimal gas compositions with H_2/CO ratios of 2 and 1, respectively.

A process flow diagram for SNG synthesis is shown in Fig. 7. Synthesizing methane requires a higher hydrogen:carbon ratio, compared with methanol synthesis, as it is evident from the key reactions:

$$CO + 3H_2 \rightarrow CH_4 + H_2O\ (-206\ \text{kJ/mol}) \tag{6}$$

$$CO_2 + 4H_2 \rightarrow CH_4 + 2H_2O\ (-165\ \text{kJ/mol}) \tag{7}$$

The process conditions are typically around 250–700°C and 10–30 bar for SNG synthesis. The reactions in Eq. 6 and Eq. 7 are favoured at lower temperature and higher pressures as seen in Di Felice and Micheli (Di Felice and Micheli, 2015). The temperature level is closely linked to the reactor choice, which can either be a water-cooled isothermal fluid bed or an adiabatic fixed/fluid bed. Adiabatic reactors are the most common ones, and three reactors or more,

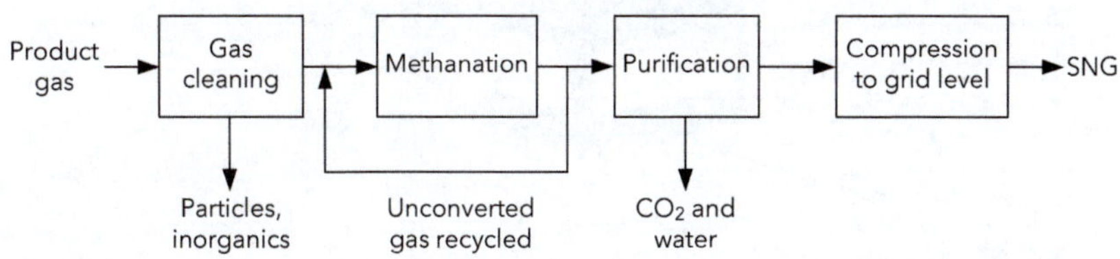

Figure 7 Typical flow diagram for SNG synthesis from product gas. Compression prior to the methanation can take place at the gasifier (fuel inlet) or at either end of the gas cleaning.

with intermediate cooling, are typically required to ensure a high conversion. The excess heat can be used to, for example, power a steam cycle for power generation. The commercial reactors are predominantly nickel-based and offer both low costs and high activity. They can also facilitate water-gas-shift reactions internally.

Methanation is a very exothermic process that has a heat release in the range of 20% of the input LHV of the reactants or ≈2 kW/Nm^3_{CH4} (≈2.6 times more than methanol on a molar basis; Rönsch et al., 2016). Therefore, heat management is essential in SNG plants and careful component control as well as plant integration are needed to ensure high efficiencies. After synthesis, a part of the gas is recycled, and the raw SNG product is purified from CO_2 and water dilution. While water is removed via cooling, pressurized gas washes such as the commercial Selexol process are mostly used (Kapetaki et al., 2015; Vitasari et al., 2011). After purification, the SNG is compressed to the local natural gas grid pressure level and directly injected into the grid.

5 Carbon balance value of by-products from thermal gasification of biomass

An overview of common process input and process output (products and by-products) from thermal biomass gasification is provided in Fig. 8.

The mass and volume fractions of by-products from biomass gasification vary substantially with the type of biomass fuel converted as well as the product portfolio from the specific process.

In general, there are three main routes to achieve carbon-negative energy production from thermal gasification of biomass, and all routes require proper

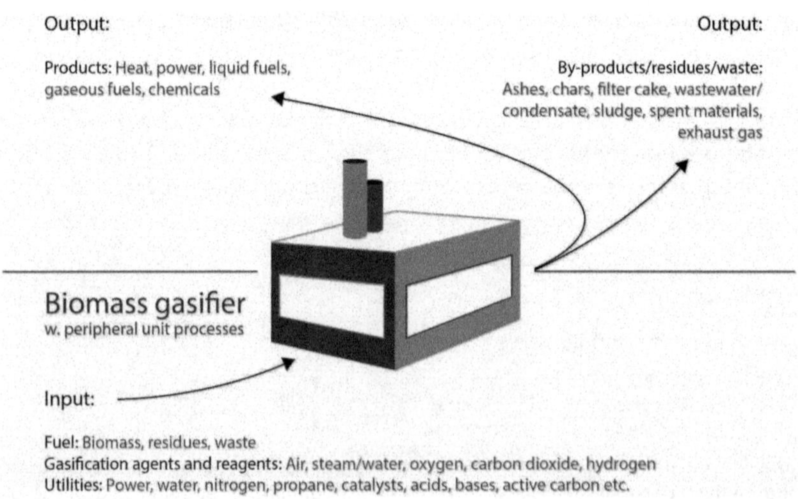

Figure 8 Illustration of common input and output from biomass gasification.

management of by-products. These routes involve production of energy services (heat, power, fuels) combined with one or several of the following by-product management strategies:

- Long-term sequestration of CO_2 captured from the off-gases;
- Long-term sequestration of recalcitrant carbon in chars and ashes; and
- Production of marketable non-energy products to substitute equivalent products on the market.

To address the influence of by-products on the carbon balance of the energy production, it is important to consider the complete carbon balance of the energy system:

$$C_{balance} = \frac{\Sigma CO_{2\,eq,\,emitted} - \Sigma CO_{2\,eq,\,stored} - \Sigma CO_{2\,eq,\,displaced} - \Sigma CO_{2\,eq,\,captured}}{\text{Net Energy Services (NES) provided}}, \quad NES > 0 \qquad (8)$$

$CO_{2\ eq,\ emitted}$ includes emissions from the exhaust as well as from biomass procurement, production of utilities, transportation and so forth. $CO_{2\ eq,\ stored}$ includes long-term sequestration of carbon in the shape of CO_2 (CCS) or recalcitrant carbon in char and ash. The temporal scope of the sequestration potential should follow the scope of the global warming potential of the assessed GHGs (IPCC, 2013). $CO_{2\ eq,\ displaced}$ includes the representative carbon intensity on the present market of marketable by-products produced from the system. $CO_{2\ eq,\ captured}$ includes the carbon captured from the atmosphere in the process, primarily via photosynthesis in the biomass growth phase. Net Energy Services provided include the net sum of all energy products provided from the biomass gasification system to society.

All elements are sums (Σ). of direct and indirect emissions for all quantified GHGs. All GHGs emitted (CO_2, CH_4, N_2O, CFCs and so forth) need to be recalculated into CO_2 equivalents using relevant impact factors and aggregated (IPCC, 2013). In addition, carbon-negative energy production requires a negative numerator and a positive denominator on the right-hand side. Systems with a negative Net Energy Service provided are not energy production systems, but may be categorized as energy conversion systems, converting low-quality energy services or carriers to high-quality energy services or carriers.

The potential for carbon-negative energy production from biomass gasification with CCS (long-term storage of CO_2 from off-gases) is substantial but struggles with economical and technical feasibility, especially with regard to retrofitting of existing plants. A recent study has indicated that equipping a biofuel production (liquid or gaseous) based on thermal gasification of biomass with CCS would double the climate benefit of the system with an additional cost of around 10% (Álamo et al., 2018; Hardisty et al., 2011).

Quantity and quality of ashes and chars from biomass gasification vary heavily with the fuel characteristics as well as the process design and operational parameters (Thomsen et al., 2016). In addition, many gasifiers will have more than one ash/char product from the process. This is illustrated in Fig. 9 with an example of a low-temperature biomass gasifier. In this design, there is a potential to extract at least three different ash/char fractions from the process which will all have very different characteristics.

For successful sequestration of carbon via ashes and chars, it is essential that the carbon in the solid residuals is stable in the environment in which it is stored. In addition, the quality of the char and ash has to be high enough for it to be feasible and attractive for the desired application. Usually, char and ash are applied to soil, and it is therefore often the carbon stability in the related terrestrial systems that are relevant for the carbon balance. Only a few studies have been conducted on the long-term stability of gasification biochar, but based on the results it can be expected that gasification biochar is highly stabile in soil systems with sequestration periods longer than a 100 years (common temporal scope for carbon balance assessments) for the majority of the carbon (Hansen et al., 2015, 2016; Shackley et al., 2012). Many more studies on the stability of carbon in pyrolysis biochar have been conducted with very positive results and expected sequestration periods from hundreds to thousands of years (Bruun et al., 2011; Budai et al., 2013; Cheng et al., 2006; Lehmann et al., 2006; Nguyen et al., 2008). The gasification system benefits

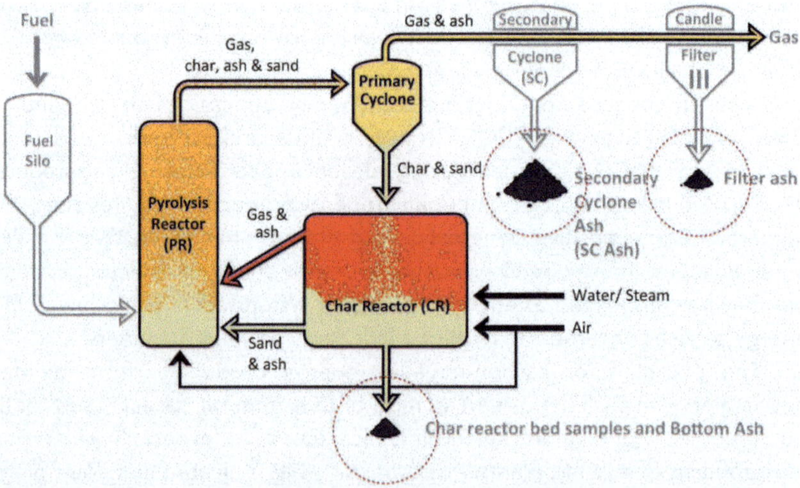

Figure 9 Low-temperature circulating fluidized-bed schematic design and ash/char products. Source: adapted with permission from Thomsen et al. (2017c).

from generally higher temperature and longer total retention time of solids in the systems, leading to lower O:C ratios which increase carbon recalcitrance, whereas pyrolysis systems often benefit from higher total carbon recovery in the solid fractions (Bai et al., 2014).

If the solids are applied in a landscaping effort, then it is also relevant if the use implies environmental risks or pollution. If the char or ash is amended to farm soil, it would often be as a fertilizer or a soil enhancer, and this would require fertilizer value and/or soil enhancement potential in the form of, for example, improvements of soil structure, water-holding capacity, cation exchange capacity and so forth. Soil enhancement effects as well as potential fertilizer value of gasification biochar and ash will vary substantially with the feedstock, design of the gasifier, the operational parameters and the soil system in which it is applied. Gasification of nutrient-rich substances such as straw or sewage sludge may provide ashes or chars with a high fertilizer value of especially potassium and phosphorous, while gasification of wood may provide valuable chars with vast soil enhancer qualities if gasifier design and end-use is properly aligned (Bruun et al., 2014; Hansen et al., 2015, 2016; Kuligowski et al., 2008, 2012; Li et al., 2017; Mackay et al., 2016; Müller-Stöver et al., 2014; Thompson et al., 2016). Two different ashes from thermal gasification of sewage sludge are shown in Fig. 10 to illustrate how different the solid residuals may be even when the fuel substrate is the same. In addition to these two samples, the complete set of solid residuals included cyclone ashes and filter ashes as well (Thomsen et al., 2016).

When producing gasification chars or ashes with a nutritional value which is marketable as a fertilizer substrate, then it is relevant to include this value in the carbon balance. This is commonly done in Life Cycle Assessment studies via System Expansion or Avoided Production and gives a reasonable credit to systems with co-production of high-quality by-products (European Commission et al., 2010). In practice, it implies subtraction of emissions from representative production of an equivalent quantity of the replaced product on the market. In addition to fertilizer value, it may also be relevant to include avoided

Figure 10 Two different ashes produced from the same sewage sludge substrate by thermal gasification in a low-temperature circulating fluidized-bed gasifier (left) and a TwoStage downdraft gasifier (right).

production of other marketable by-products including gas/water filter media, road material, metal procurement, manure treatment, BBQ charcoal, animal feeding, chemical grade pulverized active carbon (Huber, 2017; Rüegsegger and Kast, 2018).

6 System integration for polygeneration

Gasification of biomass can be an important platform for future biorefineries producing various biofuels and biochemicals. If the biorefinery also includes a heat engine or fuel cell for syngas conversion to electricity, it is possible to make a flexible biorefinery that can switch to electricity production when the electricity demand is high. Furthermore, if electricity from renewables, such as wind and solar, is supplied to the biorefinery together with the biomass input, it is possible to enable a full utilization of the biomass carbon. The electricity input can be used for water/steam electrolysis to produce hydrogen for syngas production, and oxygen for the gasifier. If solid oxide cells (SOC) are used for steam electrolysis (SOEC), these cells can also be used as fuel cells (SOFC) for electricity production, when the electricity production from renewables is low. Systems combining gasification and SOCs have been studied in numerous articles for *either* electricity production (Bang-Møller et al., 2011; Gadsbøll et al., 2017; Omosun et al., 2004; Panopoulos et al., 2006) or fuel production (Bernical et al., 2013; Clausen, 2017), but few studies look at both electricity and fuel production and investigate the potential of such polygeneration systems or flexible biorefineries (Fig. 11).

6.1 Flexible biorefineries

A flexible biorefinery integrating gasification and SOC can operate in at least three to four different modes, which makes it possible to adjust the electricity input or output to the current electricity price on the grid. The operating modes differ in gasifier operation and the SOC operation – for example, if the SOC operates in SOEC mode, the gasifier could be oxygen-blown, but if the SOC

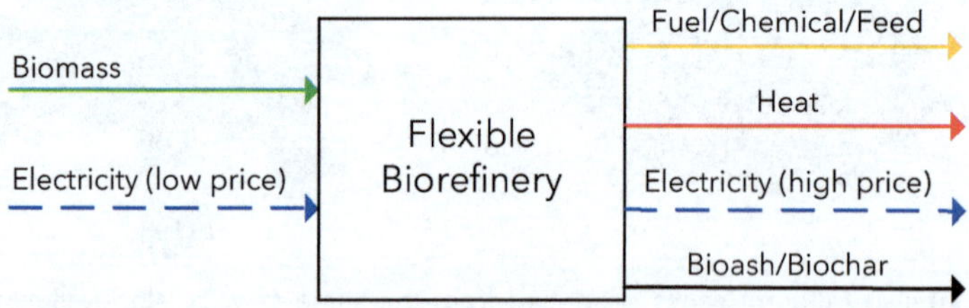

Figure 11 Diagram of a biorefinery based on thermal gasification integrated with water/steam electrolysis.

operates in SOFC mode, there is no oxygen production, and the gasifier could instead be air-blown. When the gasifier is air-blown, once-through synthesis of liquid fuels or chemicals could be feasible (Clausen et al., 2011), but production of gaseous fuels would be infeasible due to nitrogen dilution. An overview of operating modes can be seen in Table 4 and Fig. 12. Figure 12 is very simplified, and real systems would include syngas cleaning and conditioning before fuel synthesis. If, for example, methane is produced, CO_2 removal would be necessary in Fig. 12b, otherwise the produced methane would have a large CO_2 content. If a liquid fuel is produced, CO_2 removal would most likely be used in both Fig. 12a and b to lower the synthesis reactor size and reduce the syngas loss by the synthesis loop purge (Clausen, 2015).

Most of the operating modes shown in Fig. 12 have been studied previously as single-mode systems, for example, the system in Fig. 12d has been widely studied by thermodynamic modelling (Bang-Møller et al., 2011; Omosun et al., 2004; Panopoulos et al., 2006) and has also been proven experimentally (Gadsbøll et al., 2017). The experimental tests have proven the feasibility of combining gasification and SOFCs and have demonstrated biomass to electricity efficiencies of up to 43%, with a potential of increasing this to 62% when integrating a gas turbine (Gadsbøll et al., 2017). The system in Fig. 12a has so far only been studied theoretically by thermodynamic modelling (Bernical et al., 2013; Clausen, 2017). These analyses find biomass-to-biofuel efficiencies of 59–70%, depending mainly on the gasifier type and the biofuel produced. Highest efficiency is found for the system producing methane (Clausen, 2017), but this efficiency could potentially be increased to ~75% if methanol or DME is produced instead, as synthesis of these fuels is less exothermic than methane synthesis (Clausen, 2017).

Table 4 Overview of operating modes for a biorefinery based on biomass gasification and solid oxide cells

Operating mode	Gasifier	SOC	Synthesis of biofuel[a]	Electricity consumption (−)/production (+)
Electricity storage mode	Oxygen-blown	SOEC mode. Covers hydrogen demand	+++	−−−
Oxygen demand mode	Oxygen-blown	SOEC mode. Covers oxygen demand	++	−−
Once-through synthesis mode	Air-blown	SOFC mode. Unconverted gas from synthesis to SOFC	+ Only synthesis of liquid fuel	+
Electricity production mode	Air-blown	SOFC mode. Syngas directly to SOFC		++

[a] '+' is used to denote the biofuel production, '+++' is the maximum biofuel production.

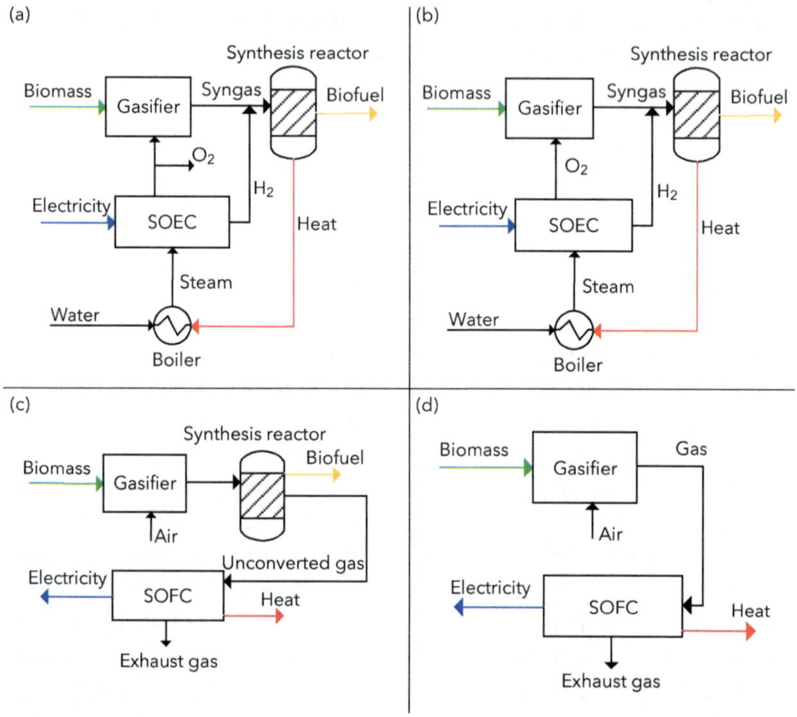

Figure 12 Simplified flowsheets of a biorefinery based on biomass gasification and solid oxide cells (SOC) operating in various modes: (a) electricity storage mode, (b) oxygen demand mode, (c) once-through synthesis mode, (d) electricity production mode. Note that the gasifier produces bioash/biochar, although not shown on the figures.

6.2 Pressurization and internal methanation in solid oxide cells (SOC)

Pressurization of both gasifier and SOC could be highly feasible for large-scale systems, and would reduce the of power consumption of power for syngas compression. Pressurized gasification is typically assumed for large-scale systems (Clausen, 2015; Larson et al., 2009) and is commercially used by coal gasifiers, but pressurized SOC operation is rarely assumed, and only few pressurized SOC experiments have been performed (Jensen et al., 2016, 2017).

If the biofuel produced is methane, there is an additional benefit of SOC pressurization, as it is possible to have internal methanation inside a state-of-the-art SOC if carbonaceous gasses are supplied to the SOC (Butera et al., 2019; Jensen et al., 2015, 2017). This can be achieved by sending humidified syngas to the SOC (Fig. 13). As the methanation synthesis reaction is highly exothermic, it is a perfect match for the endothermic H_2O and CO_2 electrolysis reactions. Thermodynamic system modelling predicts that approximately 50% of the methane is formed in

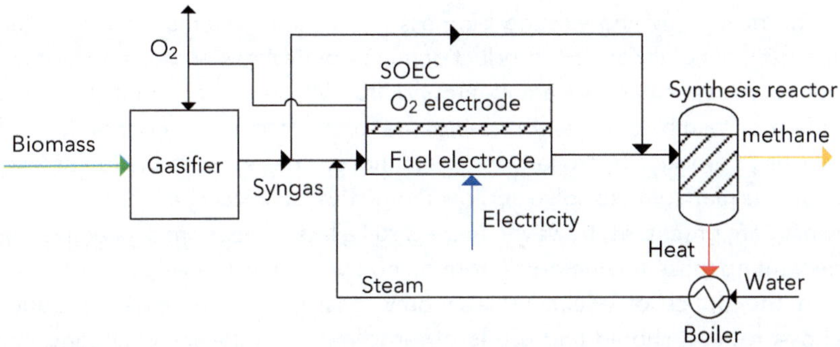

Figure 13 Simplified flowsheet of a methane-producing system based on pressurized gasification and pressurized SOC with internal methanation. Note: part of the syngas bypasses the SOEC to allow (1) a high H/C ratio in the SOEC, thereby avoiding equilibrium carbon formation (Butera et al., 2019), and (2) having the optimal ratio between H_2, CO and CO_2 for methane synthesis before the synthesis reactor (Butera et al., 2019).

the SOC; this significantly increases the output-to-input efficiency of the overall system, as heat released by the methanation reaction inside the SOC can replace a part of the electricity input to the SOEC. Another benefit of such a system is that a change in operating mode from electricity storage (Fig. 13) to electricity production (Fig. 12d) would be simpler, as the input to the SOC is always syngas, whereas the SOC changed input from steam to syngas in Fig. 12.

If other biofuels than methane are produced, it would still be possible to feed a pressurized SOEC with syngas, but it requires the use of SOECs without nickel in the fuel electrode, as nickel catalyses methane synthesis. A promising candidate is electrodes based on doped-ceria (Graves et al., 2016). Another option would be to have both gasifier and SOEC operate at atmospheric pressure; in this way, a state-of-the-art nickel containing SOEC could be used. At atmospheric pressure, thermodynamics will limit the methane production if sufficient high-temperature and high-oxygen content at the SOEC outlet is used.

7 Conclusion

Biomass gasification is a versatile and flexible process that can provide substantial contributions to modern biorefinery concepts and energy systems. The purpose of this chapter has been to give an introduction to the current state-of-the-art, recent research, development efforts and classic challenges related to integration and operation of this technology. The main parameters affecting the syngas production and composition such as gasification technologies, feedstock characteristics and operating gasification conditions have been presented. The importance of gas cleaning and conditioning depending on the downstream processes has also been highlighted.

The majority of the existing biomass gasification systems are small-scale wood gasifiers built for CHP production in integrated gas engines with power production capacities averaging around 80 kW power per unit. However, ongoing research focusses on using the product gas for Fischer–Tropsch synthesis to produce hydrocarbons, methanol synthesis or production of SNG. Research is managing to solve and overcome the technical challenges of these downstream processes; however, there is still a lack of economic incentive that is preventing these technologies from being commercially available.

In the optics of a circular economy, gasification processes' products and by-products should find use in order to minimize the waste streams and increase product and process value. Particularly, chars and ashes offer several options for being recycled in an efficient and sustainable way as active carbon or for soil amendment, remediation and carbon sequestration.

In the scope of carbon-negative bioenergy production, thermal gasification of biogenic material has substantial beneficial potentials. The potentials arise from a combination of high thermal efficiency, high fuel and product flexibility, value of by-products, possible integration with bio-CCS and SOCs and direct carbon sequestration by soil amendment of chars and ashes. In the future, this potential is expected to increase further as new concepts and solutions such as process integration schemes, combinations of different technologies and new polygeneration strategies will contribute to maximize product yield and value, increase the overall process efficiency, improve the economic viability and overcome the operational challenges that the technology is currently facing.

8 Where to look for further information

The first and second edition of the 'Handbook of Biomass Gasification' published by the BTG Biomass Technology Group BV in the Netherlands provides a good introduction to the subject for non-specialists as well as reviews of the current state of technology and a description of the latest success stories.

The IEA Bioenergy Task 33 'Gasification of Biomass and Waste' is a working group of international experts with the aim to promote the commercialization of efficient, economical and environmentally preferable thermal biomass gasification processes. On their website, it is possible to find several publications and reports regarding biomass gasification as well as a list of the most relevant conferences and workshops in the field.

9 References

Abdoulmoumine, N., Adhikari, S., Kulkarni, A. and Chattanathan, S. (2015) A review on biomass gasification syngas cleanup. *Applied Energy* 155: 294–307. DOI: 10.1016/J.APENERGY.2015.05.095.

Ahrenfeldt, J., Henriksen, U. B., Jensen, T. K., Gøbel, B., Wiese, L., Kather, A. and Egsgaard, H. (2006) Validation of a continuous Combined Heat and Power (CHP) operation of a two-stage biomass gasifier. *Energy & Fuels* 20(6): 2672–80. DOI: 10.1021/ef0503616.

Ahrenfeldt, J., Egsgaard, H., Stelte, W., Thomsen, T. P. and Henriksen, U. B. (2013a) The influence of partial oxidation mechanisms on tar destruction in TwoStage biomass gasification. *Fuel* 112. DOI: 10.1016/j.fuel.2012.09.048.

Ahrenfeldt, J., Thomsen, T. P., Henriksen, U. and Clausen, L. R. (2013b) Biomass gasification cogeneration – A review of state of the art technology and near future perspectives. *Applied Thermal Engineering* 50(2): 1407–17. DOI: 10.1016/J.APPLTHERMALENG.2011.12.040.

Álamo, G. del, Sandquist, J., Vreugdenhil, B. J., Aranda Almansa, G. and Carbo, M. (2018) *Implementation of Bio-CCS in biofuels production – IEA Bioenergy Task 33 special project*.

Asadullah, M. (2014a) Barriers of commercial power generation using biomass gasification gas: A review. *Renewable and Sustainable Energy Reviews* 29: 201–15. DOI: 10.1016/j.rser.2013.08.074.

Asadullah, M. (2014b) Biomass gasification gas cleaning for downstream applications: A comparative critical review. *Renewable and Sustainable Energy Reviews* 40: 118–32. DOI: 10.1016/J.RSER.2014.07.132.

Bai, M., Wilske, B., Buegger, F., Bruun, E. W., Bach, M., Frede, H. G. and Breuer, L. (2014) Biodegradation measurements confirm the predictive value of the O:C-ratio for biochar recalcitrance. *Journal of Plant Nutrition and Soil Science* 177: 633–7. DOI: 10.1002/jpln.201300412.

Baliban, R. C., Elia, J. A. and Floudas, C. A. (2013) Novel natural gas to liquids processes: Process synthesis and global optimization strategies. *American Institute of Chemical Engineers Journal* 59(2): 505–31. DOI: 10.1002/aic.13996.

Bang-Møller, C., Rokni, M. and Elmegaard, B. (2011) Exergy analysis and optimization of a biomass gasification, solid oxide fuel cell and micro gas turbine hybrid system. *Energy* 36(8): 4740–52. DOI: 10.1016/j.energy.2011.05.005.

Basu, P. (2013) *Biomass Gasification, Pyrolysis and Torrefraction*. 2nd edn. Dalhouse University: Elsevier Inc.

Basu, P. (2018) Design of biomass gasifiers. In: *Biomass Gasification, Pyrolysis and Torrefaction*. 3rd edn. Academic Press.

Behera, B. K. and Varma, A. (2019) *Bioenergy for Sustainability and Security*. Springer International Publishing.

Bergthorson, J. M. and Thomson, M. J. (2015) A review of the combustion and emissions properties of advanced transportation biofuels and their impact on existing and future engines. *Renewable and Sustainable Energy Reviews* 42: 1393–417. DOI: 10.1016/j.rser.2014.10.034.

Bernical, Q., Joulia, X., Noirot-Le Borgne, I., Floquet, P., Baurens, P. and Boissonnet, G. (2013) Sustainability assessment of an integrated high temperature steam electrolysis-enhanced biomass to liquid fuel process. *Industrial & Engineering Chemistry Research* 52(22): 7189–95. DOI: 10.1021/ie302490y.

Böcker-Riese, B. and Bräkow, D. (2017) Status quo from Biomass gasification CHP-plant systems in the FEE – Society for the promotion of renewable energies. In: *Task33 Workshop. Small Scale Gasification for CHP 02.-04. May 2017, Innsbruck, Berlin, 2017*. FEE – Society for the Promotion of Renewable Energies.

Boerrigter, H., Slort, D. J., Bodenstaff, H., Den Uil, H., Van Ree, R. and Veringa, H. J. (2004) *Gas Cleaning for Integrated Biomass Gasification (BG) and Fischer-Tropsch (FT) Systems Experimental Demonstration of Two BG-FT Systems ("Proof-of-Principle"). Report ECN-C-04-056*.

Brandin, J., Tunér, M. and Odenbrand, I. (2011) *Swedish Energy Agency Report Small Scale Gasification: Gas Engine CHP for Biofuels*. Växjö/Lund.

Bruun, E. W., Hauggaard-Nielsen, H., Ibrahim, N., Egsgaard, H., Ambus, P., Jensen, P. A. and Dam-Johansen, K. (2011) Influence of fast pyrolysis temperature on biochar labile fraction and short-term carbon loss in a loamy soil. *Biomass and Bioenergy* 35(3): 1182–9. DOI: 10.1016/j.biombioe.2010.12.008.

Bruun, E. W., Petersen, C. T., Hansen, E., Holm, J. K. and Hauggaard-Nielsen, H. (2014) Biochar amendment to coarse sandy subsoil improves root growth and increases water retention. *Soil Use and Management* 30(1): 109–18. DOI: 10.1111/sum.12102.

Budai, A., Zimmerman, A., Cowie, A., Webber, J. B. W., Singh, B. P., Glaser, B., Masiello, C. A., Andersson, D., Shields, F., Lehmann, J., Camps Arbestain, M., Williams, M., Sohi, S. and Joseph, S. (2013) *Biochar Carbon Stability Test Method: An Assessment of Methods to Determine Biochar Carbon Stability*. J. Lehmann.

Butera, G., Jensen, S. H. and Clausen, L. R. (2019) A novel system for large-scale storage of electricity as synthetic natural gas using reversible pressurized solid oxide cells. *Energy* 166: 738–54. DOI: 10.1016/j.energy.2018.10.079.

Cheng, C.-H., Lehmann, J., Thies, J. E., Burton, S. D. and Engelhard, M. H. (2006) Oxidation of black carbon by biotic and abiotic processes. *Organic Geochemistry* 37(11): 1477–88. DOI: 10.1016/j.orggeochem.2006.06.022.

Clausen, L. R. (2015) Maximizing biofuel production in a thermochemical biorefinery by adding electrolytic hydrogen and by integrating torrefaction with entrained flow gasification. *Energy* 85: 94–104. DOI: 10.1016/j.energy.2015.03.089.

Clausen, L. R. (2017) Energy efficient thermochemical conversion of very wet biomass to biofuels by integration of steam drying, steam electrolysis and gasification. *Energy* 125. DOI: 10.1016/j.energy.2017.02.132.

Clausen, L. R., Elmegaard, B., Ahrenfeldt, J. and Henriksen, U. B. (2011) Thermodynamic analysis of small-scale dimethyl ether (DME) and methanol plants based on the efficient two-stage gasifier. *Energy* 36(10): 5805–14. DOI: 10.1016/j.energy.2011.08.047.

Dalena, F., Senatore, A., Marino, A., Gordano, A., Basile, M. and Basile, A. (2018) Chapter 1 – Methanol production and applications: An overview. *Methanol - Science and Engineering* 3–28. DOI: 10.1360/zd-2013-43-6-1064.

Danish Technological Institute (2011) *GreenSynFuels – Economical and Technological Statement Regarding Integration and Storage of Renewable Energy in the Energy Sector by Production of Green Synthetic Fuels for Utilization in Fuel Cells*. Final Project Report, March 2011. EUDP project journal number: 64010-0011. Report Editor: Danish Technological Institute.

Di Felice, L. and Micheli, F. (2015) From biomass to SNG. In *Process Intensification for Sustainable Energy Conversion*, pp. 243–66. DOI: 10.1002/9781118449394.ch8.

E4tech (2009) *Review of Technologies for Gasification of Biomass and Wastes – Final Report*. NNFCC Project 09/008.

Energistyrelsen and Energinet.dk (2012) *Technology Data For Energy Plants*. Copenhagen.

European Commission, Joint Research Centre and Institute for Environment and Sustainability (2010) *International Reference Life Cycle Data System (ILCD) Handbook*

- *General Guide for Life Cycle Assessment - Detailed Guidance*. EUR 24708 EN. Luxembourg: Publications Office of the European Union. DOI: 10.2788/38479.

Fechete, I. and Vedrine, J. C. (2017) *Nanotechnology in Catalysis: Applications in the Chemical Industry, Energy Development, and Environment Protection*. 1st edn. Sels, B. and Voorde, M. Van De (Eds). WILEY-VCH Verlag GmbH & Co. KGaA.

FORCE Technology (n.d.) Harbooere District Heating Plant, Denmark.

Gadsbøll, R. Ø. (2018) *Biomass Gasification Polygeneration - Designing and Operating TW-Stage Systems Coupled with Solid Oxide Cell Technology*. Technical University of Denmark.

Gadsbøll, R. Ø., Thomsen, J., Bang-Møller, C., Ahrenfeldt, J. and Henriksen, U. B. (2017) Solid oxide fuel cells powered by biomass gasification for high efficiency power generation. *Energy* 131: 198–206. DOI: 10.1016/j.energy.2017.05.044.

Galadima, A. and Muraza, O. (2015) Hydroisomerization of sustainable feedstock in biomass-to-fuel conversion: A critical review. *International Journal of Energy Research* 39: 741–59. DOI: 10.1002/er.

Gao, Y. N., Liu, S., Zhao, Z., Tao, H. C. and Sun, Z. Y. (2018) Heterogeneous catalysis of CO2 hydrogenation to C2+products. *Wuli Huaxue Xuebao/ Acta Physico - Chimica Sinica* 34(8): 858–72. DOI: 10.3866/PKU.WHXB201802061.

Gatti, M., Martelli, E., Maréchal, F. and Consonni, S. (2014) Multi-objective optimization of a Rectisol® process. In: *24th European Symposium on Computer Aided Chemical Engineering*, Budapest, Hungary, pp. 1249–54. DOI: 10.1016/B978-0-444-63455-9.50043-X.

Graves, C., Martinez, L. and Sudireddy, B. R. (2016) High performance Nano-Ceria electrodes for solid oxide cells. *ECS Transactions* 72(7): 183–92. DOI: 10.1149/07207.0183ecst.

Hansen, V., Müller-Stöver, D., Ahrenfeldt, J., Holm, J. K., Henriksen, U. B. and Hauggaard-Nielsen, H. (2015) Gasification biochar as a valuable by-product for carbon sequestration and soil amendment. *Biomass and Bioenergy* 72: 300–8.

Hansen, V., Müller-Stöver, D., Munkholm, L. J., Peltre, C., Haugaard-Nielsen, H. and Jensen, L. S. (2016) The effect of straw and wood gasification biochar on carbon sequestration, selected soil fertility indicators and functional groups in soil: An incubation study. *Geoderma* 269: 99–107. DOI: 10.1016/j.geoderma.2016.01.033.

Hardisty, P. E., Sivapalan, M. and Brooks, P. (2011) The environmental and economic sustainability of carbon capture and storage. *International Journal of Environmental Research and Public Health* 8(5): 1460–77. DOI: 10.3390/ijerph8051460.

Hasler, P. and Nussbaumer, T. (1999) Gas cleaning for IC engine applications from fixed bed biomass gasification. *Biomass and Bioenergy* 16(6): 385–95. DOI: 10.1016/S0961-9534(99)00018-5.

Heeb, R. (2010) *Up-date on the Harboøre Technology May 2010 IDA Energi, Status af Danske aktiviteter* (May).

Henriksen, U., Ahrenfeldt, J., Jensen, T. K., Gøbel, B., Bentzen, J. D., Hindsgaul, C. and Sørensen, L. H. (2006) The design, construction and operation of a 75kW two-stage gasifier. *Energy* 31(10–11): 1542–53. DOI: 10.1016/j.energy.2005.05.031.

Heyne, S., Liliedahl, T. and Marklund, M. (2013) Biomass gasification - A synthesis of technical barriers and current research issues for deployment at large scale. Report No 2013:5, f3 The Swedish Knowledge Centre for Renewable Transportation Fuels and Foundation, Sweden. Available at: www.f3centre.se (April 2013). DOI: 10.13140/RG.2.1.2593.8406.

Hrbek, J. (2016) *Status Report on Thermal Biomass Gasification in Countries Participating in IEA Bioenergy Task 33*. Vienna.

Huber, M. (2017) The carbon makes the difference - Decentralized wood power plant with valuable carbon as by-product. In: *IEA Bioenergy Task 33 Workshop 'Small scale gasification CHP'*, Innsbruck, Austria, 2017, p. 23. SynCraft Das Holzkraftwerk & IEA Task 33.

IPCC (2013) *Climate Change 2013: The Physical Science Basis. Contribution of Working Group I to the Fifth Assessment Report of the Intergovernmental Panel on Climate Change*. Stocker, T. F., Qin, D., Plattner, G.-K., et al. (Eds). Cambridge and New York: Cambridge University Press.

Jensen, S. H., Graves, C., Mogensen, M., et al. (2015) Large-scale electricity storage utilizing reversible solid oxide cells combined with underground storage of CO_2 and CH_4. *Energy Environment Science* 8: 2471–9. DOI: 10.1039/C5EE01485A.

Jensen, S. H., Sun, X., Ebbesen, S. D. and Chen, M. (2016) Pressurized operation of a planar solid oxide cell stack. *Fuel Cells* 16(2): 205–18. DOI: 10.1002/fuce.201500180.

Jensen, S. H., Langnickel, H., Hintzen, N., Chen, M., Hauch, A., Butera, G. and Clausen, L. R. (2017) Pressurized reversible operation of a 30-cell solid oxide cell stack using carbonaceous gases. In: *European Fuel Cell Technology & Applications Conference - Piero Lunghi Conference*, 2017. ENEA.

Kapetaki, Z., Brandani, P., Brandani, S. and Ahn, H. (2015) Process simulation of a dual-stage Selexol process for 95% carbon capture efficiency at an integrated gasification combined cycle power plant. *International Journal of Greenhouse Gas Control* (39): 17–26. DOI: 10.1016/j.ijggc.2015.04.015.

Krylova, A. Y. (2014) Products of the Fischer-Tropsch synthesis (a review). *Solid Fuel Chemistry* 48(1): 22–35. DOI: 10.3103/S0361521914010030.

Kuligowski, K., Poulsen, T. G., Stoholm, P., Pind, N. and Laursen, J. (2008) Nutrients and heavy metals distribution in thermally treated pig manure. *Waste Management & Research* 26(4): 347–54. DOI: 10.1177/0734242X08090075.

Kuligowski, K., Gilkes, R. J., Poulsen, T. G. and Yusiharni, B. E. (2012) Ash from the thermal gasification of pig manure-effects on ryegrass yield, element uptake, and soil properties. *Soil Research* 50(5): 406–14. DOI: 10.1071/SR12075.

Larson, E. D., Fiorese, G., Liu, G., Williams, R. H., Kreutz, T. G. and Consonni, S. (2009) Co-production of synfuels and electricity from coal + biomass with zero net carbon emissions: An Illinois case study. *Energy Procedia* 1(1): 4371–8. DOI: 10.1016/j.egypro.2009.02.251.

Lehmann, J., Gaunt, J. and Rondon, M. (2006) Bio-char sequestration in terrestrial ecosystems - A review. *Mitigation and Adaptation Strategies for Global Change* 11(2): 395–419. DOI: 10.1007/s11027-005-9006-5.

Li, X., Rubæk, G. H., Müller-Stöver, D. S., Thomsen, T. P., Ahrenfeldt, J. and Sørensen, P. (2017) Plant availability of phosphorus in five gasification biochars. *Frontiers in Sustainable Food Systems* 1: 1–12. DOI: 10.3389/fsufs.2017.00002.

Liu, G., Larson, E. D., Williams, R. H., Kreutz, T. G. and Guo, X. (2011) Making Fischer-Tropsch fuels and electricity from coal and biomass: Performance and cost analysis. *Energy and Fuels* 25(1): 415–37. DOI: 10.1021/ef101184e.

Mackay, J. E., Cavagnaro, T. R., Jakobsen, I., Macdonald, L. M., Grønlund, M., Thomsen, T. P. and Müller-Stöver, D. S. (2016) Evaluation of thermally processed sewage sludge as a soil P fertiliser. *Environmental Science and Technology* Submitted.

Maitlis, P. M. (2013) *Greener Fischer-Tropsch Processes for Fuels and Feedstocks*. 1st edn. Maitlis, P. M. and Klerk de, A. (Eds). Weinheim, Germany: WILEY-VCH Verlag GmbH & Co. KGaA.

Matas Güell, B., Sandquist, J. and Sørum, L. (2012) Gasification of biomass to second generation biofuels: A review. *Journal of Energy Resources Technology* 135(1): 014001. DOI: 10.1115/1.4007660.

Molino, A., Chianese, S. and Musmarra, D. (2016) Biomass gasification technology: The state of the art overview. *Journal of Energy Chemistry* 25(1): 10–25. DOI: 10.1016/j.jechem.2015.11.005.

Molino, A., Larocca, V., Chianese, S. and Musmarra, D. (2018) Biofuels production by biomass gasification: A review. *Energies* 11(4). DOI: 10.3390/en11040811.

Müller-Stöver, D., Grønlund, M., Jakobsen, I., Ahrenfeldt, J., Holm, J. K. and Hauggaard-Nielsen, H. (2014) Ashes from low-temperature gasification as sustainable phosphorus fertilizer. *4th Sustainable Phosphorus Summit. Montpellier*, France, 1–3 September.

Narayan, V., Jensen, P. A., Henriksen, U. B., Glarborg, P., Lin, W. and Nielsen, R. G. (2016) Defluidization in fluidized bed gasifiers using high-alkali content fuels. *Biomass and Bioenergy* 91: 160–74. DOI: 10.1016/j.biombioe.2016.05.009.

Nguyen, B. T., Lehmann, J., Kinyangi, J., Smernik, R., Riha, S. J. and Engelhard, M. H. (2008) Long-term black carbon dynamics in cultivated soil. *Biogeochemistry* 89(3): 295-308. DOI: 10.1007/s10533-008-9220-9.

Nguyen, T. L. T., Hermansen, J. E. and Mogensen, L. (2013) Environmental performance of crop residues as an energy source for electricity production: The case of wheat straw in Denmark. *Applied Energy* 104: 633–41. DOI: 10.1016/j.apenergy.2012.11.057.

Omosun, A., Bauen, A., Brandon, N., Adjiman, C. S. and Hart, D. (2004) Modelling system efficiencies and costs of two biomass-fuelled SOFC systems. *Journal of Power Sources* 131(1–2): 96–106. DOI: 10.1016/J.JPOWSOUR.2004.01.004.

Panopoulos, K. D., Fryda, L. E., Karl, J., Poulou, S. and Kakaras, E. (2006) High temperature solid oxide fuel cell integrated with novel allothermal biomass gasification: Part I: Modelling and feasibility study. *Journal of Power Sources* 159(1): 570–85. DOI: 10.1016/J.JPOWSOUR.2005.12.024.

Porosoff, M. D., Yan, B. and Chen, J. G. (2016) Catalytic reduction of CO2by H2for synthesis of CO, methanol and hydrocarbons: Challenges and opportunities. *Energy and Environmental Science* 9(1): 62–73. DOI: 10.1039/c5ee02657a.

Puerari, F., Bosio, B. and Heyen, G. (2014) Energy efficiency optimisation in different plant solutions for Methanol production from biomass gasification. *Chemical Engineering Science* 37: 301-6. DOI: 10.3303/CET1437051.

Røngaard Clausen, L. (2011) *Design of Novel DME Methanol Synthesis Plants Based on Gasification of Biomass*. Technical University of Denmark.

Rönsch, S., Schneider, J., Matthischke, S., Schlüter, M., Götz, M., Lefebvre, J., Prabhakaran, P. and Bajohr, S. (2016) Review on methanation – From fundamentals to current projects. *Fuel* 166: 276–96. DOI: 10.1016/j.fuel.2015.10.111.

Rüegsegger, M. and Kast, M. (2018) *Valorization of By-Products from Small Scale Thermal Gasification Valorization of By-Products from Small Scale Thermal – White Paper elaborated 2016–2018 under IEA Bioenergy, Task 33 Thermal Gasification of Biomass and waste*. Fahrni, Switzerland: IEA Bioenergy Task 33.

Ruiz, J. A. A., Juárez, M. C. C., Morales, M. P. P., et al. (2013) Biomass gasification for electricity generation: Review of current technology barriers. *Renewable and Sustainable Energy Reviews* 18. Pergamon: 174–83. DOI: 10.1016/j.rser.2012.10.021.

Samavati, M. (2018) *Design and Analysis of Solid Oxide Electrolysis-Based Systems for Synthetic Liquid Fuels Production*. Environmental Science. DOI: 10.3390/bios6010001.

Sansaniwal, S. K., Pal, K., Rosen, M. A. and Tyagi, S. K. (2017) Recent advances in the development of biomass gasification technology: A comprehensive review. *Renewable and Sustainable Energy Reviews* 72(December 2015): 363–84. DOI: 10.1016/j.rser.2017.01.038.

Shackley, S., Carter, S., Knowles, T., Middelink, E., Haefele, S., Sohi, S., Cross, A. and Haszeldine, S. (2012) Sustainable gasification-biochar systems? A case-study of rice-husk gasification in Cambodia, Part I: Context, chemical properties, environmental and health and safety issues. *Energy Policy* 42: 49–58. DOI: 10.1016/j.enpol.2011.11.026.

Sikarwar, V. S., Zhao, M., Clough, P., Yao, J., Zhong, X., Memon, M. Z., Shah, N., Anthony, E. J. and Fennell, P. S. (2016) An overview of advances in biomass gasification. *Energy Environ. Science* 9(10): 2939–77. DOI: 10.1039/C6EE00935B.

Singh, R. N., Singh, S. P., Balwanshi, J. B. and Vishwavidhlya, D. S. (2014) Tar removal from producer gas: A review. *Research Journal of Engineering Sciences* 3(10): 16–22. DOI: 10.1002/elan.200603785.

Skoglund, N., Bäfver, L., Fahlström, J., Holmén, E. and Renström, C. (2016) Fuel design in co-combustion of demolition wood chips and municipal sewage sludge. *Fuel Processing Technology* 141: 196–201. DOI: 10.1016/j.fuproc.2015.08.037.

Stoholm, P., Nielsen, R. G., Sarbæk, L., Tobiasen, L., Fock, M. W., Richardt, K., Sander, B., Wolff, L. and Henriksen, U. B. (2002) The Low Temperature CFB gasifier - Further test results and possible applications. In: *Proceedings of the 12. European Biomass Conference*, 2002, pp. 706–9.

Stoholm, P., Nielsen, R. G., Nielsen, M. B., Sander, B., Nørholm, N., Krogh, J., Henriksen, U. B. and Qvale, B. (2007) *LT-CFB forgasser 500 kW forsøg m.v. Slutrapport (Eng.: LT-CFB gasifier 500 kW experiment and more. Final report)*. 4833. Denmark: DFBT, Anhydro, FORCE Technology, Elsam Engineering, DTU MEK.

Sun, L. and Smith, R. (2013) Rectisol wash process simulation and analysis. *Journal of Cleaner Production* 39: 321–8. DOI: 10.1016/j.jclepro.2012.05.049.

Sutton, D., Kelleher, B. and Ross, J. R. H. (2001) Review of literature on catalysts for biomass gasification. *Fuel Processing Technology* 73(3): 155–73. DOI: 10.1016/S0378-3820(01)00208-9.

Thompson, R., Thomsen, T. P. and Müller-Stöver, D. (2016) Plant phosphorus availability in thermally treated sewage sludge. In: *8th International Phosphorus Workshop (IPW8) 12–16 September 2016*, Rostock (Germany), 2016. The Leibniz ScienceCampus Phosphorus Research Rostock.

Thomsen, T. P., Hauggaard-Nielsen, H., Bruun, E. W. and Ahrenfeldt, J. (2011) *The Potential of Pyrolysis Technology in Climate Change Mitigation-Influence of Process Design and-Parameters, Simulated in SuperPro Designer Software*. Denmark. Forskningscenter Risoe. Roskilde, Denmark.

Thomsen, T. P., Ravenni, G., Holm, J. K., Ahrenfeldt, J., Hauggaard-Nielsen, H. and Henriksen, U. B. (2015) Screening of various low-grade biomass materials for low temperature gasification: Method development and application. *Biomass and Bioenergy* 79: 128–44.

Thomsen, T. P., Ahrenfeldt, J., Henriksen, U. B., Hauggaard-Nielsen, H. and Holm, J. K. (2016) *Closing the Loop - Utilization of Secondary Resources by Low Temperature Thermal Gasification*. Technical University of Denmark, Lyngby, Denmark.

Thomsen, T. P., Sárossy, Z., Gøbel, B., Stoholm, P., Ahrenfeldt, J. and Frandsen, F. J. (2017a) Low temperature circulating fluidized bed gasification and co-gasification of municipal sewage sludge. Part 1: Process performance and gas product characterization. *Waste Management* 66: 123–33. DOI: 10.1016/j.wasman.2017.04.028.

Thomsen, T. P., Hauggaard-Nielsen, H., Gøbel, B., Stoholm, P., Ahrenfeldt, J., Henriksen, U. B. and Müller-Stöver, D. S. (2017b) Low temperature circulating fluidized bed gasification and co-gasification of municipal sewage sludge. Part 2: Evaluation of ash materials as phosphorus fertilizer. *Waste Management* 66: 145–54. DOI: 10.1016/J.WASMAN.2017.04.043.

Thomsen, T. P., Sárossy, Z., Ahrenfeldt, J., Henriksen, U. B., Frandsen, F. J. and Müller-Stöve, D. S. (2017c) Changes imposed by pyrolysis, thermal gasification and incineration on composition and phosphorus fertilizer quality of municipal sewage sludge. *Journal of Environmental Management* 198: 308–18. DOI: 10.1016/j.jenvman.2017.04.072.

Tijmensen, M. J. A., Faaij, A. P. C., Hamelinck, C. N. and van Hardeveld, M. R. M. (2002) Exploration of the possibilities for production of Fischer Tropsch liquids and power via biomass gasification. *Biomass and Bioenergy* 23(2): 129–52. DOI: 10.1016/S0961-9534(02)00037-5.

Torres, W., Pansare, S. S. and Goodwin, J. G. (2007) Hot gas removal of tars, ammonia, and hydrogen sulfide from biomass gasification gas. *Catalysis Reviews* 49(4): 407–56. DOI: 10.1080/01614940701375134.

Vakkilainen, E., Kuparinen, K. and Heinimö, J. (2013) *Large Industrial Users of Energy Biomass*. Lappeenranta, Finland.

Valmet (2015) *The World's Largest Biomass Gasifier Reduces Emissions at Vaskiluodon Voima*. https://www.valmet.com/media/articles/energy/the-worlds-largest-biomass-gasifier-reduces-emissions-at-vaskiluodon-voima/.

Valmet (n.d.) *Valmet CFB Gasifier for Biomass and Waste*. https://www.valmet.com/energyproduction/gasification/.

van der Meijden, C. M., Veringa, H. J. and Rabou, L. P. L. M. (2010) The production of synthetic natural gas (SNG): A comparison of three wood gasification systems for energy balance and overall efficiency. *Biomass and Bioenergy* 34(3): 302–11. DOI: 10.1016/j.biombioe.2009.11.001.

Venvik, H. J. and Yang, J. (2017) Catalysis in microstructured reactors: Short review on small-scale syngas production and further conversion into methanol, DME and Fischer-Tropsch products. *Catalysis Today* 285: 135–46. DOI: 10.1016/j.cattod.2017.02.014.

Vitasari, C. R., Jurascik, M. and Ptasinski, K. J. (2011) Exergy analysis of biomass-to-synthetic natural gas (SNG) process via indirect gasification of various biomass feedstock. *Energy* 36(6): 3825–37. DOI: 10.1016/j.energy.2010.09.026.

Williams, R. H., Liu, G., Kreutz, T. G. and Larson, E. D. (2011) Coal and biomass to fuels and power. *Annual Review of Chemical and Biomolecular Engineering* 2(1): 529–53. DOI: 10.1146/annurev-chembioeng-061010-114126.

Woolcock, P. J. and Brown, R. C. (2013) A review of cleaning technologies for biomass-derived syngas. *Biomass and Bioenergy* 52: 54–84. DOI: 10.1016/J.BIOMBIOE.2013.02.036.

Zhang, Z., Chen, B., Chen, A. and Zhao, W. (2013a) Barriers to commercialization development of crop straw gasification technology in China and promoting policy design. *Energy Sources, Part B: Economics, Planning, and Policy* 8(3): 279–89. DOI: 10.1080/15567240903452071.

Zhang, Q., Deng, W. and Wang, Y. (2013b) Recent advances in understanding the key catalyst factors for Fischer-Tropsch synthesis. *Journal of Energy Chemistry* 22(1): 27–38. DOI: 10.1016/S2095-4956(13)60003-0.

Zwart, R., Heijden, S. Van Der, Emmen, R., Bentzen, J. D., Stoholm, P. and Krogh, J. (2010) *Tar Removal from Low-temperature Gasifiers*. ECN-E-10-008. The Netherlands: ERA-NET Bioenergy.

Chapter 2

Fast pyrolysis for biofuel production

David Shonnard, Olumide Winjobi and Daniel Kulas, Michigan Technological University, USA

1 Introduction

Pyrolysis is a thermal depolymerization and fragmentation process carried out at moderate temperatures (450–700°C) in the absence of oxygen (or air) and with a wide range of reaction residence times (1 s to minutes).[1] Pyrolysis is a process that can convert a wide range of solid materials, ranging from woody biomass and algae to waste plastics, into complex mixtures of low-molecular-weight organic molecules representative of the polymeric parent material. The thermal reactions in pyrolysis occur in stages, each corresponding to higher reaction severity, where increasing severity refers to longer reactor residence times and higher reaction temperatures. Primary reactions at low severity produce gases (CO_2, CO, H_2O), organic vapors, and liquid products. Secondary reactions act on primary products by decomposing these larger molecules into small-molecular-weight gaseous and liquid compounds as well as char (a solid containing mostly carbon). Tertiary reactions further the degradation process to produce synthesis gas (CO_2, CO, H_2O, H_2) and soot.[2] Three primary co-products are present at the pyrolysis reactor exit that include synthesis gas (CO, CO_2, H_2O), a liquid bio-oil, and solid char. The relative proportion of each of these is dependent on reaction severity.

http://dx.doi.org/10.19103/AS.2019.0027.06

Fast pyrolysis is a prominent member of a large (and growing) set of process technologies for producing liquid biofuels from biomass feedstocks. Figure 1 shows liquid transportation biofuel conversion process technologies organized into broad categories, conventional versus advanced technologies as well as biochemical and thermochemical. Fast pyrolysis is a member of advanced thermochemical biofuel process technologies and is favorably suited for conversion of lignocellulosic (woody) feedstocks. Woody biomass has advantages because it does not directly compete with food supplies in the production of biofuels, which is an often-cited criticism of biofuels. The intermediate product from fast pyrolysis is a pyrolysis bio-oil, often referred as py-oil. This intermediate, after chemical stabilizers are added and with further processing,[3] can be used either directly for combustion to generate heat or produce power[4] or upgraded through catalytic hydrotreatment to produce 'drop-in' hydrocarbon biofuel products, predominantly in the boiling range of gasoline and diesel.[5] The goal of producing a drop-in biofuel is to avoid expensive engine modifications for high biofuel blends and to enable the blending of biofuel into fossil-fuel supplies at any desired level. In pyrolysis, it is particularly advantageous if the woody feedstock is relatively dry and contains low mineral ash content. Otherwise, high moisture feedstocks require expensive and energy-intensive drying in order to process them through fast pyrolysis, and ash (alkali earth metals) often has catalytic effects resulting in lower liquid bio-oil yields and more gas production.[6,7]

Pyrolysis reactions on biomass feedstocks are carried out under a variety of temperatures, residence times of solids and vapors, and in the presence of different solvents (see Table 4 in Mohan et al. 2006).[1] *Carbonization* involves slowly heating biomass over several days at approximately 400°C. This yields charcoal as the main product, but also significant amounts of gas (CO_2, H_2O, CO).[8] *Conventional pyrolysis* involves biomass residence times of between 5 and 30 minutes, a slow heating rate, and reaction temperature up to 600°C in which the products include synthesis gas, bio-oil, and char. *Fast pyrolysis* produces mostly bio-oil by employing high heating rates, short residence times of 0.5–5 s, and temperatures of 425–650°C and produces mostly bio-oil, but small amounts of synthesis gas and char. *Ultra pyrolysis* produces mostly gaseous products and chemicals through very high heating rates, high temperature of ~1000°C, and ultrashort residence times (<0.5 s).[9,10] *Vacuum pyrolysis*, similar to fast pyrolysis, produces predominantly bio-oil at 400°C, with a 2–30 s residence time and medium heating rate.[11] *Hydro-pyrolysis* is similar to fast pyrolysis in terms of reaction temperature, heating rate, and residence time except that hydrogen is present in the vapor phase at high pressure.[12] Its advantage is in hydrogenation of free radicals to stabilize these fragments before they repolymerize and form char (Basu, 2010 - Table 3.2).[13] Thus, more bio-oil yield is normally observed. Methane is used as the reaction medium

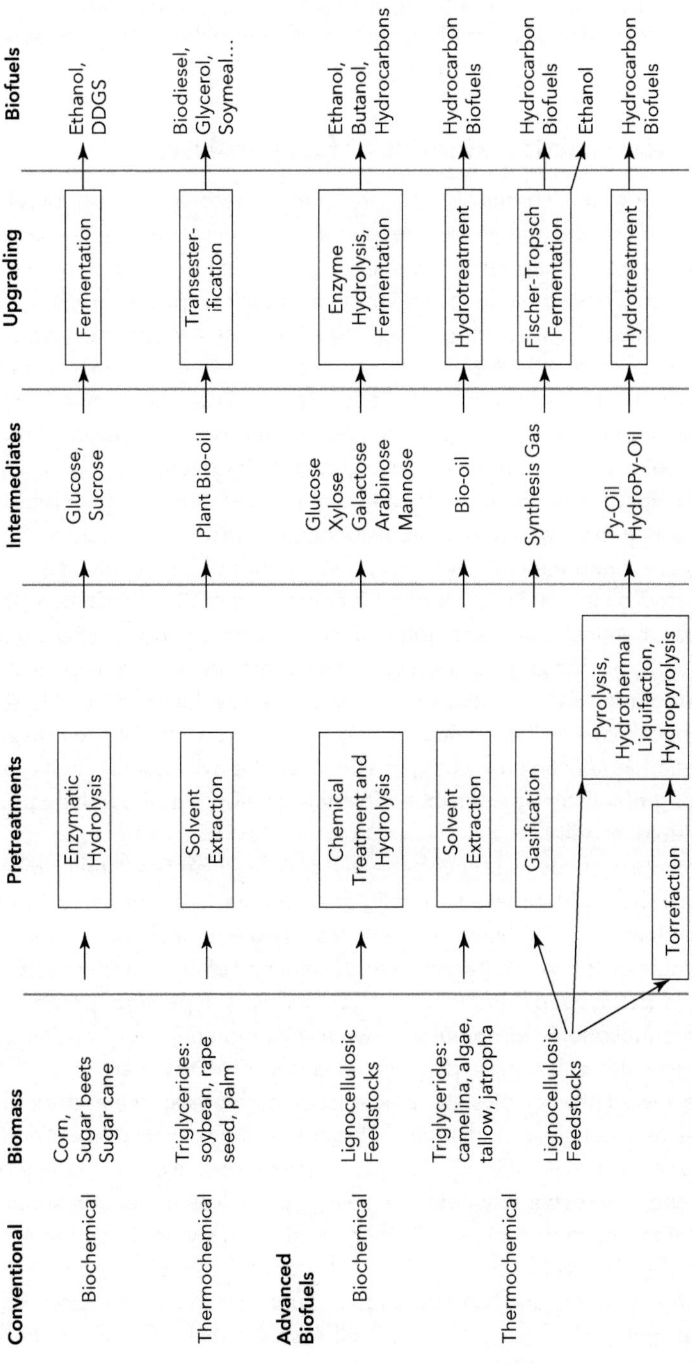

Figure 1 Conversion process technologies for production of liquid transportation biofuels from a range of biomass types.

in *methano-pyrolysis*, but the main products are chemicals (ethylene, ethane, benzene) rather than fuels, and reactor conditions include a high heating rate, a residence time of <10 s and temperatures of >700°C.[14]

2 Chemical reaction kinetics of fast pyrolysis

Understanding the chemical reaction kinetics of biomass fast pyrolysis is complicated by the many diverse phenomena occurring in the process.[15] Firstly, the reactions happen simultaneously with the transfer of energy into the biomass particle resulting in the evolution of chemical species from the solid to the vapor phase. If particle size is too large, heat transfer into the solid particle will not keep pace with chemical reaction, which in this case, would proceed from the outside of the particle to the interior. Once the evolved products are in the vapor phase, reactions continue on the primary species generated from the solid, breaking them down into smaller fragments through a series of reaction steps. In addition to these phenomena, there is a change of phase from solid to liquid from which evolution of chemical products occurs as biomass particle temperature increases. Finally, the kinetics of fast pyrolysis are further complicated by the fact that biomass is comprised of three main components - cellulose, hemicellulose, and lignin. Each of these components will respond differently to increasing temperature. For example, hemicellulose is more readily degraded than cellulose or lignin, and so evolution of volatile species from hemicellulose will occur first during fast pyrolysis. Finally, the presence of ash in the biomass can have catalytic effects on the pyrolysis reactions, leading to cracking of volatile species that affect the distribution of pyrolysis products to gas, liquid, or solid char.

A number of reviews have recently appeared on the kinetic modeling of fast pyrolysis of biomass and cellulose. Sharma et al. reviewed the status of mathematical modeling of biomass pyrolysis including models for hydrodynamics of solid particles and catalytic effects in pyrolysis.[15] The effects of pyrolysis parameters such as biomass type and pretreatments, reaction atmosphere, temperature, heating rate, and vapor residence time on lignocellulosic biomass pyrolysis was reviewed by Kan et al.[16] Wang et al. reviewed the pyrolysis characteristics of biomass main components: cellulose, hemicellulose, and lignin including recently developed experimental techniques of Py-GC-MS/FID, *in situ* spectroscopy, isotopic labeling and others, and macroscopic kinetic modeling as well as molecular simulations of elemental reaction routes.[17] Further insights on pyrolysis biomass kinetic modeling was provided by Papari and Hawboldt[18] as well as reactor models combining kinetics and computational fluid dynamics. Burnham et al.[19] reviewed global chemical kinetics of cellulose thermal decomposition noting the sigmoidal reaction character consistent with sequential, nucleation-growth, or random-scission mechanisms. The readers are referred to these thorough reviews for more details, but here we will present only the key features of kinetic modeling.

In order to demonstrate the range in complexity of biomass kinetic modeling, we briefly present results from three approaches that include first-order model, six-step sequential model, and detailed mechanistic model. Klemetsrud et al.[20] compared a global first-order kinetic model of volatiles produced from poplar biomass fast pyrolysis with a six-step sequential model. The first-order model was adapted from Shafizadeh and Chin[21] (Fig. 2a) while the six-step sequential model was from Klinger et al.[22] (Fig. 2b), where R represents solid residue produced from each of the reaction steps, P_i the volatile products and α_i stoichiometric coefficients for each volatile species in each reaction step. Integration of mass balance equations for the first-order model yield exponential functions with respect to time for each volatile species while integration of the six-step model provides sigmoidal behavior. These models were tested against poplar biomass (0.5 mg) micropyrolysis volatiles evolution data obtained using a CDS model 5200 pyroprobe coupled with GC/MS (Thermofisher GC Trace Ultra - DSQ II) at 500°C with heating rate of 1000°C/s with negligible vapor residence time (primary reaction products only). Experiments were conducted for 1, 2, 3, 5, 7, 10, 15, and 20 s to investigate volatiles production with respect to time. Key results from this study showed that (1) volatiles from hemicellulose (acetic acid) appeared earliest in the time course and achieved maximum yields within 2 s, (2) volatiles from cellulose and lignin (methyl syringol, guaiacol) appeared later and required longer times to achieved maximum yields (8 s), and (3) the six-step sequential model exhibited sigmoidal behavior and fit the volatiles data better than the first-order model. Broadbelt and coworkers (Mayes and Broadbelt; Zhou et al.)[23,24] reported a mechanistic model for fast pyrolysis

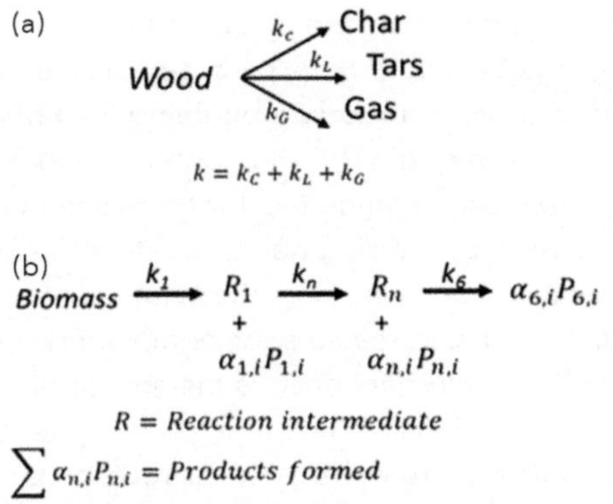

Figure 2 (a) First-order model (tars are liquid products) and (b) multicomponent six-step model.

of glucose-based carbohydrates (cellobiose, cellulose) involving 342 reactions and 103 chemical species. Their model results for molar compositions, which were compared to experimental yield data, do not decay exponentially with time but rather exhibit characteristics of a sequential reaction mechanisms.

3 Fast pyrolysis process technologies

Process technologies for fast pyrolysis vary widely depending on the objectives for converting solid materials into lower-molecular-weight products. In the area of research, the objectives are to produce data that can allow a more complete understanding of the important reaction mechanisms, validate kinetic models of these reactions, and generate kinetic parameters. Thermogravimetric analysis (TGA) is a mainstream instrument for determining the effects of increasing temperature on mass loss from solid materials. However, it is important to point out that TGA does not represent fast pyrolysis conditions due to slow heating rates and very long residence times. Micropyrolysis with gas chromatrography/mass spectroscopy is used extensively to study rates of evolution of chemical species from biomass and other solid materials as a function of reaction conditions such as temperature, residence time, gas phase atmosphere, effects of catalysts, and other factors. However, there are concerns over the very small sample size (0.1–1.0 mg) and whether micropyrolysis data represents the behavior of bulk materials that are heterogeneous. One of the advantages of micropyrolysis is that many experiments can be conducted in the same amount of time as a single experiment using larger-scale pyrolysis equipment. A recent advancement in micropyrolysis reactor design has overcome one of the limitations of standard micropyrolysis, negligible vapor residence time. Figure 3 shows a diagram of a two-stage micropyrolysis reactor (TSMR) comprised of a pyroprobe coupled to a constant-temperature tubular reactor,[25] which controls vapor residence time between 1 and 6 s, either at the same or different temperature from the pyroprobe.

Biomass fast pyrolysis reactors at the pilot, demonstration, and commercial scales include various designs, including bubbling fluid bed, circulating fluid bed, rotating cone, and rotary kiln.[16,26] The bubbling fluid bed and circulating fluid bed, together the most common reactor types commercially, operate on similar principles of using a fluidizing gas to create a mixed environment for contact between biomass particles with the heat carrier particles, normally sand. Residence times in these reactors for both vapors and solid are short, only a few seconds. The main difference is the source of energy for the fast pyrolysis reaction. In a bubbling fluid bed, hot gas from the combustion of non-condensable gases obtained from the pyrolysis reaction is the energy source. The hot combustion gases flow through heating coils inside the fluidized bed, and then are exhausted after transferring most of the stream's sensible

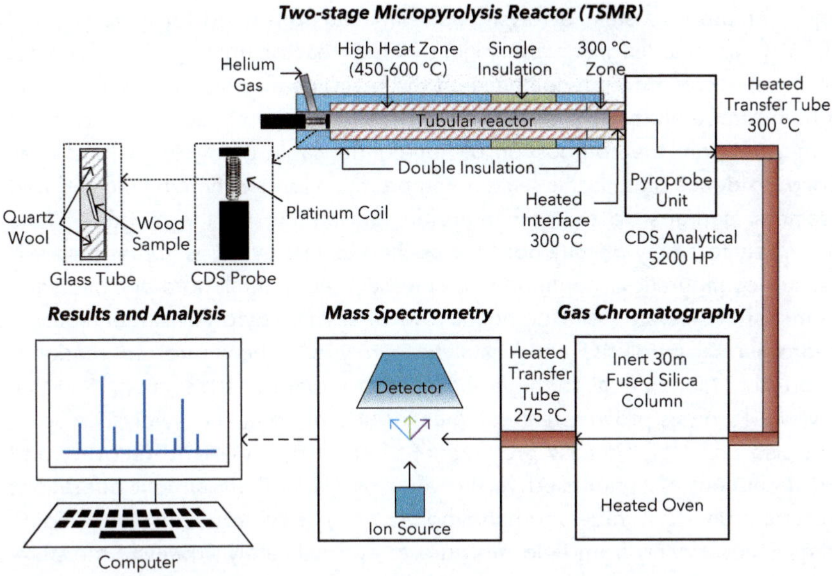

Figure 3 Experimental apparatus and methods for pyrolysis and data collection of solid samples with vapor residence time (1–6 s).

heat to the reaction zone. In a circulating fluid bed reactor (Fig. 4), a reactor type of importance commercially, sand and char are recovered together from the reaction zone and then combusted in an external hot sand recovery circuit. The hot sand, the source of heat for pyrolysis, then flows back to the reactor and combines with fresh biomass particles. Biomass particles for these reactors should be in the size range from 2 to 6 mm. Similar to fluidized beds, a rotating cone pyrolysis reactor also makes use of contact between hot sand

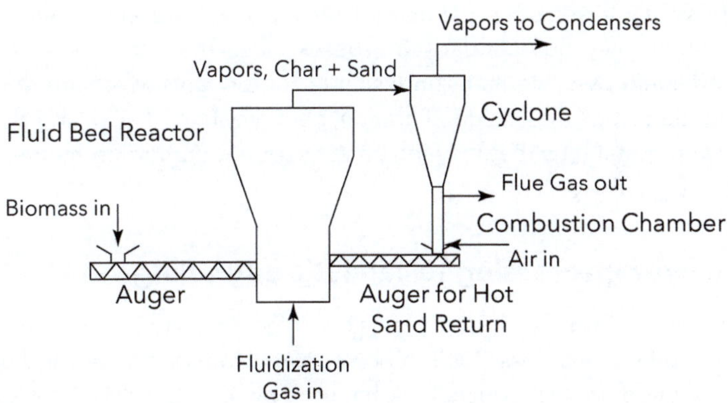

Figure 4 Schematic diagram of a circulating fluidized bed pyrolysis reactor.

and biomass to sustain pyrolysis, but eliminates the need for inert gas and downstream handling. In a rotary kiln pyrolysis reactor, heating of the biomass (20 mm) occurs through contact with the reactor wall which is maintained at a high temperature by a circulating stream of hot combustion gas. Hot gas originates from the combustion of non-condensable pyrolysis gas and char. Solid residence time in this reactor can be much longer than in fluidized bed reactors, perhaps up to 1 h,[27] therefore depending on the length of time these reactors may or may not be classified as fast pyrolysis. Other pyrolysis reactor configurations continue to be investigated at the laboratory, pilot, and demonstration scales,[16] including microwave-assisted pyrolysis, auger reactors, hydrothermal pyrolysis,[28] and catalytic pyrolysis.[29] The interested reader is referred to Mohan et al. for high-quality reactor diagrams in section 15 of that review.[1] Pyrolysis process technologies include not only the pyrolysis reactor, but also units for biomass pretreatment before pyrolysis and recovery and posttreatment of condensed vapors after pyrolysis. Pretreatment operations include biomass storage and handling, a primary size reduction step yielding large chips, drying from field moisture to approximately 10% (wt.) moisture, and a final biomass size reduction to reactor specifications.

Final size reduction is very energy intensive, contributing significantly to operating costs and to the minimum selling price of pine-derived py-oil[30] as well as to raising greenhouse gas (GHG) emissions by up to 30 g CO_2-eq./MJ py-oil.[31] Torrefaction decreases size reduction energy use by up to 90%, however this improvement comes at the cost of lower py-oil yield.[30] Downstream from the pyrolysis reactor are several unit operations for liquid py-oil recovery and liquid product stabilization. Hot vapors from pyrolysis are condensed to recover the liquid product in a single unit or in several units operated at different temperatures. The first condenser would operate at the highest condensation temperature in which only the highest-molecular-weight py-oil components are recovered. Subsequent condensers operate at progressively lower temperatures for recovery of lower-molecular-weight products until only non-condensable gases leave the last condenser. The recovered liquid py-oil contains undesirable components which are removed in stabilization steps that include filtration, neutralization of acid, and removal of dissolved metal ions that may inhibit subsequent upgrading hydrocarbon biofuel catalysts.[3]

4 From hydroprocessing to catalytic upgrading

Bio-oil produced from pyrolysis of plants has been shown to be suitable for use in boilers, co-fired with fossil fuels in power plants to provide heat and power, respectively. Red Arrow Products, USA, has used bio-oil to produce heat for over 10 years with bio-oil sprayed into a 5-MW cyclone burner through a steel nozzle and vaporized with air.[32] Co-firing of bio-oil with natural gas was demonstrated in the Netherlands, by BTG with bio-oil constituting greater than 1% of the feed in a 350 MWe natural gas-fired power station with minimal retrofitting and high system reliability.[33] However, bio-oil is not compatible to be used directly as a transportation fuel without making modifications to current vehicle engines and components. As shown in Table 1, there are several differences in the physical properties of bio-oil and crude oil that prevents bio-oil from being utilized directly as a transportation fuel. Bio-oil is highly oxygenated, resulting in lower energy density. It is also acidic and more unstable (suffers from aging) relative to fossil gasoline.

Upgrade of bio-oil to a 'drop-in' biofuel that is compatible with current pipeline and transportation infrastructure focuses on removing the oxygen in bio-oil, reducing the molecular weight to increase volatility and prevent aging. The main methods for the catalytic upgrade of bio-oil are hydroprocessing and catalytic vapor cracking (also referred to as hydrocracking). Upgrade of pyrolysis bio-oil through hydroprocessing is achieved through catalytic reaction with hydrogen to remove the oxygen in bio-oil as water. It is therefore usually referred to as hydrodeoxygenation (HDO). The process is typically at high pressure (up to 20 MPa) and moderate temperature (up to 400°C).[36,37] Precious metal catalysts on supports that are more stable in water environment have been shown to perform better than catalysts such as sulfide cobalt molybdenum supported on alumina for HDO of bio-oils.[38,39] Recent studies have also reported that HDO of bio-oils carried out in multiple sequential processes with increasingly severe conditions results in better yield than a one-step HDO step.[38,40] Catalytic vapor cracking of bio-oil over zeolites removes oxygen mainly as CO_2 (decarboxylation) and water

Table 1 Properties comparison between bio-oil and crude oil

Physical property	Bio-oil[1]	Crude oil[34]
Moisture content (wt %)	15–30	0.1
pH	2.5	–
Specific gravity	1.2	0.86
Elemental analysis		
C	54–58	83–86
H	5.5–7.0	11–14
O	35–40	<1
N	0–0.2	<1
S	0.02[35]	<4
Ash	0–0.2	0.1
HHV (MJ/kg)	16–19	44

from pyrolysis bio-oil at atmospheric pressure.[36] The loss of carbon during hydrocracking results in lower carbon efficiency for hydrocracking relative to HDO. However, hydrocracking avoids the use of hydrogen and high pressure as required in HDO. Due to the similarity of these two upgrade approaches to the operations in the production/refining of gasoline, the upgrading of bio-oil can be achieved using current infrastructure for fossil-petroleum refining. This will be beneficial to the capital expenditure of a pyrolysis-based biofuel system. Hydrodesulfurization (HDS) like HDO also utilizes hydrogen, however for the removal of sulfur in refining gasoline while the reduction of high boiling, high-molecular-weight fractions of crude oil into valuable gasoline and other products is achieved through fluid catalytic cracking (FCC). It should be noted that for these two upgrade pathways, different reactions such as decarbonylation, decarboxylation, and hydrogenation also take place during the upgrade of bio-oils.

In general, the HDO pathway is preferable because the high-molecular-weight fractions from the pyrolysis of plants are much lower than the high molecular weights found in crude oil, therefore, cracking of bio-oil can lead to the production of hydrocarbons lower than the gasoline range and gases. Also, catalytic vapor cracking is highly susceptible to the formation of coke that deactivates the surfaces of the catalyst utilized resulting in short catalyst lifetimes. The considerable amount of water produced during HDO results in the instability of the supports of catalysts such as sulfide cobalt molybdenum or nickel molybdenum on alumina.[36] Precious metal catalyst on more stable supports such as Pd/C, Ru/TiO_2 have been shown to perform better than the traditional catalysts for HDS.[41,42]

However, fully deoxygenating pyrolysis bio-oil using HDO requires a significant amount of hydrogen, which will considerably impact the economics and environmental footprint of the pyrolysis-based biofuel system if the hydrogen is fossil-derived. Recent studies have investigated ways of minimizing fossil-hydrogen by steam-reforming portions of the pyrolysis bio-oil to produce hydrogen.[43-45] A recent study by Sorunmu et al. suggested the use of electrochemical deoxygenation (EDOx) in the upgrade of bio-oil to minimize or entirely offset the use of fossil-hydrogen relative to HDO.[46] The EDOx process uses catalytic electrode membranes to generate the hydrogen required for the deoxygenation reaction at the cathode and oxygen at the anode.[46] The EDOx can accomplish the upgrade of bio-oil as a 'stand-alone' upgrade step (total EDOx) or integrated with HDO as the mild HDO step in the sequential HDO set-up (partial EDOx). Further description of the EDOx approach can be found in the literature.[46] The fossil-hydrogen requirement for the upgrade of bio-oil can also be offset by using an integrated hydropyrolysis-hydroconversion pathway.[47-49] The integrated hydropyrolysis-hydroconversion pathway developed by GTI and usually referred to as IH² pathway steam reforms the low molecular hydrocarbons (C1–C3) from the process to produce the hydrogen required for the HDO of bio-oil. A detailed discussion of HDO treatment of bio-oil as well as reaction mechanisms and kinetics of HDO can be found in the literature.[34, 50-52]

5 Co-products and uses in the market

Pyrolysis thermally decomposes biomass to produce three products that include a solid product called biochar, a liquid product called bio-oil, and a non-condensable gaseous product termed syngas. The bio-oil contains various molecular-weight compounds and can be catalytically upgraded to a drop-in biofuel while biochar and syngas are co-products. The composition and yields of these products are effected by many parameters, including but not limited to feedstock, residence time, particle size, and temperature. Slow pyrolysis at lower temperatures favors the formation of biochar while fast pyrolysis favors the formation of bio-oil. Table 2 shows typical product yields for slow and fast pyrolysis.

Focusing on the co-products, biochar is a porous, carbonaceous solid residue created from the thermal decomposition of the biomass.[54] It is generally considered resistant to chemical and biological decomposition making it potential source for carbon sequestration.[55] There are many different uses for biochar, which are summarized below. The end use for the gas product of pyrolysis is typically as an energy carrier and is combusted on-site for process energy.

5.1 Biochar uses in the market

In 2015, the US Department of Energy's Bioenergy Technologies Office (BETO) recommended that a focus be put on the development of co-product uses to enable a more sustainable production of biofuels. These co-products should be an environmentally favorable direct replacement for a petroleum-based product and capitalize on the inherent structure of biomass while adding an additional revenue source to the pyrolysis biofuel process.[56]

Table 2 Table of typical yields from pyrolysis of woody biomass[53]

	Liquid (%)	Char (%)	Gas (%)
Fast pyrolysis	50–70	10–25	10–30
Intermediate pyrolysis	35–45	30–40	20–30
Slow pyrolysis	20–50	25–35	20–50

Conventional pyrolysis systems burn the co-products for energy, displacing fossil fuels in the process while achieving climate-change mitigation benefits. The highly carbonaceous biochar, containing 60–90% carbon, has an energy content in the most favorable case comparable to high-quality coals.[54] Due to the comparable energy content along with close similarity in properties (ultimate and proximate analysis, grindability) biochar can be burned in a co-fired power plant along with coal.[57] Displacing coal in the power plants with biochar provides GHG emission savings to the electricity generation sector which, according to current biofuel regulations in the United States, can help liquid transportation biofuel meet emission reduction targets in the Renewable Fuel Standard (RFS, https://www.epa.gov/renewable-fuel-standard-program). In addition to being an energy source, biochar has other higher value applications with the potential to improve biofuel production sustainability.

5.1.1 Soil amendment

Biochar may be applied to farm fields as a soil amendment for sustainable soil quality improvement and carbon sequestration.[54,58-60] The internal pores of biochar developed during pyrolysis increase the capacity of soil to hold water and nutrients, reducing the need for fertilizers. A study by Brunn et al. looked into the effects of slow and fast pyrolysis biochar on soil C and N dynamics after soil application.[61] Independent of pyrolysis method, biochar has the potential to sequester C when incorporated as a soil amendment. Biochar from both pyrolysis methods, especially fast, shows potential to immobilize N in the soil. By immobilizing N, biochar from fast pyrolysis has been found to decrease the amount of fertilizer necessary to apply on fields.[62] Several studies have also found that applying biochar as a soil amendment reduces N_2O soil emissions by 40–80%.[63,64] One uncertainty with soil amendments is the optimal soil amendment application rate and how long the benefits will last. Studies have shown crop improvements from biochar application rates between 5 and 50 ton/ha.[65,66] Most soil amendment environmental studies report that the fertilizer and emission benefits will last anywhere between 1 and 10 years[67,68] before reapplication of biochar.

5.1.2 Activated carbon

Biochar can be processed into activated carbon through physical or chemical activation. Both activation methods increase the carbon's pore structure, resulting in a high degree of porosity and a large internal surface area.[69] Physical activation uses a gas, such as CO_2 or steam, along with an oxidizing agent at high temperatures to develop the pores.[70] Common chemical activation agents are potassium hydroxide, phosphoric acid, and zinc chloride.[71] Activated carbon has a higher market value than biochar, making its production and sale economically favorable to the biofuels plant.[72] Activated carbon made from non-renewable materials such as coal also has a heavy carbon footprint and involves the use of heavy metals for activation.

5.1.3 Other uses

The Ithaka Institute identified 55 potential uses for biochar, highlighting its versatility.[73] Preliminary research shows that biochar has the potential to displace materials that are toxic, expensive, or with a large carbon footprint in many different fields. Research conducted by Yan et al. found that preparing pervaporation membranes using biochar nanoparticles improved the performance of the membrane when separating ethanol from water.[74] A study by Peterson found that renewable biochar has the potential to be a substitute for carbon black in flexible, low filler rubber composite applications, showing improved tensile strength, elongation, and toughness.[75] The one limitation found was that the biochar needs to be low ash to be used in carbon black. Other studies have looked into adding biochar into concrete recipes to help decarbonize another high carbon footprint material.[76,77] These studies found that biochar can improve humidity control and weight in concrete[73] while also absorbing immobilizing toxic radioactive waste that it comes into contact with.[76] Research by Huggins et al. found that biochar could also be used as a sustainable electrode material for electricity production in microbial fuel cells.[78] Biochar microbial fuel cell electrodes showed comparable performance to current materials while having economic and environmental benefits.

In addition to the above uses, biochar can also have cascading uses, where it is used multiple times. A typical proposed cascade usage is filtering waste water from food or drink production and then putting the biochar back in the soil, with the organic nutrients it trapped during filtration, as a soil amendment and fertilizer.[73] Overall biochar can be used in a variety of different ways to aid in the sustainability of biofuel production.

6 Techno-economic analysis (TEA) and life cycle assessment (LCA) review

For pyrolysis-based biofuels to be a qualified renewable transportation fuel in the United States and other nations, they must meet the requirements set by environmental agencies such as the US RFS and be cost competitive in the transportation fuel market. LCAs are used to evaluate the GHG emissions of a pyrolysis-based conversion process and pathway, thus quantifying some of the environmental impacts of biofuels. Likewise, TEAs are used to quantify the economic impacts by conducting a preliminary economic analysis on the pyrolysis-based conversion process.

As highlighted in previous sections, fast pyrolysis of biomass maximizes liquid yields while slow pyrolysis maximizes the yield of solids. Very few TEA and LCA studies looking at biofuel from slow pyrolysis of biomass were found in literature relative to fast pyrolysis of biomass. This may be because of the focus on alternative liquid biofuel production that is favored by fast pyrolysis in comparison to slow pyrolysis that favors solid biofuel production. This section summarizes TEA and LCA studies on the fast pyrolysis pathway with the objective of producing preferably 'drop-in' hydrocarbon transportation biofuel.

6.1 Life cycle assessment (LCA)

Some studies have investigated the life cycle emissions of pyrolysis-based conversion pathways. However, direct comparisons of the results of these studies is made challenging due to some variability in their life cycle assessment (LCA) methods such as differences in the functional unit, allocation methods, feedstock type, final pyrolysis product, and so on. With the objective of being an alternative to fossil fuels, some of the studies, in line with the US EPA, utilized energy as the functional unit in their studies.

For pyrolysis bio-oil combusted in boilers as an alternative to heavy fuel oil, Steele et al. determined GHG emissions of 32.3 g CO_2-eq/MJ for bio-oil with the co-product biochar used internally to provide electricity required by the process while excess electricity not utilized in the process was modeled to displace electricity from the grid.[35] Winjobi et al. calculated GHG emissions of 36.26 g CO_2-eq/MJ for bio-oil also combusted for heat from a pyrolysis-only process referred to as a one-step pathway in their study.[31] Some of the studies looked at using the bio-oil produced from pyrolysis for electricity generation. Fan et al. reported GHG emissions of about 3.89 g CO_2-eq/kWh of electricity generated using a gas turbine combined cycle (GTCC) with heat recovery for bio-oil generated from the pyrolysis of waste wood integrated with the GTCC unit. In addition, emissions of 104.12 g CO_2-eq/kWh of electricity generated were reported when the bio-oil is combusted in a coal-fired plant using US average grid electricity for pyrolysis processing.[4] Co-products from their study were utilized internally to provide process heat. Dang et al. reported GHG emissions as low as 260 g CO_2 eq/kWh relative to 1050 g CO_2-eq/kWh from existing bituminous coal power plants when the heavy ends of the bio-oil are blended and co-fired (60:40 wt.) with bituminous coal while the biochar co-product is utilized for soil amendment.[79]

The upgrade of bio-oil to hydrocarbon transportation fuel in most of the studies that have investigated pyrolysis-based bioenergy system is through HDO using hydrogen. Hsu estimated potential emissions from the pyrolysis of forest residue to be as low as 25 g CO_2-eq/MJ biofuel when the electricity used in the process is biomass-derived.[80] For some of these LCA studies it was observed that the most significant contribution to the overall life cycle emissions is from the production of hydrogen required for HDO through steam-reforming of natural gas. Hsu reported this contribution for hydrogen production to be about 51% of the overall life cycle emissions.[80] Therefore, more recent studies have looked at either eliminating or partially reducing the fossil-derived hydrogen. Snowden-Swan et al. in their study reported emissions of about 31.5 to 36.8 g CO_2-eq/MJ renewable gasoline when forest residue and poplar undergoes pyrolysis respectively with a cut of the bio-oil reformed for hydrogen production and the biochar utilized internally for steam and electricity generation.[81] Dang et al. reported emissions as low as −18.14 g CO_2-eq/MJ when only the pyrolytic portion of the bio-oil (about 38% of the bio-oil) was upgraded to biofuels while the remaining portion of the bio-oil, the aqueous phase, was steam-reformed to produce the hydrogen required for the hydrotreatment.[43] The remaining hydrogen not used internally in their study, displaced hydrogen produced from conventional pathways. Han et al. in their study investigated three different pathways with the required hydrogen produced through (1) reforming fuel gas/natural gas, (2) reforming a portion of the pyrolysis oil with biochar combusted for electricity, and (3) reforming a portion of the bio-oil and using biochar for soil amendment.[82] However, reforming the bio-oil and utilizing the char for soil amendment gives the most significant reduction of about 112% relative to fossil fuels.[82] Winjobi et al. estimated GHG emissions of about 38.4 g CO_2-eq/MJ biofuel for transportation fuel produced from the pyrolysis of pine residues for a heat-integrated pyrolysis-only bioenergy system with a partial reduction in natural gas for hydrogen production achieved by reforming light hydrocarbon gases from the upgrade step.[83] Biochar in their study was modeled to displace coal while process heat was obtained from the combustion of natural gas. Fan et al. and Winjobi et al. also reported emissions of about 12.75 and 12.42 g CO_2-eq/MJ fuel blend, respectively, for the integrated hydropyrolysis-hydroconversion pathway (IH^2®) developed by GTI that fully eliminated fossil natural gas for hydrogen production through the steam-reforming of light hydrocarbon gases from the hydroconversion step.[48,49] Small credits for the displacement of ammonia were included in the analyses and biochar was internally used to provide the electricity required by the IH^2® process.

Upgrade of bio-oil to transportation biofuels in the LCAs discussed above was mostly achieved through HDO. Some LCAs have also looked at either modification to the HDO step or alternatives to HDO for biofuel production. Vinescu et al. examined different upgrade approaches that included single-stage hydrotreating and hydrocracking, esterification or ketonization followed by hydrotreating and hydrocracking, and two-stage hydrotreating and hydrocracking. They reported the lowest emission of about 2240 g CO_2-eq/

kg of upgraded fuel for the single-stage hydrotreating and hydrocracking but believe that more favorable/lower net emissions can be obtained from other upgrade configurations if the CO_2 absorbed by the corn stover is also included in their analyses.[84] Sorunmu et al. in their study used LCA to investigate an alternative approach, EDOx instead of HDO for the upgrade of pyrolysis oil to renewable gasoline and diesel.[85] In their study, a full EDOx replaces the typical two-stage hydrotreaters in HDO while for the partial EDOx, only the mild hydrotreater is replaced. For both the partial and full EDOx processes, GHG emissions of about 88% to 95% lower than conventional gasoline and diesel were determined.

Some studies considered land as a scarce resource that can be used to cultivate more than one feedstock for bioenergy systems. Therefore, they considered the land required for plant cultivation as their functional unit. Kauffman et al. in their studies believe a portion of land that is used for corn cultivation can be utilized for bioenergy systems using the corn grain as well as the corn stover.[86] Using LCA, they evaluated net emissions of about 4.66 ton CO_2-eq/ha (−2.99 ton CO_2-eq/ha for stover fast pyrolysis pathway and 7.65 ton CO_2-eq/ha for grain ethanol production) with biochar used as a soil amendment.[86]

The use of the co-product biochar in multiple applications has the most significant impact of all the co-products in most of the analyses. This is because biochar is a carbon-rich co-product with a substantial amount of energy. Beyond the use of biochar for soil amendment and combustion for process heat examined by most of these studies, Kulas et al. further investigated its use for making activated carbon. In their study, they reported the use of biochar for activated carbon to be favorable relative to soil amendment and combustion for process heat with potential GHG emissions of about 15.8 g CO_2 eq/MJ biofuel for transportation fuel produced from the pyrolysis of poplar residues for a heat-integrated pyrolysis-only bioenergy system in their study.[72] It is noteworthy that despite the differences in the scope and functional units in these studies, a common trend in all these analyses is that the treatment/utilization of the co-products from the bioenergy system strongly influences the ability of the system to be a carbon-negative bioenergy system. A summary of these LCAs published in open literature for pyrolysis biofuel pathways is shown in Table 3.

6.2 Techno-economic analysis (TEA)

Techno-economic analysis (TEA) is a tool that uses process models to quantify the technical and economic performance of a biorefinery. The model generates a simple and inexpensive estimate of the financial return from the capital investment necessary for a biorefinery. The absence of commercial-scale

Table 3 Summary of Life Cycle Assessments of biofuel and bio-oil from fast pyrolysis conversion pathway

Main product	Feedstock	Co-products	Functional unit	Capacity	Emissions (g CO_2-eq)	Additional notes (Allocation method)
Bio-oil[35]	Southern pine	-	1 MJ of bio-oil	2000 t/day	32.3	Char utilized internally. Credits from excess energy displaced electricity from the grid (System expansion)
Biofuel[84]	Corn stover	Biochar	1 kg of biofuel	2000 MT/day	2240	Co-products/by-products used internally. Energy avoided from producing by-products through other routes not considered in analysis
Bio-oil[31]	Pine	Biochar	1 MJ of bio-oil	1000 MT/day	32.26	Credits from displacing some of the coal used in a coal power plant (System expansion)
Biofuel[83]	Pine	Biochar	1 MJ of biofuel	1000 MT/day	38.4	Credits from displacing some of the coal used in a coal power plant (System expansion)
Biofuel[72]	Poplar	Activated carbon	1 MJ of biofuel	1000 MT/day	15.8	Credits from displacing bituminous coal used in making activated carbon (System expansion)
Bio-oil[4]	Forest waste	-	1 kWh electricity generated	400 MT/day	104.12	Char utilized internally for process heat
Biofuel[82]	Corn stover	Biochar	1 MJ of biofuel	2000 MT/day	11.28	Emissions estimated from the percentage reduction stated by the authors (88%). Electricity from biochar and fuel gas combustion. Surplus electricity exported to the grid (Energy allocation)
Bio-oil[79]	Corn stover	Biochar	1 kWh electricity generated	2000 MT/day	260	Bio-oil co-fired with coal (60:40 wt. %). Biochar sequestered into the soil. (Mass allocation)
Biofuel[82]	Corn stover	Biochar	1 MJ of biofuel	2000 MT/day	−11.28	Emissions estimated from the percentage reduction stated by the authors (112%). Biochar used as a soil amendment to sequester carbon. N in biochar also displaced conventional N fertilizer (System expansion)

(*Continued*)

Table 3 (*Continued*)

Main product	Feedstock	Co-products	Functional unit	Capacity	Emissions (g CO_2-eq)	Additional notes (Allocation method)
Biofuel & Ethanol[86]	Corn & corn stover	Biochar	1 ha of corn cultivation	-	4.66 ton	Ethanol produced from corn and biofuel from corn stover. Biochar used as a soil amendment to reduce N_2O emissions and sequester carbon (System expansion)
Biofuel[80]	Forest residue	-	1 MJ of biofuel	2 kT/day	25	No co-products produced in this analysis. Biomass-derived electricity assumed using a portion of the feed
Biofuel[81]	Forest residue	-	1 MJ of biofuel	-	31.5	Biochar utilized internally for power and heat
Biofuel[85]	Forest residue	Oxygen	1 MJ of biofuel	300 MT/day	5–11	Emissions estimated from percentage reduction stated by the authors. Credits for displacing oxygen from conventional production (System expansion)
Biofuel[48,49]	Forest residue	Ammonia	1 MJ of biofuel	500 MT/day	13	IH2® process. Ammonia displaced conventional ammonia for fertilizers. (System expansion)
Biofuel[43]	Corn stover	Biochar	1 MJ of biofuel	2000 kg/day	−18.14	Electricity credit from biochar utilization. Also credits from surplus hydrogen from reforming aqueous phase of bio-oil displacing fossil-hydrogen (System expansion)
Biofuel[81]	Poplar	-	1 MJ of biofuel	-	36.8	Biochar utilized internally for power and heat

[a] Biofuel implies hydrocarbon biofuel.

cellulosic biorefineries makes TEAs necessary to compare the economic performance of the pyrolysis pathway to fossil-fuel pathways.[87] One downside of the TEA is the high uncertainty in the results due primarily to the assumptions needed to simplify the complex process, however this uncertainty can be quantified through Monte Carlo analysis using methods pioneered by Brown and Wright.[88] Nonetheless, TEAs can be used to calculate useful economic metrics like the minimum selling price (MSP) of bio-oil/biofuel or either the internal rate of return (IRR) or net present value (NPV) of a biorefinery. The process inputs (capital investment, operating costs, and feedstock costs) are typically either discounted or annualized in the cash flow. MSP calculations are done at a desired IRR and can be used to break down the economic burdens of each unit operation onto the liquid product price. This allows for a quick snapshot of what process steps are the most expensive, either in terms of equipment or operational costs, and may lead to energy efficiency and other improvements. IRR or NPV calculations use a biofuel selling price equal to the equivalent energy commodity price (i.e., price of gasoline). This gives investors a preliminary analysis of the economic favorability of the biorefinery. A summary of the recent TEAs published in open literature for pyrolysis biofuel pathways is shown in Table 4.

Recent TEA studies indicate that the MSP of pyrolysis-based liquid hydrocarbon biofuel is between \$2.20 and \$4.00 gallons of gasoline equivalents (gge) and for pyrolysis bio-oil is between \$0.94 and \$1.73/gallon. The wide variation in prices is mostly due to the many different assumptions that can be made in a TEA. A study by Brown found that capacity, Lang factors (i.e., the ratio of total project investment to total purchased equipment cost), the percentage of hours per year that the biorefinery is operational, feedstock cost, and biorefinery lifespan are rarely identical even across analyses for the same pathway.[87] As shown in Table 4 there are many potential feedstocks that can be used, all having different feedstock costs and the pyrolysis yields. Sensitivity analysis is typically used to uncover the effects of these assumptions. Research shows that the economic analyses are most sensitive to product yield, IRR, feedstock cost, and capital investment.[72,89-91] Brown et al. found that the assumption of biorefinery lifetime also greatly influences the biofuel MSP.[87]

7 Commercialization and demonstration of biomass fast pyrolysis

Pyrolysis of biomass has enjoyed some commercial success through the actions of a small number of global firms producing a range of products in multiple international locations. Since 1989 the Canadian firm Ensyn designed and licensed a number of fast pyrolysis units for the Kerry Group in Rhinelander, WI, of 30–40 tons/day capacity producing food and flavor ingredients as well

Table 4 Summary of techno-economic analysis of biofuel and bio-oil from fast pyrolysis conversion pathway

Main product	Year of study	Feedstock	Capacity	Total project investment	IRR (%)	Plant life (years)	Minimum selling price (gge)
Biofuel[92]	2010	Corn stover	2000 MTPD	$200 million	15[a]	20	$2.92
Biofuel[90]	2009	Hybrid poplar	2000 MTPD	$329 million	10	20	$2.21
Biofuel[44]	2010	Stover	2000 MTPD	$217 million	10	20	$2.29
Biofuel[91]	2013	Red oak	2000 MTPD	$210 million	11.40[a]	20	$2.82
Biofuel[93]	2013	Stover	2000 MTPD	$429 million	10	20	$2.57
Biofuel[94]	2012	Cryptomeria residue	10,000 t/yr	$4 million	17	20	$21.76/GJ
Biofuel[89]	2015	Pine	1000 MTPD	$361 million	10	20	$4.01
Biofuel[72]	2017	Poplar	1000 MTPD	$392 million	10	20	$3.72
Biofuel[95]	2017	Red oak	2000 MTPD	$67.5 million	10	30	$2.48
Bio-oil[96]	2015	Wood furniture sawdust	20–30 dm^3/day	5.1 million Thai Baht (THB)	10	20	15 THB/dm^3 ($0.45/L)
Bio-oil[97]	2010	Pine	100 MTPD	$6 million	10	10	$0.94/gallon
Bio-oil[30]	2013	Pine	1000 MTPD	$327 million	10	20	$1.04/gallon[b]
Bio-oil[30]	2013	Pine	1000 MTPD	$132 million	10	20	$1.32/gallon
Bio-oil[98]	2000	Rice husk	24 MTPD	$0.4 million	10	10	$1.73/gallon
Gasoline and diesel blendstock[89]	2010	Woody biomass	2000 MTPD	$700 million	10	10	$3.39

[a] Cash flow table solved for IRR.
[b] With torrefaction.

as renewable heating fuels from wood residues. The Ensyn technology is based on a circulating fluidized bed design termed Rapid Thermal Processing (RTP™). In Renfrew, Ontario, Canada, Ensyn operates a 3-million gallon/year RTP facility for production of renewable fuel oil (RFO (pyrolysis bio-oil)) for US clients in building heat applications. Production from this facility also provides RFO for testing in petroleum refineries as a co-processed input to the FCC unit of a petroleum refinery. The company's website claims to be developing several new RFO facilities in Canada, the United States, Brazil, and Malaysia at various stages of design and construction. In the Netherlands, the company BTG-BTL operates two fast pyrolysis facilities based on the rotating cone reactor technology in Hengelo, the Netherlands with a capacity of 20 million L py-oil/year and also in Malaysia processing palm empty fruit bunches with a capacity of 2 tons/day. In the city of Joensuu, Finland, the Fortum-Valmet combined heat and power (CHP) facility provides 95% of the heat requirements for the city's 75 000 inhabitants with an electricity generation capacity of 54 MW. The CHP facility augments the city's energy demand with fast pyrolysis bio-oil generated using a fluid bed reactor from forest residues, wood chips, and sawdust at a capacity of 50 000 tonnes bio-oil/year. At the Karlesruhe Institute of Technology (KIT) facility in Karlsruhe, Germany a fast pyrolysis pilot plant of capacity 500 kg/h operates with a twin-screw auger reactor to convert dry biomass into a liquid bio-oil product, char, and non-condensable pyrolysis gas (https://www.bioliq.de/english/64.php).[99] This demonstration facility provides a bio-oil intermediate for upgrading to gasoline and heavy fraction hydrocarbon products.

8 Conclusion and future trends

This chapter has provided a concise review of the wide range of technical, economic, environmental, and commercialization issues with fast pyrolysis of biomass. Fast pyrolysis is a versatile technology that is able to process a wide range of biomass types from energy crops, forest and agricultural residues, municipal solid waste, and algae. Pyrolysis converts low energy-dense biomass into a higher energy-dense carrier exhibiting a diverse range of properties and chemical species with potential uses in energy and chemical synthesis. Pyrolysis is a simple process conceptually, adding heat in a controlled fashion to cause the degradation of solid polymeric materials in wood to low-molecular-weight chemicals. Despite this conceptual simplicity, pyrolysis remains extremely difficult to model in terms of chemical kinetics despite decades of research, while still recognizing significant advancements in this area have been made. At the process scale, many advances in reactor designs have been made as well as in downstream separations of potentially valuable components and in the diverse energy and chemical applications of pyrolysis bio-oil. One of the most promising applications of fast pyrolysis of biomass is in the production of hydrocarbon drop-in biofuels by catalytic hydrotreatment of intermediate pyrolysis bio-oil. Techno-economic analyses of pyrolysis bio-oil and of hydrocarbon biofuels have shown promising results despite not yet achieving cost parity with fossil fuels on an energy equivalent basis at the current crude oil price of approximately \$60–70/barrel. Environmental LCAs have mostly focused on GHG emissions and cumulative energy demand and in the studies reviewed here demonstrate significant GHG emission reductions for pyrolysis bio-oil, electricity generation, and for hydrocarbon biofuels. In both the TEA and LCA studies, effects of different process configurations, energy minimization, co-product uses, biomass type, and pretreatments have been investigated. Commercial applications of biomass fast pyrolysis have achieved some initial successes in the global energy market, with prospects of more success in the future. As the world's population continues to increase, thus placing higher demands on current energy resources, this presents opportunities for renewable alternatives such as biomass fast pyrolysis to take hold and grow.

9 Where to look for further information

Future research in fast pyrolysis of woody biomass or fiber fractions of municipal solid waste for production of liquid transportation biofuels will focus on reducing costs of production to achieve parity with fossil fuels such as petroleum-derived gasoline, diesel, and jet. To reach this goal, improvements in reactor design and operation are needed to better control vapor residence time and enhance rates of heat transfer to the solid biomass during fast pyrolysis. In some cases, enhancing heat transfer could be accomplished by introducing just enough oxygen into the pyrolysis reactor to generate the energy required internally, thus overcoming heat transfer resistance common in current reactor designs. Improving catalyst reactivity and stability could further reduce the hydrotreatment cost component of biofuels. Source of hydrogen in the hydrotreatment will continue to be an intense research area in order to lower GHG emissions, especially when low-cost conversion of renewable feedstocks are developed such as solar- and biomass-derived hydrogen. Future research will also investigate value-added applications of biofuel co-products, especially biochar, to improve overall economics. TEA and LCA studies will continue to improve our understanding of future commercial production of biofuels and co-products through fast pyrolysis of biomass. These and other research issues are being investigated by top research universities and national laboratories across the globe, in addition to emerging bioenergy companies and well-established petroleum industry firms.

10 References

1. Mohan, D., Pittman, C. U. and Steele, P. H. Pyrolysis of wood/biomass for bio-oil: a critical review. *Energy and Fuels* 2006, 20(3), 848–89.
2. NSF. *Breaking the Chemical and Engineering Barriers to Lignocellulosic Biofuels: Next Generation Hydrocarbon Biorefineries*. Ed. Huber, G. W. University of Massachusetts Amherst, National Science Foundation, Chemical, Bioengineering, Environmental, and Transport Systems Division, Washington DC, 2008, 180p.
3. Mihalek, M. J. Life cycle assessments (LCAs) of pyrolysis-based gasoline and diesel from different regional feedstocks: corn stover, switchgrass, sugar cane bagasse, waste wood, guinea grass, algae, and albizia. Master's Thesis, Michigan Technological University, 2014.
4. Fan, J., Kalnes, T. N., Alward, M., Klinger, J., Sadehvandi, A. and Shonnard, D. R. Life cycle assessment of electricity generation using fast pyrolysis bio-oil. *Renewable Energy* 2011, 36(2), 632–41.
5. Jones, S., Meyer, P., Snowden-Swan, L., Padmaperuma, A., Tan, E., Dutta, A., Jacobson, J. and Cafferty, K. *Process Design and Economics for the Conversion of Lignocellulosic Biomass to Hydrocarbon Fuels: Fast Pyrolysis and Hydrotreating Bio-oil Pathway*. National Renewable Energy Laboratory (NREL), Golden, CO, 2013.
6. Yildiz, G., Ronsse, F., Venderbosch, R., van Duren, R., Kersten, S. R. and Prins, W. Effect of biomass ash in catalytic fast pyrolysis of pine wood. *Applied Catalysis B: Environmental* 2015, 168, 203–11.
7. Ukaew, S., Schoenborn, J., Klemetsrud, B. and Shonnard, D. R. Effects of torrefaction temperature and acid pretreatment on the yield and quality of fast pyrolysis bio-oil from rice straw. *Journal of Analytical and Applied Pyrolysis* 2018, 129, 112–22.
8. Ronsse, R., Nachenius, R. O., Prins, W. Carbonization of biomass. In: *Recent Advances in Thermochemical Conversion of Biomass*. Eds. Pandey, A., Bhaskar, T., Stöcker, M. and Sukumaran, R.. Elsevier, Amsterdam, 2015, p. 504.
9. Graham, R. G., Bergougnou, M. A., Mok, L. K. S. and de Lasa, H. I. Fast pyrolysis (ultrapyrolysis) of biomass using solid heat carriers. In: *Fundamentals of Thermochemical Biomass Conversion*. Eds. Overend, R. P., Milne, T. A. and Mudge, L. K.. Springer Netherlands, Dordrecht, 1985, pp. 397–410.
10. Demirbas, A. Pyrolysis of ground beech wood in irregular heating rate conditions. *Journal of Analytical Applied Pyrolysis* 2005, 73(1), 39–43.
11. Garcia-Perez, M., Chaala, A., Pakdel, H., Kretschmer, D. and Roy, C. Vacuum pyrolysis of softwood and hardwood biomass: comparison between product yields and bio-oil properties. *Journal of Analytical Applied Pyrolysis* 2007, 78(1), 104–16.
12. Resende, F. L. Recent advances on fast hydropyrolysis of biomass. *Catalysis Today* 2016, 269, 148–55.
13. Basu, P. *Biomass Gasification and Pyrolysis: Practical Design and Theory*. Academic Press, Amsterdam, 2010, p. 365.
14. Bridgwater, A. V. and Bridge, S. A. A review of biomass pyrolysis and pyrolysis technologies. In: *Biomass Pyrolysis Liquids Upgrading and Utilization*. Eds. Bridgwater, A. V. and Grassi, G.. Springer Netherlands, Dordrecht, 1991, pp. 11-92.
15. Sharma, A., Pareek, V. and Zhang, D. Biomass pyrolysis–a review of modelling, process parameters and catalytic studies. *Renewable and Sustainable Energy Reviews* 2015, 50, 1081-96.

16. Kan, T., Strezov, V. and Evans, T. J. Lignocellulosic biomass pyrolysis: a review of product properties and effects of pyrolysis parameters. *Renewable and Sustainable Energy Reviews* 2016, 57, 1126–40.
17. Wang, S., Dai, G., Yang, H. and Luo, Z. Lignocellulosic biomass pyrolysis mechanism: a state-of-the-art review. *Progress in Energy and Combustion Science* 2017, 62, 33–86.
18. Papari, S. and Hawboldt, K. A review on the pyrolysis of woody biomass to bio-oil: focus on kinetic models. *Renewable and Sustainable Energy Reviews* 2015, 52, 1580–95.
19. Burnham, A. K., Zhou, X. and Broadbelt, L. J. Critical review of the global chemical kinetics of cellulose thermal decomposition. *Energy and Fuels* 2015, 29(5), 2906–18.
20. Klemetsrud, B., Klinger, J., Ziv, E. B. and Shonnard, D. A kinetic study of the fast micro-pyrolysis of hybrid poplar. *Journal of Analytical and Applied Pyrolysis* 2017, 128, 353–62.
21. Shafizadeh, F. and Chin, P. P. *Thermal Deterioration of Wood*. ACS Symposium Series, American Chemical Society, 1977.
22. Klinger, J., Bar-Ziv, E. and Shonnard, D. Unified kinetic model for torrefaction-pyrolysis. *Fuel Processing Technology* 2015, 138, 175–83.
23. Mayes, H. B. and Broadbelt, L. J. Unraveling the reactions that unravel cellulose. *The Journal of Physical Chemistry A* 2012, 116(26), 7098–106.
24. Zhou, X., Nolte, M. W., Mayes, H. B., Shanks, B. H. and Broadbelt, L. J. Experimental and mechanistic modeling of fast pyrolysis of neat glucose-based carbohydrates. 1. Experiments and development of a detailed mechanistic model. *Industrial and Engineering Chemistry Research* 2014, 53(34), 13274–89.
25. Gracida-Alvarez, U. R., Mitchell, M. K., Sacramento-Rivero, J. C. and Shonnard, D. R. Effect of temperature and vapor residence time on the micropyrolysis products of waste high density polyethylene. *Industrial and Engineering Chemistry Research* 2018, 57(6), 1912–23.
26. Bridgwater, A. V. Review of fast pyrolysis of biomass and product upgrading. *Biomass and Bioenergy* 2012, 38, 68–94.
27. Phounglamcheik, A., Babler, M. U., Donaj, P., Amovic, M., Ljunggren, R. and Engvall, K. Pyrolysis of wood in a rotary kiln pyrolyzer: modeling and pilot plant trials. *Energy Procedia* 2017, 105, 908–13.
28. Knezevic, D., van Swaaij, W. and Kersten, S. Hydrothermal conversion of biomass. II. Conversion of wood, pyrolysis oil, and glucose in hot compressed water. *Industrial and Engineering Chemistry Research* 2009, 49(1), 104–12.
29. Kelkar, S., Saffron, C. M., Li, Z., Kim, S.-S., Pinnavaia, T. J., Miller, D. J. and Kriegel, R. Aromatics from biomass pyrolysis vapour using a bifunctional mesoporous catalyst. *Green Chemistry* 2014, 16(2), 803–12.
30. Winjobi, O., Shonnard, D. R., Bar-Ziv, E. and Zhou, W. Techno-economic assessment of the effect of torrefaction on fast pyrolysis of pine. *Biofuels, Bioproducts and Biorefining* 2016, 10(2), 117–28.
31. Winjobi, O., Shonnard, D. R., Bar-Ziv, E. and Zhou, W. Life cycle greenhouse gas emissions of bio-oil from two-step torrefaction and fast pyrolysis of pine. *Biofuels, Bioproducts and Biorefining* 2016, 10(5), 576–88.
32. Oy, G. F. N. Bio-oil applications. Available at: http://www.greenfuelnordic.fi/bio-oil_applications (accessed July 06, 2018).

33. Lehto, J., Oasmaa, A., Solantausta, Y., Kytö, M. and Chiaramonti, D. Fuel oil quality and combustion of fast pyrolysis bio-oils. *VTT Technology* 2013, 87, 79.
34. Mortensen, P. M., Grunwaldt, J.-D., Jensen, P. A., Knudsen, K. and Jensen, A. D. A review of catalytic upgrading of bio-oil to engine fuels. *Applied Catalysis A: General* 2011, 407(1–2), 1–19.
35. Steele, P., Puettmann, M. E., Penmetsa, V. K. and Cooper, J. E. Life-cycle assessment of pyrolysis bio-oil production. *Forest Products Journal* 2012, 62(4), 326–34.
36. Bridgwater, T. Challenges and opportunities in fast pyrolysis of biomass: part II. *Johnson Matthey Technology Review* 2018, 62(2), 150–60.
37. Elliott, D. and Baker, E. *Hydrotreating Biomass Liquids to Produce Hydrocarbon Fuels*. Pacific Northwest Lab., Richland, WA, 1986.
38. Elliott, D. C. and Hart, T. R. Catalytic hydroprocessing of chemical models for bio-oil. *Energy and Fuels* 2008, 23(2), 631–7.
39. UOP and Ensyn to form joint venture to offer second generation biomass technology. Available at: https://www.uop.com/?press_release=uop-and-ensyn-to-form-joint-venture-to-offer-second-generation-biomass-technology (accessed July 13, 2018).
40. Vispute, T. Pyrolysis oils: characterization, stability analysis, and catalytic upgrading to fuels and chemicals. Open Access Dissertation, University of Massachusetts Amherst, 2011.
41. Wildschut, J., Arentz, J., Rasrendra, C., Venderbosch, R. and Heeres, H. Catalytic hydrotreatment of fast pyrolysis oil: model studies on reaction pathways for the carbohydrate fraction. *Environmental Progress and Sustainable Energy* 2009, 28(3), 450–60.
42. Elliott, D. C., Hart, T. R., Neuenschwander, G. G., Rotness, L. J. and Zacher, A. H. Catalytic hydroprocessing of biomass fast pyrolysis bio-oil to produce hydrocarbon products. *Environmental Progress and Sustainable Energy* 2009, 28(3), 441–9.
43. Dang, Q., Yu, C. and Luo, Z. Environmental life cycle assessment of bio-fuel production via fast pyrolysis of corn stover and hydroprocessing. *Fuel* 2014, 131, 36–42.
44. Wright, M. M., Daugaard, D. E., Satrio, J. A. and Brown, R. C. Techno-economic analysis of biomass fast pyrolysis to transportation fuels. *Fuel* 2010, 89, S2–10.
45. Snowden-Swan, L. J., Spies, K. A., Lee, G. and Zhu, Y. Life cycle greenhouse gas emissions analysis of catalysts for hydrotreating of fast pyrolysis bio-oil. *Biomass and Bioenergy* 2016, 86, 136–45.
46. Sorunmu, Y., Billen, P., Elangovan, S. E., Santosa, D. and Spatari, S. Life-cycle assessment of alternative pyrolysis-based transport fuels: implications of upgrading technology, scale, and hydrogen requirement. *ACS Sustainable Chemistry and Engineering* 2018, 6(8), 10001–10.
47. Tan, E. C., Marker, T. L. and Roberts, M. J. Direct production of gasoline and diesel fuels from biomass via integrated hydropyrolysis and hydroconversion process–a techno-economic analysis. *Environmental Progress and Sustainable Energy* 2014, 33(2), 609–17.
48. Fan, J., Gephart, J., Marker, T., Stover, D., Updike, B. and Shonnard, D. R. Carbon footprint analysis of gasoline and diesel from forest residues and corn stover using integrated hydropyrolysis and hydroconversion. *ACS Sustainable Chemistry and Engineering* 2015, 4(1), 284–90.
49. Winjobi, O., Tavakoli, H., Klemetsrud, B., Handler, R., Marker, T., Roberts, M. and Shonnard, D. Carbon footprint analysis of gasoline and diesel from forest residues

and algae using integrated hydropyrolysis and hydroconversion plus fischer-tropsch (IH2 Plus cool GTL). *ACS Sustainable Chemistry and Engineering* 2018, 6(8), 10766–77.
50. Elliott, D. C. Historical developments in hydroprocessing bio-oils. *Energy and Fuels* 2007, 21(3), 1792–815.
51. Furimsky, E. Catalytic hydrodeoxygenation. *Applied Catalysis A: General* 2000, 199(2), 147–90.
52. Zhang, Q., Chang, J., Wang, T. and Xu, Y. Review of biomass pyrolysis oil properties and upgrading research. *Energy Conversion and Management* 2007, 48(1), 87–92.
53. Brownsort, P. A. *Biomass Pyrolysis Processes: Review of Scope, Control and Variability*. UK Biochar Research Centre, University of Edinburgh, Newcastle University, 2009, p. 38.
54. Lee, Y., Park, J., Ryu, C., Gang, K. S., Yang, W., Park, Y.-K., Jung, J. and Hyun, S. Comparison of biochar properties from biomass residues produced by slow pyrolysis at 500°C. *Bioresource Technology* 2013, 148, 196–201.
55. Crombie, K., Mašek, O., Sohi, S. P., Brownsort, P. and Cross, A. The effect of pyrolysis conditions on biochar stability as determined by three methods. *GCB Bioenergy* 2013, 5(2), 122–31.
56. Bailey, A., Leong, G. J. and Fitzgerald, N. *Bioproducts to Enable Biofuels Workshop Summary Report*. Office of Energy Efficiency and Renewable Energy (EERE), Washington DC, 2015.
57. Abdullah, H. and Wu, H. Biochar as a fuel: 1. Properties and grindability of biochars produced from the pyrolysis of mallee wood under slow-heating conditions. *Energy and Fuels* 2009, 23(8), 4174–81.
58. Sohi, S. P. Carbon storage with benefits. *Science* 2012, 338(6110), 1034–5.
59. Woolf, D., Amonette, J. E., Street-Perrott, F. A., Lehmann, J. and Joseph, S. Sustainable biochar to mitigate global climate change. *Nature Communications* 2010, 1, 56.
60. Lehmann, J., Gaunt, J. and Rondon, M. Bio-char sequestration in terrestrial ecosystems–a review. *Mitigation and Adaptation Strategies for Global Change* 2006, 11(2), 403–27.
61. Bruun, E. W., Ambus, P., Egsgaard, H. and Hauggaard-Nielsen, H. Effects of slow and fast pyrolysis biochar on soil C and N turnover dynamics. *Soil Biology and Biochemistry* 2012, 46, 73–9.
62. Steiner, C., Glaser, B., Geraldes Teixeira, W., Lehmann, J., Blum, W. E. and Zech, W. Nitrogen retention and plant uptake on a highly weathered central Amazonian Ferralsol amended with compost and charcoal. *Journal of Plant Nutrition and Soil Science* 2008, 171(6), 893–9.
63. Yanai, Y., Toyota, K. and Okazaki, M. Effects of charcoal addition on N_2O emissions from soil resulting from rewetting air-dried soil in short-term laboratory experiments. *Soil Science and Plant Nutrition* 2007, 53(2), 181–8.
64. Van Zwieten, L., Singh, B., Joseph, S., Kimber, S., Cowie, A. and Chan, K. Y. Biochar and emissions of non-CO_2 greenhouse gases from soil. In: *Biochar for Environmental Management: Science and Technology*. Eds. Johannes, L. and Stephen, J.. Earthscan Publications Ltd, Washington DC, 2009, vol. 1, pp. 227–50.
65. Alhashimi, H. A. and Aktas, C. B. Life cycle environmental and economic performance of biochar compared with activated carbon: a meta-analysis. *Resources, Conservation and Recycling* 2017, 118, 13–26.

66. Blackwell, P., Riethmuller, G. and Collins, M. Biochar application to soil. In: *Biochar for Environmental Management: Science and Technology*. Eds. Johannes, L. and Stephen, J. Earthscan Publications Ltd, Washington DC, 2009, vol. 1, pp. 207–26.
67. Roberts, K. G., Gloy, B. A., Joseph, S., Scott, N. R. and Lehmann, J. Life cycle assessment of biochar systems: estimating the energetic, economic, and climate change potential. *Environmental Science Technology* 2009, 44(2), 827–33.
68. Cowie, A. L., Cowie, A. J. and Solutions, R. C. Life cycle assessment of greenhouse gas mitigation benefits of biochar. *Report to IEA Bioenergy Task* 2013, 38, 136–45.
69. Azargohar, R. and Dalai, A. Steam and KOH activation of biochar: experimental and modeling studies. *Microporous and Mesoporous Materials* 2008, 110(2–3), 413–21.
70. Byrne, J. and Marsh, H. *Porosity in Carbons: Characterization and Applications*. Edward Arnold, London, 1995, pp. 2–7.
71. Lozano-Castello, D., Lillo-Rodenas, M., Cazorla-Amoros, D. and Linares-Solano, A. Preparation of activated carbons from Spanish anthracite: I. Activation by KOH. *Carbon* 2001, 39(5), 741–9.
72. Kulas, D., Winjobi, O., Zhou, W. and Shonnard, D. Effects of coproduct uses on environmental and economic sustainability of hydrocarbon biofuel from one- and two-step pyrolysis of poplar. *ACS Sustainable Chemistry and Engineering* 2018, 6(5), 5969–80.
73. Schmidt, H.-P. and Wilson, K. 55 uses of biochar. *The Biochar Journal* 2014. Available at: https://www.biochar-journal.org/en/ct/2.
74. Lan, Y. and Wang, W. Application of tree biochar in PDMS pervaporation membranes. *Advances in Polymer Technology* 2017, 37(6), 1979–86.
75. Peterson, S. C. Utilization of low-ash biochar to partially replace carbon black in styrene-butadiene rubber composites. *Journal of Elastomers and Plastics* 2013, 45(5), 487–97.
76. Laili, Z., Yasir, M. S., Wahab, M. A., Mahmud, N. A. and Abidin, N. Z. Evaluation of the compressive strength of cement-spent resins matrix mixed with biochar. *Malaysian Journal of Analytical Sciences* 2015, 19(3), 565–73.
77. Gupta, S. and Kua, H. W. Factors determining the potential of biochar as a carbon capturing and sequestering construction material: critical review. *Journal of Materials in Civil Engineering* 2017, 29(9), 04017086.
78. Huggins, T., Wang, H., Kearns, J., Jenkins, P. and Ren, Z. J., Biochar as a sustainable electrode material for electricity production in microbial fuel cells. *Bioresource Technology* 2014, 157, 114–19.
79. Dang, Q., Mba Wright, M. and Brown, R. C. Ultra-low carbon emissions from coal-fired power plants through bio-oil co-firing and biochar sequestration. *Environmental Science and Technology* 2015, 49(24), 14688–95.
80. Hsu, D. D. Life cycle assessment of gasoline and diesel produced via fast pyrolysis and hydroprocessing. *Biomass and Bioenergy* 2012, 45, 41–7.
81. Snowden-Swan, L. J. and Male, J. L. *Summary of Fast Pyrolysis and Upgrading GHG Analyses*. Pacific Northwest National Laboratory (PNNL), Richland, WA, 2012.
82. Han, J., Elgowainy, A., Dunn, J. B. and Wang, M. Q. Life cycle analysis of fuel production from fast pyrolysis of biomass. *Bioresource Technology* 2013, 133, 421–8.
83. Winjobi, O., Zhou, W., Kulas, D., Nowicki, J. and Shonnard, D. R. Production of hydrocarbon fuel using two-step torrefaction and fast pyrolysis of pine. Part 2: life-cycle carbon footprint. *ACS Sustainable Chemistry and Engineering* 2017, 5(6), 4541–51.

84. Vienescu, D., Wang, J., Le Gresley, A. and Nixon, J. A life cycle assessment of options for producing synthetic fuel via pyrolysis. *Bioresource Technology* 2018, 249, 626–34.
85. Sorunmu, Y., Billen, P., Elangovan, S., Santosa, D. M. and Spatari, S. Life cycle assessment of alternative pyrolysis-based transport fuels: implications of upgrading technology, scale and hydrogen requirement. *ACS Sustainable Chemistry and Engineering* 2018, 6(8), 10001–10.
86. Kauffman, N., Hayes, D. and Brown, R. A life cycle assessment of advanced biofuel production from a hectare of corn. *Fuel* 2011, 90(11), 3306–14.
87. Brown, T. R. A techno-economic review of thermochemical cellulosic biofuel pathways. *Bioresource Technology* 2015, 178, 166–76.
88. Brown, T. R. and Wright, M. M. Techno-economic impacts of shale gas on cellulosic biofuel pathways. *Fuel* 2014, 117, 989–95.
89. Winjobi, O., Shonnard, D. R. and Zhou, W. Production of hydrocarbon fuel using two-step torrefaction and fast pyrolysis of pine. Part 1: techno-economic analysis. *ACS Sustainable Chemistry and Engineering* 2017, 5(6), 4529–40.
90. Jones, S. B., Valkenburg, C., Walton, C. W., Elliott, D. C., Holladay, J. E., Stevens, D. J., Kinchin, C. and Czernik, S. *Production of Gasoline and Diesel from Biomass via Fast Pyrolysis, Hydrotreating and Hydrocracking: A Design Case*. Pacific Northwest National Laboratory, Richland, WA, 2009.
91. Zhang, Y., Brown, T. R., Hu, G. and Brown, R. C. Techno-economic analysis of monosaccharide production via fast pyrolysis of lignocellulose. *Bioresource Technology* 2013, 127, 358–65.
92. Brown, T. R., Wright, M. M. and Brown, R. C. Estimating profitability of two biochar production scenarios: slow pyrolysis vs fast pyrolysis. *Biofuels, Bioproducts and Biorefining* 2011, 5(1), 54–68.
93. Brown, T. R., Thilakaratne, R., Brown, R. C. and Hu, G. Techno-economic analysis of biomass to transportation fuels and electricity via fast pyrolysis and hydroprocessing. *Fuel* 2013, 106, 463–9.
94. Ning, S.-K., Hung, M.-C., Chang, Y.-H., Wan, H.-P., Lee, H.-T. and Shih, R.-F. Benefit assessment of cost, energy, and environment for biomass pyrolysis oil. *Journal of Cleaner Production* 2013, 59, 141–9.
95. Badger, P., Badger, S., Puettmann, M., Steele, P. and Cooper, J. Techno-economic analysis: preliminary assessment of pyrolysis oil production costs and material energy balance associated with a transportable fast pyrolysis system. *BioResources* 2010, 6(1), 34–47.
96. Jaroenkhasemmeesuk, C. and Tippayawong, N. Technical and economic analysis of a biomass pyrolysis plant. *Energy Procedia* 2015, 79, 950–5.
97. Dang, Q., Wright, M. M. and Li, W. Technoeconomic analysis of a hybrid biomass thermochemical and electrochemical conversion system. *Energy Technology* 2018, 6(1), 178–87.
98. Islam, M. N. and Ani, F. N. Techno-economics of rice husk pyrolysis, conversion with catalytic treatment to produce liquid fuel. *Bioresource Technology* 2000, 73(1), 67–75.
99. Flash pyrolysis of dry biomass. Available at: https://www.bioliq.de/english/64.php (accessed December 07, 2018).

Chapter 3

Producing biofuels with torrefaction

Donald R. Fosnacht, Natural Resources Research Institute – University of Minnesota, USA

1 Introduction

With carbon-capture and sequestration technology still in development, many coal-based power plants are looking for alternative ways to reduce CO_2 emissions. Some utilities have started mixing coal with materials such as woody biomass, which could help them meet the expected environmental targets. Co-firing with wood and coal is becoming a viable 'bridge strategy' for increasing the use of renewable resources while reducing atmospheric CO_2. Worldwide, over 200 test burns have been completed for co-firing wood with coal at large-scale coal-fired power plants to show the feasibility of this technique to reduce CO_2 in plant emissions.[1,2]

http://dx.doi.org/10.19103/AS.2019.0027.08

Compared with fossil fuels, raw biomass has not been widely utilized in the electric power generation industry due to its relatively low energy density. Biomass pretreatment technologies have therefore been developed to densify biomass into forms that can be stored and handled in a manner consistent with coal use at power generation operations.[3] The biomass industry is currently focusing on biomass pretreatment technologies for either pelletizing raw biomass fuels or pelletizing torrefied biomass fuels. The wood pelletizing process for the production of wood fuel pellets is a well-developed technology worldwide.

The torrefied wood industry, however, is in a 'development stage' in that many torrefaction processes are being researched and refined, with no one technology perfected or preferred as yet. Still progress is being made, especially at the bench and pilot scale. IEA Bioenergy reports that about a dozen different technologies, from oscillating bed to rotary drum, are in use around the world. Nine facilities are in the United States, while 23 other facilities span the globe from Nova Scotia to New Zealand to Europe. Projects are ongoing to scale up the production of stable and high-quality pellets or briquettes at hundreds of thousands of tons per year.

The global electric power industry is seeking 'refined' renewable fuel products as its fuel source to partially or fully replace coal and to therefore reduce carbon and other significant emissions. 'Refining' is a generic term for different fuel-processing technologies including steam explosion, torrefaction, and hydrothermal carbonization (HTC) (also called wet torrefaction). Torrefaction, 'a partial pyrolysis process', significantly improves the suitability of biomass for co-firing in coal-fired power plants while providing the potential to increase the percentages of biomass versus using untreated wood pellets. The quality of the torrefaction process depends on the balance between temperature and residence time to preserve a maximum of energy density to achieve certain fuel properties like grindability and moisture resistance.[4] In addition, torrefaction and HTC also can be used as pretreatment technologies to produce biomass-based synthetic gas and concentrated carbon products such as char and activated carbon. The type of biomass also impacts which process is used. Woody materials have been the main source for torrefaction processes, whereas steam explosion and HTC can employ other non-woody biomass materials without significant moisture reduction prior to processing.

While the lignin content in wood is usually enough to bind wood pellets, other forms of biomass require binders to strengthen them. Starch, sugars, guar gum, fenugreek, paraffin oils, or lignin can be added to make the pelletized biomass more durable.[5] Water repellent pellets or briquettes are required to produce an acceptable coal replacement product that can be shipped in bulk in open containers and stored in a manner similar to coal. As of 2018, emerging biomass torrefaction companies have significantly improved their ability to produce high-quality products with pellets of comparable durability to conventional wood pellets. But key areas of work remain, including:

- densifying with and without binders to enhance the bulk density of the produced fuels,
- developing moisture resistance regimes to allow outdoor storage, and
- optimizing the shape and size of the fuel products, and the degree of pretreatment required to reduce ash content and to achieve the desired fuel values in the products.

In addition, the magnitude of thermal treatment performed on the biomass material has a great impact on the ease of densification. More thoroughly treated materials become very difficult to pelletize using conventional wood pellet processing techniques and can favor alternative technologies such as ram or rotary briquetting.

1.1 Use of torrefied fuels in power plants[6]

Various tests on the use of torrefied biomass have been conducted. The largest demonstrations have occurred because of a program sponsored by the Consortium for Advanced Wood to Energy Solutions (CAWES) which was funded by the US Endowment for Forestry and Communities and the US Forest Service of the US Department of Agriculture.

1.1.1 Southern Company's Plant Scholz

Prior to the start of the CAWES program, a significant test was also conducted by Southern Company, at its Gulf Power subsidiary, which successfully tested the use of conventional pellets that had been torrefied in a mobile torrefaction facility. Approximately 500 tons of white wood pellets were converted and used in this test program to achieve 100% coal substitution. The company concluded that the use of torrefied materials was a straightforward path to substituting increasing amounts of coal in power generation.

1.1.2 Minnesota Power's Taconite Harbor Plant

During the CAWES program, Minnesota Power conducted a test program using approximately 120 tons of torrefied pellets produced by New Biomass at their Quitman, Mississippi, facility. Minnesota Power summarized their observations for the torrefied fuel trial at their Taconite Harbor facility in Minnesota. Significantly more material is needed to establish the quantitative impact of using torrefied fuel at their facility. However, the trial, conducted in September

2016, aimed to substitute 25% of the western sub-bituminous coals in use. The biomass came from a stockpile of product that had an energy content of approximately 22 050 kJ/kg (9500 BTU/lb.). The material was not a prime product that had been freshly made by New Biomass and was tested over a 2-day period. In the first case, one pulverizer was used at 100% torrefied pellets with three pulverizers on coal, and in the second case, two pulverizers operated at 50% torrefied fuel and 50% coal, while also operating two more pulverizers at 100% coal. Basic findings from Minnesota Power are as follows:

- The torrefied fuel was found to be an acceptable fuel substitute for coal, but changes in operating conditions were needed to accommodate the new fuel.
- Some pulverization issues in emptying the pulverizer were found, with under-torrefied or 'white' wood being too fibrous for easy comminution by the pulverizers, and it therefore collected in the system.
- The material was dustier to handle, but they used various techniques to control the dust levels.
- The sample size was useful to gauge the utility of using the torrefied fuel for the future, but larger trials were deemed necessary to fully measure environmental and operating impacts.

1.1.3 Portland General Electric's (PGE) Boardman, Oregon Plant

A much larger test was held during the CAWES program at Portland General Electric's Boardman, Oregon, Power Station. Various suppliers of torrefied fuels were used in a series of test burns to assess needed coal plant modifications. This work was coordinated by Oregon Torrefaction and involved securing production of torrefied products using equipment in Oregon and obtaining additional materials from New Biomass and other sources. Approximately 5000 tons of torrefied material was used at the plant. During the test period, the bulk of the material tested was produced at the New Biomass plant in Quitman, Mississippi. PGE reported: 'This series of test burns were performed from December 2016–February 2017 on the B pulverizer both to practice safe operations with biomass fuel and to determine the necessary pulverizer modifications required for biomass processing. These modifications were needed due to the poor grindability of the biomass fuel, which is not as friable as coal. In addition, practice modification was required for safe operations with a low-moisture, high-volatile biomass fuel that was very different in combustion behavior than coal. An additional complication was the fact that Boardman's pulverizers were designed for grinding coal to a very high fineness (75% or more passing a 200 mesh screen (74 μm)), so certain changes were required in order to 'open up' the mills for a biomass fuel that not only had a much larger fuel particle size, but was 50–70% less dense than coal. Significant pulverizer modifications were made to accommodate the torrefied fuel.'

Several single-pulverizer test burns were conducted, each one yielding greater and greater success, and concluded with a highly successful burn on February 7, 2017, where the B pulverizer was able to maintain 45 455 kg/h (100 000 lbs./h) of fuel flow. After the single-mill testing, pulverizers B, C, F, and G were selected to process biomass for a full-unit test burn, and each of these pulverizers was modified using the lessons learned from the single-mill test. After pulverizer modification, a major test using the four pulverizers was conducted. During the multiple biomass test burns, there were events that occurred where specific equipment and/or processes did not function as predicted. The lessons learned from this process resulted in a steady process improvement in the test burn program and were responsible for the successful full-unit test burn. On February 23, 2017, Boardman successfully burned 1653 tons of torrefied biomass at complete substitution for coal. After completion of the primary test program, the remaining fuel pellets were consumed over the course of time at approximately 25% substitution level.

1.1.4 Comparison of power plant conversion costs

Ontario Power in Canada has converted two plants in Western Ontario to completely use biomass materials. In one case, they modified the power plant to utilize conventional pellets, and the capital costs for this modification were estimated to be C$170 000 000. In the second case, the power plant decided to use advanced wood pellets produced from steam explosion processing methods (Zilkha or Arbaflame), and the capital cost to allow use of the materials was only C$5 000 000. The capital cost reduction illustrated that the advanced wood pellets could be used like coal in that second plant example.

1.1.5 Challenges and changes

Production of torrefied fuels is still in its infancy. Compared to the production of conventional pellets, very little production capacity exists to routinely produce commercial quantities of this type of fuel product. The successful use of torrefied fuels at the three plants noted, and the parallel work to improve the physical quality, energy density, and moisture resistance of the fuels is an important enabler in achieving products in the future that can meet the needs of power plants. At Plant Scholz, it was found that the material can lead to spontaneous heating like sub-bituminous coal. Practices developed at the various power plants to routinely control the problems associated with handling coal were used to handle the torrefied biocoal as well. As illustrated by PGE, modifications of the grinding system were made to accommodate the efficient use of the material. Under-torrefied materials appear to have fibrous constituents from the original plant structure that can accumulate in the pulverizer, and this can lead to buildup of material in the system that can be a fire hazard.

The key is to break down this fibrous structure in the torrefaction treatment so that good grindability like coals is obtained. The physical properties for the products needed at a given end-user are dictated by location and environmental conditions at the plant location. The situation at PGE differed significantly from that at Plant Scholz in Florida, as the Boardman location is an arid climate compared to the gulf coast of Florida. Significant breakdown of pellets occurred with the first trial at Plant Scholz, and this was likely due to the type of processing that was used to produce the fuel at this location (conventional pellets were converted to torrefied pellets) and the atmospheric conditions at the Florida location. At Boardman, various snow and rain events did occur with the stored fuels, but the pellets–even though dustier than handling coal–largely remained intact. The outer few inches of the piles showed some breakdown, but the interior of the pile did not seem to be impacted by the inclement conditions. The bulk of the materials at Boardman were supplied with torrefied fuels made from processed biomass and not converted from conventional pellets. The learning from these trials is that the physical integrity and energy content of the fuel products still needs to be optimized in terms of moisture resistance, elimination of fibrous residuals in the produced fuel, and effective physical strength to reduce dust formation.

1.1.6 Economic considerations

A European economic analysis indicates that, considering all aspects of potential fuel use, advanced wood pellets compared to conventional pellets have a significant economic advantage when logistics and actual cost of use at the power plant are considered. This reflects the higher energy content per unit volume of torrefied materials compared to raw biomass and conventional wood pellets and the avoidance of significant capital expense at current coal-fueled power plants that are required to use conventional wood pellets compared to the torrefied materials.

2 Biomass as a sustainable energy source

On a worldwide basis there are various drivers for the use of biomass as a substitution for coal in power plants. The primary consideration is the reduction of greenhouse gases and other environmental issues associated with coal combustion in existing power plants. This is driven by commitments by various nations to reduce their emissions as signatories to the Kyoto protocol or through other commitments at a state or local level. In the United States, Renewable Portfolio Standards (RPS), also referred to as Renewable Electricity Standards (RES), are policies designed to increase the generation of electricity from renewable resources. A detailed listing of state-level power which includes both the RPS and performance standards that apply to a given state is available. This information can be reviewed to understand the local situation for a given power plant application.[7]

These policies require or encourage electricity producers within a given jurisdiction to supply a minimum share of their electricity from designated renewable resources. Generally, these resources include wind, solar, geothermal, biomass, and some types of hydroelectricity, but may include other resources such as landfill gas, municipal solid waste, and tidal energy. RPS, are employed in the United States to facilitate the use of biomass as a renewable fuel source to partially replace coal as the primary fuel source in US power plants.

Biomass generation is forecasted to increase in the United States nearly fourfold from 2010 to 2035 driven by two main factors[8]:

- Federal requirements to use more biomass-based transportation fuels which leads to increased electricity generation as a coproduct from liquid fuel facilities such as cellulosic ethanol refineries.
- Co-firing of biomass with coal increases over the projection period, induced partially by state-level RPS as well as favorable economics in regions with significant forestry residues (see Fig. 1).

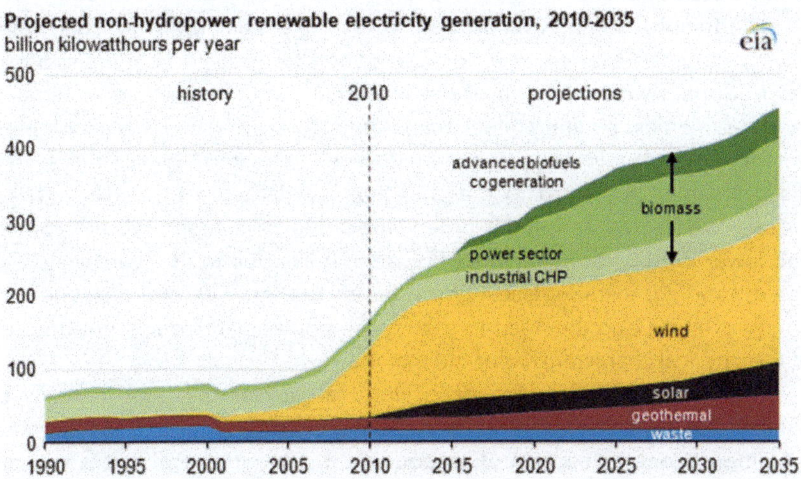

Figure 1 Projected non-hydro renewable electricity generation 2010-2035 in the United States.[8]

The market in the rest of the world is being driven by actions to reduce the carbon footprint of fuels used to produce electricity and in industrial applications. A good review of the use of wood pellets around the world is presented by the International Energy Agency in their Task 40 report.[2] Substantial conventional and advanced biomass-based fuels are expected to continue to grow in the years ahead.

Carbon-capture and sequestration technology is still under development; so many coal plants are looking for alternative ways to reduce the CO_2 from larger-scale fossil-fueled power plants. Some utilities have started mixing their coal with a cheap material such as woody biomass that could help them meet the expected environmental targets.

Until adequate methods of sequestering and storing and/or utilizing CO_2 from coal-fired power plant stack gases are developed and implemented, co-firing with wood and coal is a viable 'bridge strategy' for increasing the use of renewable resources while reducing atmospheric CO_2. Worldwide, over 200 test burns for co-firing wood with coal at large-scale coal-fired power plants have been conducted to show the feasibility of this technique to reduce CO_2 in plant emissions.[1] Co-firing with biomass is a 'ready-to-deploy' method of transitioning to a decarbonized electric power grid.

Biomass co-firing offers the potential to solve multiple problems. The current fleet of lower-cost, coal-fired, base load electricity generators are producing over 50% of the US power supply. With the 1990 Clean Air Act Amendments (CAAA) requiring reductions in emissions of acid rain precursors such as sulfur dioxide (SO_2) and nitrogen oxides (NO_x) from utility power plants, co-firing biomass at existing coal-fired power plants is one of many possible compliance options.

In addition, co-firing with biomass fuels from sustainably grown, dedicated energy crops are viewed as a possible option for reducing net emissions of CO_2. Coupled with the need of the industrial sector to dispose of biomass residues, biomass co-firing offers the potential for solving multiple problems at potentially modest investment costs. However, untreated biomass has several disadvantages compared to the coal that it is targeted to displace. These include low-energy density for green wood due to its high moisture content and lower fixed carbon content, poor grindability due to the tough nature of lignin bonding, extra logistical costs due to the inherent moisture and low-energy content, and the need to modify the power plant equipment because of the physical characteristics of the materials.

Production of dried, pelletized wood addresses the issue partially by concentrating the energy value in the fuel product, but it still does not meet the inherent energy content of an equivalent amount of coal that is typically employed. This limits the amount of coal displacement that can be employed at the power plant without derating the boiler system. In addition, as will be noted later, significant capital investment must be made at existing plants to accept even pelletized woody, dried biomass at high levels. Further pretreatment of dried biomass using one of several treatment methods (i.e. dry torrefaction, steam explosion, HTC (wet torrefaction)) can overcome the limitations of the other forms of biomass, and this can lead to reduced capital costs in implementing a biomass co-firing strategy[9] and reduced logistical costs in material movement and material storage.

The energy content of the fuel is usually a design criterion for a given boiler system. Plants using US Powder River Basin sub-bituminous coal types have energy contents that are much lower than bituminous coals from eastern US states. This can vary by coal type, but generally the Western coals have energy contents around 19728 kJ/kg (8500 BTU/lb.) versus 27850 kJ/kg (>12000 BTU/lb.) for the Eastern coals. This difference is a key factor in the pretreatment of biomass to make its fuel content more like the coal that it is displacing.

Compared with fossil fuels, biomass has relatively low-energy density and requires major modifications to the equipment. This low-energy density results in prohibitively large transportation costs combined with inconvenient biomass storage and handling issues. The plants using biomass on a co-firing basis generally are smaller in overall capacity (less than 50 MW) in the United States. A database of plants with their plant capacities is available and can be found summarizing the situation as of 2016.[10]

There are great benefits for co-firing biomass with coal, and these are summarized by Future Metrics below[11]:

- 'The environmental benefits are immediate and quantifiable. To lower carbon emissions by 10% requires a ratio of wood pellets to coal of about 11.24% pellets and 88.76% coal.
- The power generation assets that are fueled with pulverized coal gain significant new value. At a co-firing rate that results in a 10% CO_2 reduction, the increase in cost of generation is estimated to be less than one penny per kilowatt-hour.
- The coal producers have a long-term market for their product with a certainty for demand over the next several decades. Co-firing with biomass is not possible with gas turbines.
- The wood pellet producers have a new and gradually increasing market also with known demand.'[11]

Others note that biomass use in co-firing offers the potential to solve other key problems:

- 'The current fleet of low-cost, coal-fired, base load electricity generators are producing over 50% of the nation's power supply. With the 1990 Clean

Air Act Amendments (CAAA) requiring reductions in emissions of acid rain precursors such as sulfur dioxide (SO_2) and nitrogen oxides (NOx) from utility power plants, co-firing biomass at existing coal-fired power plants is viewed as one of many possible compliance options.

- In addition, co-firing using biomass fuels from sustainably grown, dedicated energy crops are viewed as a possible option for reducing net emissions of carbon dioxide (CO_2).
- Coupled with the need of the industrial sector to dispose of biomass residues, biomass co-firing offers the potential for solving multiple problems at potentially modest investment costs.
- Torrefied biomass can lead to reduced capital costs in implementing a biomass co-firing strategy (by making the biomass similar to the coal it is displacing)'.[12]

There are limits on how much co-firing can be achieved without significant plant modification. The European experience for biomass co-firing by premixing and co-milling has been summarized and, in general, indicates that 5–10% on a heat input basis of biomass can be employed without major equipment modification. The key constraints that are noted include[13]:

- Availability of suitable biomass;
- The limitations of on-site biomass reception, storage, and handling facilities;
- The limitations associated with the ability of the coal pulverizer mills to co-mill the biomass materials; and
- Safety issues associated with bunkering and milling of the mixed biomass-coal blends.

Other comments made based on the European experience include:

- Co-firing in pulverized coal combustion (PCC) boilers is easy to achieve at low rates using wood chips or other biomasses simply by adding the material to the coal feed to the existing mills; fly ash quality is not usually an issue.
- Wood pellets can be used at higher co-firing rates in PCC boilers using more extensive modifications–some employ 100% biomass but at significant capital expense.
- Modified fire safety systems are essential.
- Deposition and corrosion are containable.
- Efficiencies are not very greatly reduced.
- Importance of ensuring sustainability of biomass production, consistency, and biomass product standards is fully recognized by the utilities.

Some plants have converted to 100% biomass, but to accomplish this, they spend significant capital to modify the plant to accommodate the materials. Examples are the Schiller Station in Portsmouth, New Hampshire which converted one of its three 50 MW lines to burn wood chips, but the line required installation of a new fluidized bed boiler at a conversion cost of $70 million.[13] Another example is the Drax Power Plant in the United Kingdom that has converted two of its six plants to burn 100% wood pellets.[14]

3 The development of biomass pretreatment technologies

3.1 Wood pelleting–dried wood processed into a uniform shape and size

Various biomass pretreatment technologies have been or are being developed to densify biomass into forms that can be stored and handled in a manner consistent with coal usage at power generation operations without the need for massive equipment modifications and new storage facilities. In reviewing the status of biomass pretreatment technologies, it is well known that the biomass industry is focusing on biomass pretreatment technologies for pelletizing raw biomass fuels or otherwise densifying torrefied biomass fuels. The wood pelletizing process for the production of conventional pellets is a well-developed technology and turnkey plants are now implemented. US examples include Enviva Partners, Highland Pellets, New England Wood Pellet, Fram Renewable Fuels, Georgia Biomass, Green Circle, New Biomass, and others.

3.2 Concentration of energy beyond dried wood and enhancing fuel properties

The pretreatment wood industry, by contrast, is in a 'development stage' in that many pretreatment processes (torrefaction, steam explosion, and HTC) are being researched and refined, with an uncertainty of which technologies will prevail to commercialization. US examples include Zilkha, New Biomass Energy, River Basin Energy, and others. The matrix in Table 1 summarizes the current state of developments in North America. The world situation is reviewed later.

The industry is seeking 'refined' renewable fuel products to partially or fully replace coal as its fuel source to reduce carbon and other emissions (e.g. mercury, sulfur). 'Refining' is a generic term for different fuel pretreatment processing technologies including steam explosion, torrefaction, and HTC (sometimes referred as wet torrefaction). All these technologies are employed to produce a fuel that is closer to coal in overall properties. Raw pelletized biomass products are not easily milled like coal, and therefore are not easily used as fuel sources feeding typical powder coal injection (PCI) burners in

Table 1 Torrefaction developments in North America (www.energyjustice.net/map/nationalmapbiomass)

		Additional info				Comments		
		Plant 1						
Company name	Country	Type (0 = lab scale, 1 = pilot, 2 = commercial scale)	Status (0 = planned, 1 = in operation, 9 = inactive)	Plant location	Capacity	Major feedstock	Technology	Product shape
HM3 Energy	USA	1	1	Troutdale, OR	454.4 kg/h	Wood biomass	HM3 Energy	Briquettes
Torrefaction Process Inc.	USA	2	0	Carson City, NV		Woody biomass	Proprietary Design	Pellets
CNFBiofuel, Inc	USA	1	0	Salida, CA	3 t/d	Woody biomass	CNFbiofuel, Inc./Inventive Resources	Pellets
Renewable Fuel Technologies	USA	1	9	San Mateo, CA	1 t/d	Woody biomass	Mobile Torrefaction Equipment,	pellets
River Basin Energy, WRI	USA	1	1	La ramie, WY	6000 kg/h		Rotary Kiln	Pellets
AEON Energy Solutions/Konza Renewable Fuels	USA	1	1	Healy, KS	1 t/h		Rotary Drum	
Earth Care Products	USA	1	1	Independence, KS	20,000 t/a		Rotary Drum	
Torrsys	USA	1	9	Minneapolis, MN	1.44 t/d	Woody biomass	Torrsys	

Biomass Energy Resources LLC	USA	1	1	Frisco, TX				
New Biomass Energy	USA	2	1	Quitman, MS	30,000 t/a	Waste wood and bio	Oil heated screw	Pellets
AgriTech Producers-SC, LLC	USA	1	0	Allendale, SC	13,000 t/a		Screw reactor	
NCSU, North Carolina State University	USA	2	0	Columbia, SC	5 t/h	Woody Biomass	Screw reactor	
Integro Earth Fuels	USA	1	1	Greenville, SC	11,000 t/a	Woody Biomass	Multiple hearth	
Terra Green Energy	USA	1	1	McKean, PA	12 t/d	Wood Feedstock	Multiple hearth	
Biomass Secure Power Inc	USA	2	0	Nalchitoches, LA	340,000t/a	Woody biomass		Pellet
American Biocarbon	USA	2	1	White Castle, LA	250,000 t/a	Sugar plant waste	Rotary Dryer and Kiln	Briquettes
Natural Resources Research Institute	USA	2 Demo	1 & 0	Coleraine, MN	2 to 5 t/d	Woody biomass	Rotary Kiln, Moving Bed	Pellets, briquettes, extrudates, char
Terradyne Energy Inc.	Cana da	2	0	Saint Stephens, NB		Woody biomass	'heat lock technology' cited	pellets
Airex Energy	Cana da	2	1	Becancour, QC	2 t/h	Woody biomass	Cyclonic Bed	Pellets, char

(Continued)

Table 1 (Continued)

		Additional info				Comments		
		Plant 1						
Company name	Country	Type (0 = lab scale, 1 = pilot, 2 = commercial scale)	Status (0 = planned, 1 = in operation, 9 = inactive)	Plant location	Capacity	Major feedstock	Technology	Product shape
Manitoba Hydro	Cana da	1	0	Elie, MB		Woody biomass	Preseco Oy	Pellets
Global Bio-Coal Energy	Canada	2	0	Lumby, BC		Woody biomass	Turbo-Dryer Thermal Unit	
Diacarbon	Canada	1	0	Merritt, BC	35,000-40,000 t/a	Woody Biomass	Diacarbon	pellets, char
Biomass Secure Power Inc.	Canada	2	0	Lake Cowichan, BC	340,000 t/a		Not Described	pellets

power plant boiler systems. In general, the energy density is higher for the thermal-processed biomass.[15]

4 Biomass densification technologies

Densification of biomass improves the bulk density of the material by mechanical means to improve logistics and storage. Mechanical type presses, for example, are used to produce biomass pellets, tablets, and/or cubes. Raw biomass fibers can be converted from a density of 80 kg/m^3 (5 lbs/ ft^3) to a density of 640 kg/m^3 (40 lbs/ft^3) in the densification process.[17]

In general, biomass densification focuses on pelleting, but briquetting is considered as well. Good-quality pellets can be produced without additional binders being used. If no binders are used the process window product quality that meets logistics and end-use targets can be very tight. Good-quality control of pelleting process conditions is essential, especially in the case of producing torrefied biomass pellets. For torrefied materials, temperature control and particle sizing are important parameters that impact the ease of densification. In addition, since fuels are manufactured in the pretreatment process, special attention to safety issues is essential to reduce self-heating and dust explosion problems.[18] The degree of torrefaction conversion significantly impacts the ease of densification using pelletizing, as the binding properties of lignin are significantly altered by employing a higher temperature and/or increasing the treatment time. Both factors lead to higher material losses through removal of increasing amounts of volatile matter from the biomass.

4.1 Biomass densification pelleting/briquetting process

Biomass densification processes are divided into two broad categories, closed and open die compaction. Biomass pellets are formed by a continuous extrusion process through an open die. Biomass briquettes are formed in a roll press machine utilizing closed cup compaction on at least one roller.[19] Wood pellets and torrefied wood (TW) pellets currently dominate biomass renewable fuel sources, and wood pellets dominate the home heating marketplace.

Loose biomass material having an average density of 80 kg/m^3 (5 lbs./ ft^3) typically has high moisture (30–50% H_2O), non-uniform particle size, and is susceptible to spoilage through biological reaction with bacteria and mold. The pelletizing/briquetting process then converts low-density biomass into high-density pellets/briquettes with an average density of 640 kg/m^3 (40 lbs./ ft^3). This pelletized/briquetted product typically has lower moisture, uniform size, is easy to store in covered storage facilities, and can be transported long distances.[17]

Biomass pelleting processes manufacture products having the following advantages as compared to loose biomass products:

- Uniform size (10–12mm × 6mm), density, and moisture content;
- Moisture content, (6–8% H_2O);
- Easy to transport, convey, and feed using existing systems;
- High heating value ~18.5 GJ/t (15.9 MMBTU/ton);
- Multiple uses such as power generation, domestic heating, biofuels production, and animal bedding; and
- High export value.

Many factors impact the strength and durability of pellets. Strength refers to the compressive and impact resistance, while durability refers to the friability or abrasive resistance. Parameters impacting strength and durability include additive and/or binder addition, moisture content, particle size, pelletizing temperature, pelletizing pressure, steam conditioning (steam explosion processing), torrefaction processing, and other factors (i.e. species processed).[19]

Additives in biomass pellet production can increase the strength and durability of wood pellets. The composition of biomass materials directly influences bonding characteristics. High lignin content biomass can readily be formed into pellets using contained lignin as natural binders. Biomass with lower lignin content may need binding agents to help develop improved bonding.[20] Torrefaction reduces the effective binding properties of the lignin through thermal modification.

Moisture content has a large effect on pelleting and the strength of the pellets produced. During the pelleting process, some water is necessary for the development of intermolecular forces. However, too much moisture can lead to poor pellet quality. Li and Liu[38] suggest that the optimal moisture content for the densification of woody biomass is between 6% and 12%. Optimal moisture content for biomass pelleting reported in the literature spans a broad range, suggesting that the effect of moisture on the densification process is not well understood.[19]

Particle size also plays a key role in both the pelleting process and the subsequent use of the fuel product in the power generation. For pelleting, it is a key variable determining the strength of the pellets produced. When the feedstock is compacted, the distance between the particles is reduced and the intermolecular attractive forces play a role in bonding. Commonly, the attractive force is increased as the particle size is decreased.[20]

Temperature is another key variable that affects the pelletizing process. Higher temperatures produced during pelleting can partially melt the biomass constituents, which results in the formation of solid bridges between particles as pellets are cooled down.[20]

Pressure applied to biomass materials during the pelleting process is also a key factor in producing strong, durable pellets. Mechanical pressure developed in pellets mills is one of the major factors that influence particle bonding effects. Higher pelleting pressure will result in an improved bonding effect.[20]

5 Pretreatment technologies: an overview

The Natural Resources Research Institute is engaged in work to develop demonstration-level production of solid biofuel densified products that can be stored outside, have high bulk densities for ease of logistical transport, have good handling characteristics that minimize dust generation, possess grinding characteristics similar to coal used in power plants, and have fuel contents that match or exceed sub-bituminous coal levels. During the work, two pretreatment technologies have been investigated for concentrating the energy content of raw biomass. These include torrefaction using an indirectly fired rotary kiln process or a moving bed system at the demonstration level and HTC technology at the bench and pilot scale. The Institute has also collaborated with Syngas Technologies on a moving bed, directly heated steam-based process at the pilot scale.

A key factor in showing the full technical feasibility of using the pretreatment technologies is to demonstrate that the produced particulate fuel products can be densified to a level that allows good logistical and handling practices to be routinely attained. It has been found that hydrothermally carbonized materials can be agglomerated using a variety of densification devices including pelleting and briquetting in a repeatable and practical manner using commercial densification equipment with and without the use of binders.

However, torrefied materials have proven to be much more difficult to densify using a variety of densification equipment, especially as the degree of torrefaction increases. Uniformly torrefied materials at a high energy level appear to be especially difficult to densify but have the attributes of high fuel value and good grindability, with very little residual fiber content compared to less-torrefied material or steam-exploded biomass.

Therefore, the work undertaken and explained in the following discussion illustrates that that highly torrefied materials can be satisfactorily densified to produce high-energy-content products that have good physical properties, possess acceptable moisture resistance, low ash, sulfur and mercury content, and have bulk densities that can lead to improved logistics. The densification practices involve optimizing overall process conditions on an integrated systems basis and include moisture level, densification pressure, mix preparation pressure, and the use of appropriate binders when required. The densification system that seems to show the greatest promise for the highly torrefied materials is briquetting although recent work with a new biomass-based additive also improves the ability to form good pellets. Work will continue in examining other densification options and in improving the conditions used and discussed in this report. Steam explosion can also be used to produce fuel products with good physical properties and moisture resistance but possesses less energy per unit volume than well-torrefied materials, and the resultant product is very hard and difficult to grind compared to torrefied or hydrothermally carbonized materials. Hydrothermal carbonization is a useful technology for producing advanced energy products, but its development lags behind that of the other two technologies.

6 Steam explosion[21]

Steam explosion increases the calorific value of biomass due to the removal of moisture and volatiles and provides for the thermal degradation of hemicelluloses present within the biomass particles. The removal of hemicelluloses also changes the mechanical properties of the biomass. Steam exploded biomass has more 'coal-like' properties compared to untreated biomass creating more favorable grinding and combustion characteristics.[21]

The hygroscopic nature of biomass is high due to the presence of hydroxyl (OH) groups within hemicellulose and cellulose (see Fig. 2). These (OH) groups provide active bonding sites for water molecules. Depending on the severity of the steam explosion process, the number of available hydroxyl groups is often drastically reduced.[21] As a result, the carbon content of the biomass increases as oxygen and hydrogen atoms present in the hydroxyl groups are removed during

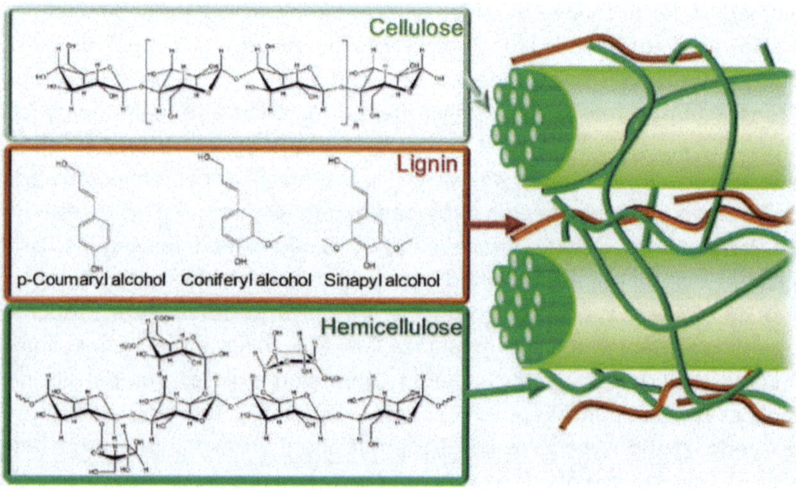

Figure 2 Relationship between cellulose lignin and hemicellulose.[16]

the steam explosion process. Typical materials processed for biomass preatment are illustrated in Fig. 3. The type of material will directly impact the processing conditions needed and the type of process required. In consolidation of biomass materials, various technologies can be used and pelleting is illustrated in Fig. 4. An example of a typical steam conditioning device is shown in Fig. 5.

The changes in the physical appearance of wood pellets treated with increasing steam explosion treatment times are shown in Fig. 6. The reduction of fibrous particles resulting from steam explosion pretreatment is shown in

Figure 3 The types of biomass and their composition of biomass materials directly influences bonding characteristics in conventional pellets.[19]

Figure 4 Mechanical pressure applied within pellet mills is one of the major factors that influences particle bonding within biomass pellets.[19]

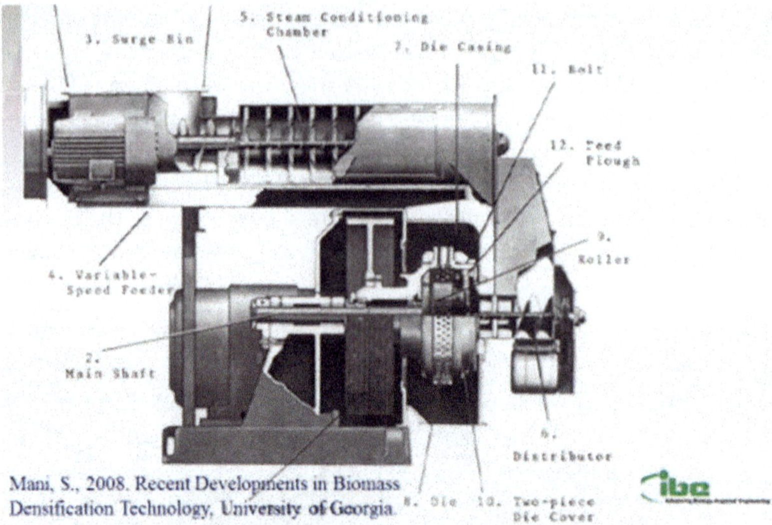

Figure 5 Biomass steam conditioning equipment design.[17]

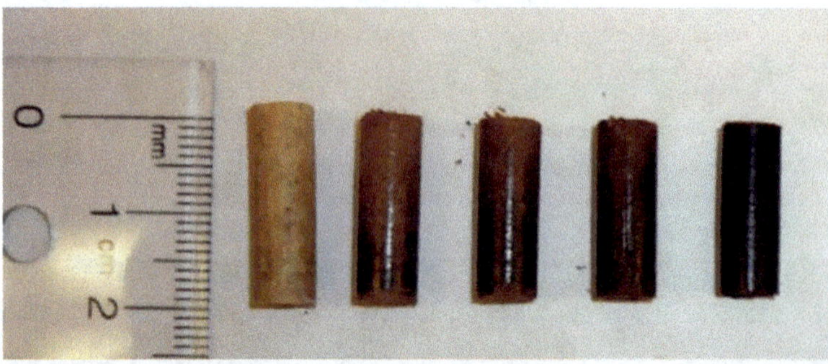

Figure 6 Physical appearance of wood pellets treated at different steam-explosion conditions.[20] (From left to right: untreated, 200°C for 5 min, 200°C for 10 min, 220°C for 5 min, and 220°C for 10 min. Pellets are 6.62 mm in diameter and about 18 mm in length.)

Fig. 7.[22] Two companies that extensively use the steam explosion concept are Zilhka and Arbaflame.[23] The removal of the hydroxyl groups from the biomass structure greatly improves the ability of the materials to shed water and remain intact through water submersion. This is illustrated in Fig. 8. This can reduce storage costs associated through use of outside storage. The 'white pellets' shown on the left glass basically have soaked up all the water in the glass and have disintegrated in structure, while the steam-exploded pellets have absorbed much moisture and remain intact as shown in the glass on the right of Fig. 8.

Steam explosion biomass pretreatment technology can be applied prior to torrefaction pretreatment to improve pellet processing as illustrated in the following pellet process flow diagram (Fig. 9).

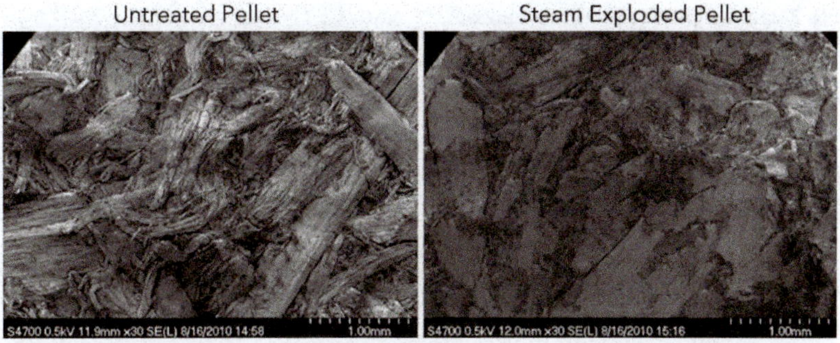

Figure 7 SEM photos of the cross-section of pellets made from (left) untreated and (right) steam-exploded (220°C for 5 min) feedstock at low magnification (30×). Untreated pellets show the stack of fibrous particles. The fibrous structure is not visible in the treated sample.[22]

Figure 8 Water resistance properties shown by steam-exploded wood pellets above on the right versus water soaked untreated wood pellets shown on the left.[23]

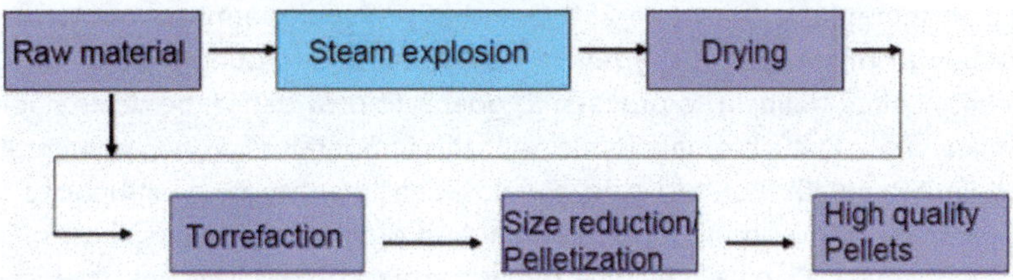

Figure 9 Use of the steam-explosion process prior to torrefaction to improve the biomass pelletizing process.[17]

Table 2 Biomass pellet quality comparison[17]

Properties	Saw dust	Wood pellets	Steam-exploded pellets
Moisture (%)	40	7-8	2-3
Energy content (MJ/kg)	10	19	20
Bulk density (kg/m³)	180	650	800
Energy density (GJ/m³)	1.8	12.4	16
Moisture uptake	High	High	Low

The rationale for combining these approaches is to enhance both the energy content and the grindability of the final product. Steam-exploded material has a higher energy density than typical conventional pellets but less than that obtained after a torrefaction treatment. In addition, it is harder to grind than torrefied material. The torrefaction treatment can enhance the grindability and the energy content of the final product relative to either 'white pellets' or steam-exploded material alone. Work at the University of Minnesota's Natural Resources Research Institute (NRRI) has shown that adding some portion of steam-exploded material to torrefied materials can greatly aid in the densification of the final products. In this case, the steam explosion process can be viewed as one way to create natural binders that allow enhanced processing for torrefied materials that have enhanced energy content relative to other pretreatment regimes. Similar findings have been found in using hydrothermal carbonized materials as an additive to torrefied materials.[31]

In order to provide a comparison of pellet properties produced by various biomass pretreatment techniques, Table 2 is shown illustrating the differences in biomass pellet properties.

7 Dry torrefaction

Torrefaction of biomass can best be described as a partial pyrolysis processing technique. The biomass is chipped to a uniform size and heated to a temperature of between 250°C and 320°C under atmospheric pressure in an atmosphere free of oxygen. It is similar to the roasting process used to process coffee beans. The processing goal is to provide for the destruction of hemicellulose in the biomass particles. In this temperature range, carbonization and limited volatization of lignin, cellulose, and hemicellulose structures take place. Depending on the input material and process conditions chosen, 30% of the dry mass is typically devolatized as torrefaction gas. This gas product can be burned as fuel in a separate combustion chamber to evaporate the moisture out of the incoming biomass feed and/or to provide heat for the torrefaction process. Typically, about 90% of the lower heating value (LHV) of the raw biomass material is contained in the torrefied product, which generally amounts to 70% of the original dry mass.[24] If we consider raw biomass at roughly 40% moisture, this would imply that for 1 unit of torrefied biomass produced, one would need approximately 2.4 units of raw biomass feed material. The volatile matter evolved during processing consists of various organic and inorganic molecules as shown in Table 3.[25]

The advantages of applying the torrefaction pretreatment process to biomass are listed as follows[4,35]:

1 Significant cost reductions in transport and handling of the product;
2 Broader feedstock basis–geographically plus types of raw material;
3 Almost zero biodegradation of product when stored (little interaction with biological agents);
4 Large variety of applications;
5 Reduction in both capital and operating expenses (CAPEX and OPEX) at the end user (can utilize like coal);
6 Immediate use in existing coal-fired plants;
7 Improved grindability and water resistance characteristics;

Table 3 Volatile matter from torrefaction treatment[25]

Species	Atomic composition
Acetic acid	$C_2H_4O_2$
Glyoxal	$C_2H_2O_2$
Propanal	C_3H_6O
Propanedial	$C_3H_4O_2$
5-Hydroxymethyl-furfural	$C_6H_6O_3$
Levoglucosan	$C_6H_{10}O_5$
Xylose monomer	$C_5H_8O_4$
Paracoumaryl alcohol	$C_9H_{10}O_2$
Phenol	C_6H_6O
Sinapaldehyde	$C_{11}H_{12}O_4$
Hydrogen	H_2
Carbon monoxide	CO
Carbon dioxide	CO_2
Methane	CH_4
Formaldehyde	CH_2O
Methanol	CH_4O
Ethylene	C_2H_4
Acetaldehyde	$C_2H_4O_2$
Ethanol	C_2H_6O
Water vapor	H_2O

8 Cleaner combustion (less ash, mercury, sulfur than coal);
9 Tailored products (can vary in the degree of torrefaction);
10 Uniform fuel value; and
11 Less smoke and reactive volatile matter (removal of oxygenated hydrocarbons through the torrefaction regime).

As a result of applying the torrefaction pretreatment process to biomass, a significant improvement in the suitability of biomass for co-firing in coal-fired power plants is produced while providing the potential to enable higher co-firing percentages at reduced rates versus using untreated wood pellets. The quality of the torrefaction process depends on the balance between temperature and residence time to attain maximum energy density and to achieve certain fuel properties like grindability and hydrophobicity. For densification (pelleting or briquetting) of the product, it is important to keep an amount of lignin as the natural binder in the biomass in order to avoid the need for additional binders.[23] Retained weight percentages of the various components of wood as it passes through increasing temperatures stages during the torrefaction process are shown in Fig. 10.[26]

Some representative coal heating values and chemistry compared to TW pellets heating value and chemistry values are shown in Table 4.[26] The reductions in ash, sulfur, nitrogen, and chlorine content because of torrefaction

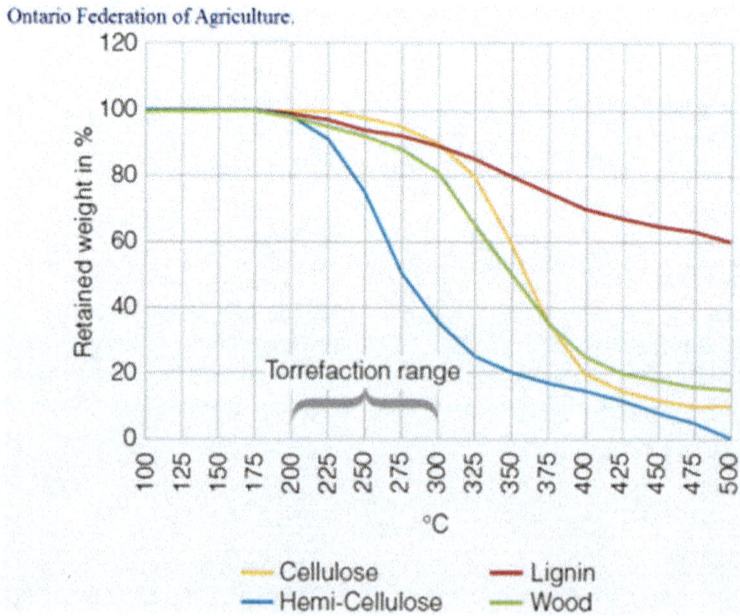

Figure 10 Changes in weight percentages of the main biomass components.[26]

Table 4 Coal heating value and chemistry compared to torrefied wood pellets

Item	Coal (can vary by type)	Torrefied pellets (varies by degree of torrefaction)
Heating value	25 GJ/T	22 GJ/T
Ash content	10%	3%
Sulfur content	3%	0.1%
Nitrogen content	1.5%	0.2%
Chlorine content	0.05%	0.01%

demonstrates the reduction in toxic gas emissions that can be obtained in using torrefied fuel products. As noted previously, the degree of torrefaction can greatly alter the ease of densification and make pellet formation through pelleting processing very difficult. In addition, trials at actual power plants of torrefied materials containing lightly torrefied material or untorrefied materials (used to help form pellets) can cause grinding difficulties in conventional coal pulverization systems. The formed products under these conditions also have reduced moisture resistance. This has led to the development of alternative densification methods for the highly torrefied materials produced by the Natural Resources Research Institute, University of Minnesota. Rotary briquetting, ram briquetting, and extrusion are actively being developed as an alternative to the standard pelletizing methods currently employed for white wood pellets.

The benefits of torrefaction processing of biomass are shown as follows with the processes' ability to produce a more homogeneous output product:

1 Uniformity of both physical and chemical properties;
2 Allows sourcing of different types of woody biomass in a single device, and therefore the economics of densification can be improved;
3 Potential for utilizing different types of local woody biomass for energy use in a single end user combustion unit, and therefore improving fuel availability, supply reliability, and reduces fuel costs; and
4 Reduced handling and storage costs through concentration of energy values and potentiality in lowering storage costs.

Kumar et al. have evaluated various pathways for producing torrefied pellets.[27] They describe the pathways as illustrated in Fig. 11.

When torrefied biomass pellets are produced, two different production pathways can be followed as listed below:

1 Pathway 1: Biomass is first dried and torrefied, then it is ground and then pelletized.
2 Pathway 2: Biomass is first dried , ground, pelletized, and then torrefied.

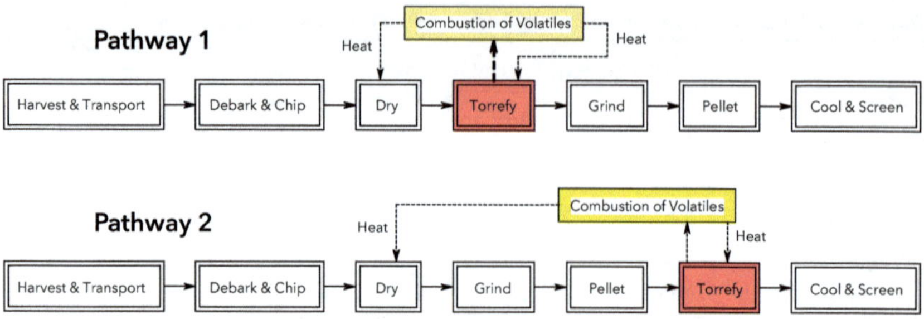

Figure 11 Alternative production pathways for torrefied pellets.

Table 5 Overall energy/mass balance for the two torrefaction pathways[27]

Product	Unit energy required MJ/t
White pellets	1873
Pathway 1 torrefied pellets	3300
Pathway 2 torrefied pellets	2939

In both cases, the volatile matter from the torrefaction step is used to supplement external fuel for drying and for torrefaction of the biomass.

The two pathways have distinctly different requirements. Pathway 1 often requires high energy and/or use of binders to densify the torrefied biomass into strong, durable pellets. Pathway 2 requires maintenance of physical compaction during and after torrefaction and requires that pellets be torrefied as efficiently as wood chips. The uniformity of feeding white wood pellets to the torrefaction reactor also can help improve overall process efficiency, but the physical properties of the torrefied product are less robust compared to the product made from Pathway 1. Construction of an energy and mass balance of the two pathways gives valuable insight into the energy requirements of each torrefaction pathway as shown in Table 5[28]:

The results in Table 5 indicate that potentially less energy is required in processing via the second pathway. In addition, the bulk density of the materials from Pathway 2 is less than that obtained from Pathway 1. Work at NRRI has shown an interesting result in that the processing by Pathway 2 does result in a product that when simply crumbled has an enhanced bulk density and improved hydrophobic character relative to the uncrumbled product from Pathway 2. The crumbled product results in a very coarse particulate material that may be a good fuel product. In addition, fines produced from the crumbling treatment can be a good supplement in white wood preparation to produce energy-enhanced conventional pellets.

Torrefied biomass can be densified into a range of sizes as shown in Fig. 12:

Figure 12 Torrefaction products can be densified into a range of sizes (Andritz, AG).

Torrefaction significantly improves heating values and the energy density of the fuel products as summarized in data from the IEA Bioenergy Task Report shown in Table 6.[28]

A table showing the growing list of torrefaction initiatives around the world as of 2015 is shown in Table 7.[29] There is no consensus on the type of technology that will emerge as the best for creating uniform, high-density fuel products using torrefaction treatments. Very few of the development efforts have reached the production size generally employed by the current white pellet industry. A comparison of the advantages and disadvantages of the various technologies is analyzed in Table 8.

As can be seen each technology has certain advantages in terms of flexibility, technical maturity, and production capabilities. The actual choice on which technology to select for commercialization will be dependent on the size of the plant and the variety of products targeted for production. For example, an oil-heated screw reactor will have an upper limit on the temperatures that can be used for processing due to the limits imposed by oil used as the energy carrier. A properly configured, indirectly heated rotary kiln can increase to much higher temperatures depending on the materials of construction and the overall design. In this case, it can likely be used to produce higher temperature chars, activated carbon, and other carbon products besides its use in torrefaction. Fluidized beds will require preconditioning of the materials to a specific smaller particle size requirement before they enter the reactor, whereas, some of the other technologies can process coarser chips directly. Moving bed systems may possess enhanced

Table 6 Comparison of fuel properties[28]

Item	Wood	Wood pellets	Torrefaction pellets	Charcoal	Coal
Moisture content (wt%)	30-45	7–10	1-5	1-5	10–15
Lower heating value (MJ/kg)	9–12	15–18	20–24	30–32	23–28
Volatile matter (% db)	70-75	70-75	55-65	10–12	23.28
Fixed carbon (% db)	20-25	20-25	28-35	85-87	50-55
Bulk density (kg/l)	0.2–0.25	0.55–0.75	0.75–0.85	~0.2	0.8–0.85
Energy density (GJ/m^3)	2.0–3.0	7.5–10.4	15.0–18.7	6-6.4	18.4–23.8
Dust	Average	Limited	Limited	High	Limited
Hydroscopic Properties	Hydrophilic	Hydrophilic	Hydrophobic	Hydrophobic	Hydrophobic
Biological degradation	Yes	Yes	No	No	No
Grindability	Poor	Poor	Good	Good	Good
Handling	Special	Special	Good	Good	Good
Quality variability	High	Limited	Limited	Limited	Limited

Table 7 Various torrefaction initiatives around the world[29]

Developer	Technology	Location	Production capacity (t/y)
Clean Electricity Generation	Oscillating Bed	Derby (UK)	30000
Horizon Bioenergy	Oscillating Belt Conveyor	Steerwijk (NL)	45000
New Biomass Energy	Screw Reactor	Quitman (USA/MS)	30000
Topell Energy (Blackwood Energy)	Torbed Reactor	Duiven (NL)	60000
Torr Coal B. V.	Rotary Drum	Dilsen-Stokkem (BE)	30000
Airex	Cyclonic Bed	Becancour (Can/QC)	16000
Agri-Tech Producers LLC	Screw Reactor	Allendale (USA/SC)	13000
Andritz	Rotary Drum	Frohleiten (AT)	10000
Andritz	Moving Bed	Stenderup (DK)	10000
BioEndev	Screw Reactor	Holmsund, Umea (SWE)	16000
CMI Nesa	Multiple Hearth	Seraing (BE)	Undefined
Earth Care Products	Rotary Drum	Independence (USA/KS)	20000
Grupo Lantec	Moving Bed	Urnieta (SP)	20000
Integro Earth Fuels, LLC	Multiple Hearth	Greenville, (USA/SC)	11000
LMK Energy	Moving Bed	Mazingarbe (FR)	20000
Teal Sales Inc	Rotary Drum	White Castle (USA/LA)	15000
Torrec	Moving Bed	Mikkeli (FI)	10000
Airless Systems	Unknown	Latvia	40000
Atmosclear	Rotary Drum	Latvia, New Zealand, USA	50000
Bioenergy Dev & Production	Fluidized Bed	Nova Scotia (Can)	Undefined
BioEnergy Development North AB	Rotary Drum	O-vik (SWE)	25000
BioLake B. V.	Screw Conveyor	Eastern Europe	5000-10000
EBES AG	Rotary Drum	Frohnleiten (AU)	10000
ECN	Moving Bed	Stenderup (DK)	10000
HM3 Energy	Torr B Technology	Gresham (USA/OR)	3500
Integro Earth Fuels, LLC	Turbo Dryer	Foxboro (USA/NC)	80000
Renergy/4Energy Invest	Rotary Drum	Amel, Ham (BE)	38000
River Basin Energy	Fluidized Bed	Laramie (USA/WY)	48000
Rotawave, Ltd.	Microwave Reactor	Terrace (CAN, BC)	110000
Thermya/Grupo Lantec	Moving Bed	Umieta (SP)	20000
Thermya/LMK Energy	Moving Bed	Mazingarbe (FR)	20000
WPAC	Unknown	Unknown	35000

Table 8 Comparison of reactor types and advantages and disadvantages

Reactor type	Percent of developers using	Typical demonstration capacity (t/h)	Advantages	Disadvantages
Rotary Drum	25.0	3–11	Well established for drying. Relatively inexpensive; potentially can go to higher temperatures for other value-added products.	Indirectly heated can lead to lead to uneven heat transfer; higher capacities will require significant equipment investment and space.
Screw Reactor	12.5	1–5	Can handle variety of particle sizes, good heat transfer possible with both internal and external heating.	Oil heating can limit temperatures that can be used and not allow higher value products to be attained; potential for product attrition due to friction in the system; multiple reactors may be required for high production rates.
Multiple Hearth	9.4	1–5	Variety of particle sizes can be treated; proven technology; relatively compact and inexpensive.	Uneven heat transfer depending on particle uniformity.
Moving Bed	21.9	2–7	Simple and low cost; high reactor capacity and high heat transfer especially with superheated steam as media.	Potential for high pressure drop and particle bridging; non-uniform temperature across bed if using indirect heating.
Fluid Bed/Torbed	9.4	4–9	Good heat transfer and scalability; very fast heating rates.	Requires fluidizable particle sizes and may require inert fluidizing gas.
Cyclonic/Oscillating Bed	9.4	2–5	Variety of particle sizes can be handled; well-proven technology for drying, limited higher temperature capabilities for more advanced products.	Scalability for higher production levels and potential for belt clogging and higher maintenance.
Microwave	3.0	12	Direct, internal heating of the particles due to coupling with moisture; likely can be scaled similar to commercial baking ovens.	New technology and electrical conversion energy efficiency may reduce efficiency.

Figure 13 Demonstration scale torrefaction system at NRRI.

efficiency compared to other methods but may also be more problematic due to potential for material bridging in the reactor. Ultimately, pretesting of material to be processed by a given combination of material preparation and processing regimes is necessary before implementing a commercial plant equipment configuration. The Natural Resources Research Institute, University of Minnesota has invested in demonstration-level equipment which can help developers make the correct equipment and processing choices before implementing a commercial facility. The demonstration-scale system implemented at NRRI is shown in Fig. 13.

8 Pelleting/briquetting

Some products have better densification properties than others after pretreatment. This is largely due to the retained lignin amount and lignin form that remains after processing. Specific conditions for densification will vary by processing temperature employed in pretreatment of the biomass and subsequently the temperature employed in the densification step. This directly links the two processes and if the materials are densified without cooling to ambient conditions, binding agents may or may not be needed. For example, if the materials are torrefied at the lower temperatures employed during this processing, enough soft lignin material may be available to act as a binding agent that will result in a strong and stable compact after the densification step. If higher temperature and more rigorous torrefaction conditions are employed, the lignin appears to crystallize and lose its binding capability and this makes densification without a binder more difficult. Under these conditions, an external binding agent is added to facilitate the densification process.

8.1 Binder types

Binders are divided by their function into matrix type, film type, solvent type, and chemical binders[30]:

- Matrix type binders embed the particles into a substantially continuous binder phase. The properties of the briquettes, therefore, are largely determined by the properties of the binder.
- Film type binders are like glues and usually depend upon the evaporation of water or some solvent to develop their strength.
- Solvent type binders are sometimes used, even though the material can be briquetted with pressure alone, as lower pressures can be employed and briquettes with a more porous structure can be made this way.
- Chemical binders can be either film or matrix type.

8.2 Natural binders

The pelleting of biomass is assisted by the presence of lignin that melts and flows to act as a binder. The addition of a small fraction of wax is often practical. While the lignin content in wood is usually enough to bind pellets, other forms of biomass require special conditioning to strengthen them. Sometimes binders such as starch, sugars, paraffin oils, or lignin must be added to make the biomass malleable.[5] The natural binders in biomass can be activated (softened) under high pressures in the presence of moisture (e.g. water soluble carbohydrates) and in some cases increased temperature (e.g. lignin, protein, starch, and fat). When pressure is removed and the binder cools, it hardens or 'sets up' forming bridges or bonds between particles, which has the effect of binding them together and making the resulting product more durable.[22]

Activating (softening) the natural binding components through moisture and temperature in the range of glass transition is essential to produce highly durable briquettes and pellets. The glass transition temperature is the minimum temperature required for activation of natural binders to produce durable densified products. The measured temperatures of roll-press briquettes and pellets ranged from 51°C to 81°C, which is well within the range of glass transition temperatures of corn stover and switchgrass (i.e. 50–113°C). Therefore, by providing enough moisture and a temperature in the range of the glass transition for the biomass materials, natural binding components can be fully activated to produce higher amounts of natural binders from the biomass cells to enhance their binding functionality.[22]

Common binders in biomass pellet production include bonding agents such as starch, molasses, natural paraffin, plant oil, lignin sulfate, lignosulfonates, and synthetic agents.[20]

8.3 Binders to improve mechanical durability

Some of the most common binders tested so far are lignosulfonates, starches (potato peel, potato flour, maize starch, corn starch, wheat starch), vegetable and mineral oils, sodium carbonate, urea, glycerol, and various forms of lignin. Potato flour is one of the most cost effective and contains a relatively small amount of alkali-minerals such as potassium.[26] Many of these binders, however, are hydrophilic and will dissolve if the densified material is exposed to moisture. This will cause the material to potentially fall apart.

8.4 Binder additives used in biomass pellet production

A detailed description of common binders used in biomass pellet production is summarized below[20]:

Starch–Bonding agents (from corn or rice) can also be used to decrease abrasion. A small amount of starch, less than 2% by mass, increases pellet strength. The most common starches are derived from potatoes and corn. *Lignin*–Lignin acts as a binder in-situ in the feed material, although it can also be added as a binder and is obtained as a by-product of the pulp and paper industry. At elevated temperatures, lignin softens and helps the binding process. There are limits on how much can be used and depending on the type of lignin variable moisture resistance may not be attained. *Fiber*–Fiber can be classified as water-soluble and water-insoluble fibers. Water-soluble fibers increase the viscosity of the feed and positively affect the structural integrity of the pellets. Water-insoluble fibers may entangle and fold between particles or fibers. Other potential additives to improve pellet quality include hydrated lime and pea starch. Fibers, however, can also be a problem in grinding the materials to meet the particulate size targets for PCI using conventional pulverization systems. This was the case in several of the biomass trials that have occurred to date.

Lime–Biomass combustion appliance manufacturers often recommend the addition of lime (CaO) to reduce clinker formation and slagging. Limestone creates a chemical compound such as $CaSO_4$ which has a higher melting temperature, thus stays in the bottom ash. Limestone also has the added benefit of reducing HCl formation. It can also interact with any molasses addition to form a binding structure in a densified product.

Molasses–Molasses alone or in combination with hydrated lime is a common industrial binding agent used in various briquetting applications.

8.5 The need for water-resistant binders in biomass production

Conventional pellets are neither waterproof nor water repellant. Lignin, while present in significant quantity in torrefied wood, is largely blocked from coming to the surface during pelleting. Generally, the amount of lignin that finds its way to the surface does not bind well to the charred surface of torrefied particles. Addition of water-resistant binders is required to produce a water-repellent torrefied pellet or particle.[22]

Although torrefied wood particles are hydrophobic, their actual water repellency is very low. Pelleting into a highly water repellent pellet or briquette is required for the torrefied wood industry to produce an acceptable coal replacement product that can be shipped in bulk in open containers. Various efforts are underway to develop a natural binder that can impart moisture resistance to the densified materials.

Mississippi State University (MSU) has developed a low-cost binder to produce a high-density pellet with high water resistance. The pyrolysis oil binder is produced from biomass and has a neutral pH of 6.2, such that the pellet does not produce corrosive gases during their combustion. No chemical cross-linking agents, such as formaldehyde, are required for making this binder, which is very environmentally friendly. At 2% loading, biomass contained in the binder adds roughly 10% to the higher heating value (HHV) of the pellets that it binds.

The MSU binder is suitable to produce densified highly water-repellent pellets with many different biomass feedstocks. These products include densified switchgrass pellets, char, torrefied wood, and powdered coal/torrefied wood pellets. The binder utilized has provided increased density and water repellence for various products. This study reports only on the results of producing pellets with torrefied wood. The utilization of this binder is similar to the utilization of any carbon-based biomass in the sense that combustion will have a neutral influence on global carbon balance.[22]

The Natural Resources Research Institute, University of Minnesota, has shown that both steam-exploded woody material and hydrothermally processed northern wood materials possess very excellent binder characteristics and can be added to torrefied material to greatly facilitate strong, sound densified compacts that possess moisture resistance. In addition, the Institute has identified another biomass-based material that is a by-product of alcohol production that seems to have excellent binding properties and imparts moisture resistance. This work is not yet reported in the literature but seems to hold significant promise for future biofuel application.[31]

Figure 14 Hydrothermal carbonization pilot plant at NRRI.

9 Hydrothermal carbonization (wet torrefaction)

HTC (or wet torrefaction) is a technique that processes biomass under pressure and temperature in an autoclave type of environment. Typical process conditions include pressures up to 50 bar (735 psi) and temperatures between 200°C and 250°C. Under these conditions, raw biomass will break down to produce a material similar to dry torrefied products, but with excellent binding properties. Work conducted at the bench-scale and pilot-scale at the NRRI has shown that the solid material produced has both good cohesion and high-energy value. The slurry from HTC reactor is processed to separate the solid and liquid components using centrifuging technology. The separated solids still retain some moisture but have a mud-like appearance. This 'energy mud' can easily be densified into solid compacts using standard equipment and possess energy contents similar to the material produced via dry torrefaction. In addition, the products appear to resist moisture degradation largely due to the better bindability of the material and the loss of oxygenated components to the liquid phase during processing.[29,30] Briquettes or other shapes can also be made from the collected solids with low-pressure densification methods. Researchers have also found that product from the HTC process can be blended at various levels with fuels produced using torrefaction to produce densified, stable, moisture resistant compacts that have high-energy value ~20 890 kJ/kg (>9000 Btu/lb.). Other researchers have found that HTC material shows similar results.[14] Materials beyond woody biomass have also been processed with similar results and these include switch grass, rice hulls, miscanthus, and corn stover.[31,32] An example of an HTC system that has reached the pilot stage is shown in Fig. 14.

10 Advantages and limitations of biomass pretreatments

As noted in the previous sections, steam explosion, torrefaction, and HTC can be used to concentrate the energy content of raw biomass. This takes the raw biomass and makes it more compatible with solid fuels like coal and facilitates its use at power plant facilities without major changes in power plant equipment. Experience in using either raw biomass or conventional pellets indicate that major changes in equipment are necessary to utilize these latter materials at high substitution rates. In addition, a key issue with biomass use is logistics. Concentrating the energy content so that less material must be handled and transported to the user site lowers the costs associated in these areas. The costs associated with material storage can also be improved if the upgraded materials can be stored in a manner like coal. Steam-exploded and HTC materials possess good moisture resistance, but torrefied materials with good moisture resistance still appear to be a developing area.

Also, as noted previously, the polar components of raw biomass appear to be significantly altered using the three treatments noted and this improves the hydrophobic character of the products. However, when the treatment regime exceeds a certain degree for torrefaction, the resultant materials are difficult to densify without a binder; and compaction then requires use of binding materials that may not result in hydrophobic properties for the resultant products. If this is the case, then covered storage may be necessary to avoid material breakdown.

The pretreated materials also weaken the lignocellulosic structure of the produced biomass fuels and this can lead to better grinding characteristics of the fuel products produced relative to raw biomass or conventional pellets. This then can allow co-milling of the processed biomass with coal without significant equipment modification. The use of the three techniques should be viewed as a facilitation step in increasing the potential utilization of biomass in established coal-based power plants and industrial applications that currently utilize coals. The degree of energy compatibility will be dependent on the original boiler or process design and the type of coal used in designing the original facility. For example, if the facility is designed to use Western coals from the United States such as Powder River Basin coals, all three pretreatment techniques can produce energy products with similar energy content to the coal employed. If the coals used are from bituminous characteristics, dry torrefaction appears to be the technology that will allow products that more closely match the desired energy characteristics.

Another key attribute for use of the pretreated materials is their potential positive impact on reducing emissions from power generation. Woody biomass has inherently lower sulfur and mercury compared to coal. Increasing use of these biomass fuels in the coal blend will have a positive impact on the emissions profile at the end user and likely will reduce overall emissions control costs. As the pretreated biomass is more compatible with the coals that are to be displaced, this allows for greater substitution levels at the user facilities and may allow attainment of desired levels of pollutants to aid the facility in meeting current or anticipated environmental standards.

10.1 Limitations of pretreated biomass

Currently, the main limitation of refined biomass products is the uncertainty concerning availability and quality. At this stage, there are only a few facilities for refined biomass that can deliver the substantial amounts required by the end users. For most of those facilities, the big challenge remains to produce a constant biomass quality due to the variability in raw biomass specifications and the complexity of process control.

Except for the HTC and the steam explosion processes, the presence of inorganic constituents in the biomass is usually not influenced by the biomass refining process. That means if high alkali and chloride-containing biomass is used for the process, the resulting refined fuel will typically keep this high amount of chloride and alkali. This can potentially result in the well-known slagging, fouling, and corrosion issues with boiler operations or with industrial applications using refractory lining materials and this will also restrict the co-firing ratio.

Because part of the organic material in the biomass is evaporated, the internal surface area of the biomass fuel increases significantly through the refining process. This usually leads to a higher reactivity and certain safety issues. Though some tests revealed explosion characteristics comparable to coal, other tests have shown explosion characteristics that exceed that of coal.

Because of the currently experienced variation in quality of the pretreated material (torrefaction process control and performance) and of the densified biomass (durability after pelleting or briquetting), the products may tend to have higher dustiness levels compared to the coals that they displace. This implies that specific countermeasures must be taken to avoid problems at the user handling and storage sites.

The hydrophobicity of the refined biomass must also be understood. At this stage of technology development–it strongly depends on the pretreatment process control and performance. Stability of torrefied pellets over several months has been observed, but in other cases the storage environment can lead to disintegration of products after 2 days depending on the conditions at the storage location. Significant work is underway to develop enhanced water resistance properties for the pretreated materials, and as noted previously, this will depend on the degree of pretreatment and the linkage of the densification process to the energy concentration method employed to process the biomass materials. If drying alone is used, no hydrophobicity develops. If the pretreatment method drives out the polar components in the original biomass, then the produced material does develop water resistance properties. The final characteristics of the produced fuel product will then depend on what technology is used to gain effective densification.

10.2 Recent technology developments

As of 2018, some important progress is noted:

- Torrefaction technology has been proven on pilot-scale and several demonstration and (semi)commercial facilities have been realized.
- The companies involved have significantly improved their ability to produce high-quality products, with pellets of comparable durability to conventional wood pellets. Other densification methods have also been demonstrated.
- The torrefied pellets exhibit comparable supply costs; however, for the end user it provides superior handling and combustion characteristics.
- Total cumulative production figures are estimated at 70 000–120 000 t of torrefied product to date, but various new plants are under development to increase processing capabilities.
- The product has been used in coal plants, gasifier(s), and non-industrial facilities, although in very few cases for some extended periods.
- Some developers, however, have re-focused on the market for torrefied material: they consider smaller domestic or industrial markets more promising than large-scale utilities.
- Some producers are adding torrefied biomass to conventional pellets to enhance the pellet energy levels (increase the MJ/t) of conventional pellet products shipped. An example of this is the New Biomass plant in Quitman, Mississippi.

11 Use of torrefied biomass in power plants[6,30-34]

11.1 Testing of torrefied biomass at Southern Corporation's Gulf Power Company Scholtz Power Plant[6,34]

Tests were run at Southern Company's Plant Scholz (Fig. 15) to test the potential utility of using up to 100% torrefied fuel as a substitute for coal. The Scholz Power Plant contained a 40 MW wall-fired boiler system with a maximum load of 49 MW. The TW was in the form of pellets produced by torrefying conventional wood pellets near the plant site. The pellets were torrefied using a portable torrefaction process as shown in Fig. 16.

The resultant pellets were brittle and fractured when placed in outside storage. The heating value of the pellets after torrefaction was approximately ~24 800 kJ/kg (10 700 BTU/lb.). This was compared to the original pellets

Figure 15 Overview of Gulf Power's Plant Scholz.

Figure 16 Portable torrefaction kiln used to process the conventional pellet materials.

before torrefaction at ~20 640 kJ/kg (8894 btu/lb.) and the bituminous coal normally employed at 30 565 kJ/kg (13 169 btu/lb.). The explosiveness (as measured by KST value) for the torrefied pellets was comparable to PRB sub-bituminous coal.

Figure 17 Material appearances while in storage (dry on left and wet on right).

Various safety measures were employed during the test program and these included the installation of a dust suppression chemical system; water sprays at various transfer points and at walls; observation ports in mills; the use of cold air in starting the grinding mills and the mills were generally started and stopped using 100% coal addition. In addition, all personnel were cleared during loading operations. The torrefied materials were stored outside the plant in piles in a similar manner to coal (Fig. 17). During storage, the torrefied materials exhibited self-ignition with some pile fires after 3–4 weeks of storage. The remedy for this problem was to reshape the pile and compact it based on procedures generally used for Powder River Basin sub-bituminous coal from the western United States. During use, rains caused the original moisture of the torrefied materials to increase from 4.5% to approximately 25%. In addition, the piles contained significant fine particles.

Use of various levels of torrefied fuel was planned for the three mills employed at the plant. Single mill tests were planned at 0%, 20%, 50%, 75%, and 100% torrefied addition levels and 70–75% and 100% using all three milling operations. The tests were targeted to measure boiler efficiency and emission levels. Some notable results during the test showed that the maximum load decreased as the amount of torrefied materials increased and the boiler efficiency decreased from 86.1% to 85.1% at 100% substitution. The main reason for the drop was the very high moisture level of the materials after they gained moisture in the piles. The emission results indicated significant drops in NO_x and SO_2 as the amount of torrefied material increased. The plant opacity levels were not changed in using the materials. During the program not all the planned usage levels were employed. This is summarized in Table 9. The results showed that the use of torrefied material did allow large amounts of renewable energy to be generated in a coal-fired power plant in a 'relatively easy manner' even with materials containing high levels of fines with high moisture. The demonstration indicated up to 100% of the torrefied materials could be used in a pulverized coal boiler. Precautions like those used in handling sub-bituminous

Table 9 Results for power generation tests using various torrefied fuel amounts

Test	Power (MW) gross	%TW (wt)	%TW (Energy)	TW (MW)
1	49	0	0	0
2	41	0	0	0
3	40	6	4	2
4	41	17	13	5
5	40	25	19	8
6	36	29	22	8
7	41	0	0	0
8	32	0	0	0
9	28	0	0	0
10	38	75	69	26
11	42	67	60	25
12	28	67	60	17
13	21	100	100	21

coal could be employed with the torrefied fuel. In addition, even though boiler efficiency is reduced as the amount used increases, the emission levels were greatly reduced in using the torrefied materials.

11.2 Comparison of experience in using wood pellets versus advanced wood fuels at Ontario Power Generation[36,37]

The province of Ontario in Canada has banned the use of coal in their power plants. This has caused Ontario Power Generation (OPG) to convert to various renewable fuel options. In the western part of the province are two power plants. The western most plant is in Atikokan, Ontario and the other is in Thunder Bay, Ontario. Both plants are operated as 'peaker' plants and are operated on a variable basis as electricity production is required and both use biomass as the fuel. At Atikokan, a 200 MW facility, the decision was made to utilize conventional pellets as the fuel. This necessitated significant conversion activities at the plant site to accommodate the use of the pellets including installation of storage facilities for the pellets (10000 t of storage capability with two 5000 t capacity silos) (Fig. 18), a new fuel handling and storage system for ensuring safety, various modifications to the furnace and distributed control systems, and new trucks receiving areas and a transfer tower for the materials.

The total cost of the conversion was C$170000000. In contrast to this approach, the decision was made to use advanced wood energy pellets at the Thunder Bay plant (~200 MW capacity) to avoid the large capital costs associated with the Atikokan conversion. The plant was modified (largely safety

Figure 18 Pellet storage facilities at the Atikokan Power Plant of OPG.

features were added) at a cost of approximately C$5 000 000 and the plan was to use torrefied or steam-exploded materials as the fuel for the modified facility. The plant has been operated with various steam-exploded products on a demonstration basis after the conversion was completed. The plant was able to store the product outside even with the normal Canadian winter conditions. The conversion cost comparison was C$33/kW for Thunder Bay versus C$500/kW for Atikokan.

11.3 Taconite Harbor Energy Center (THEC) torrefied wood test burn–September 2016[6]

The Taconite Harbor Energy Center (THEC) is located near Schroeder, Minnesota, on the North Shore of Lake Superior, and has a generation capability of 225 MW. There are three units located at THEC, although THEC Unit 3 was retired from operation in June 2015 and reduced THEC's generation capability to approximately 150 MW. The three 75 MW units were originally purchased by Minnesota Power from bankrupt LTV Steel Mining Co. in 2001. Investment was made to bring the units up to utility industry standard, and they were restarted in 2002. The three generating units are housed in a single building with shared electrical and heating infrastructure and a single control room for unit operations. The facility is located at an active shipping port on Lake Superior and receives coal shipments for its operations.

As part of its 2015 Integrated Resource Plan (IRP), Minnesota Power considered various remission options at THEC to remain a cost-effective generation resource. Refueling with TW as the primary fuel is one option. The benefit of TW is that it is easy to handle and could be used at THEC with minimal expected modifications to the facility.

But TW, although like coal, does have different material characteristics compared to coal. Therefore, to gain knowledge on the fuel's characteristics, a test burn was conducted to identify potential issues with burning the fuel. Based on fuel availability, fuel handling logistics, and safety, a predetermined blend of TW (25%) and coal (75%) was planned for the burn. This level of use was deemed sufficient to determine if burning TW as a primary fuel is physically feasible at THEC without major equipment modifications. The test burn was not intended to identify every issue associated with utilizing TW as a primary fuel including the economic feasibility. The main objectives of the test burn were to validate that the fuel can be safely and reasonably handled, pulverized, and combusted.

This test burn was conducted in two separate tests on Unit 1 at THEC on September 21 and 22, 2016. Fuel was ground through four coal mill pulverizers before it was fed into the boiler. The first test was a short duration test (approximately 3 h) and was intended to identify how a single mill handles 100% TW for pulverizing. The other three mills remained 100% coal to get the 25% TW/75% coal blend. The mills are designed for grinding and drying a fuel that has approximately 25% moisture versus the TW, which contains very little moisture (6–7%). The mills are intended to operate with hot primary air for both drying and conveying the pulverized fuel to the boiler. For this test, the amount of hot air was reduced and supplemented with cold air to maintain conveying. Some of the handling characteristics of the material during the plant test are shown in Fig. 19.

The second test was to burn a 25% TW/75% coal blend, accomplished by mixing a 50%:50% blend in two of the four mills. The TW was blended with the coal in the coal yard before being sent to the bunkers. Due to the relative

Figure 19 Coal handling at outside storage and within power plant building.

dryness of the blended fuel, there were dust and explosion safety concerns. To mitigate the hazard, low pressure <207 kPa (< 30 psi) steam was injected into the bunker prior to loading the TW/coal mixture to ensure there is no potential for explosions.

Load was stabilized and maintained between 45 MW and 50 MW net for both tests. Both days tested approximately 25% wood and 75% coal. Day 1 test: 100% wood through 1A mill, 100% coal through 1B, 1C, 1D mills for approximately 3 h from 08:05–11:05. Day 2 test: 50% wood/coal through 1A and 1B mills, 100% coal through 1C and 1D mills for approximately 8.5 h from 08:00–16:30.

Basic findings

- TW is a viable fuel for THEC; it was successfully handled, pulverized, and combusted in the boilers with no major system modifications during the test burn.
- To safely utilize TW as a primary fuel, modifications were made to the fuel handling and pulverizing system to accommodate the lower moisture contents of TW compared to the traditional fuel for THEC sub-bituminous coal.
- If the TW feedstock was properly produced to be truly hydrophobic, it could be stored in the fuel yard like coal, with no need to keep it covered.

12 Case study: use of torrefied wood pellets at Portland General Electric's Boardman Facility[6]

The largest-scale test was conducted at the 617 MW coal-based power plant of Portland General Electric (PGE) outside of Boardman, Oregon. This plant consumed approximately 5000 tons of torrefied pellets made by various suppliers and included around 4000 tons from New Biomass Energy, Quitman, Mississippi, other materials supplied by a Canadian supplier, and the rest made in Oregon, using equipment available near the plant location. The overall program of procurement was coordinated by Oregon Torrefaction with funding from the US Endowment for Forestry and Communities and the USDA Forest Service.

As noted in previous discussions, the key piece of equipment in the power plant system is the pulverizer. This device comminutes the coal to the desired size for combustion purposes at the boiler. At Boardman, eight pulverizers in total exist at the plant. Coal is brought into the pulverizers with hot air at 371°C (700°F). Air enters at the bottom of the pulverizers and undersize particles are swept into the combustion burners, with oversize particles returning to the pulverizers via a cyclone. Four combustion burners are fed by each pulverizer. One of the pulverizers is shown in Fig. 20.

Figure 20 Typical pulverizer.

The coal typically employed at the plant comes from the Power River Basin in the western United States and is brought to the plant by unit train. The typical coal properties employed are shown in Table 10. An example of a grinding wheel is shown in Fig. 21. The plant employs two banks of four pulverizers,

Table 10 Typical coals used at the Boardman Plant

Item	Value
Coal source	Wyoming, Northern Montana (three mines)
Mine names	Belair, Buckskin, North Antelope
Energy content	19,600 kJ/kg (8,450 BTU/lb.)
Fixed carbon	51%
O_2 in minerals	10.5%
Ash	5%
Hydrogen	3.4%
H_2O	24%
Nitrogen	0.8%
Sulfur	0.3%

Figure 21 Example of grinder in pulverizer.

Figure 22 One of two pulverizer banks.

and these units feed 32 combustion burners at the boiler. Figure 22 illustrates these features. The pulverizers grind the coal to the required size and feed the material through burners into the fire box of the boiler. The firebox is shown in Fig. 23 with the burners illustrated in Fig. 24.

A detailed test plan was laid out in a step-wise manner to conduct the various aspects of the work at the Boardman facility. It started cautiously to gain a better understanding of the fuel that would be used to replace the Powder River Basin Coal and then proceeded as modifications to the equipment were made to accommodate the new fuel. Various modifications to the pulverizers were implemented based on this learning. In addition, changes to air flow rate and air temperature were modified to allow high levels of torrefied fuels to be used. The properties of the fuel were monitored during program execution, as

Figure 23 Boiler with Fire Box View Port.

Figure 24 Multiple burners in boiler system.

was the performance of the system. A target fuel rate of 36,364 kg/h (80,000 lb./h) was employed for the plant, and the results show that this rate of fuel use could be attained as things progressed at the plant.

12.1 Preliminary test

On December 6, 2016, at 10:24 am, the Boardman Power Plant commenced its initial attempt at deploying one of six pulverizers fueled on 100% torrefied biomass. This was the first time a Boardman pulverizer operated on 100% torrefied biomass fuel, and it was a prelude to the Plant attempting a full load test using 100% biomass on all deployed pulverizers. For all tests, the high-level goals were to demonstrate that torrefied biomass fuel can (1) grind well; (2) burn well; and (3) the ash behavior was acceptable.

12.2 The pulverizer plugged using torrefied biomass

The one-pulverizer test revealed that the torrefied biomass fuel did not grind as well as hoped. This was shown by a gradually increasing Pressure Delta (DP) approaching the 7.47 kPa (30-inches-of-water) alarm level. This DP level warns of material buildup inside the pulverizer that, in turn, increases the potential to plug the pulverizer and eventually force its shutdown. This buildup of biomass inside the pulverizer limited the fuel infeed to 22 727 kg/h (50 KPPH (thousand lbs./h), although 27 273 kg/h (60 KPPH) was achieved for a brief period. To try to get a better handle on the nature of the DP increase, the test was conducted twice over a period of 8 h. On the second trial, the DP also increased, although it appeared to hold steady for a longer period time. Decreasing the infeed rate also decreased the DP, which suggested a way to control the plugging problem.

12.3 Torrefied biomass yielded a good burn

Combustion consultants and plant operators reported that the pulverized fuel delivered to the boiler yielded a very good burn. Subsequent testing of the torrefied biomass–in this case, woody material–has shown that it does not pulverize to particles that are as small as those typically seen during the burning of coal at the plant. The higher reactivity of the biomass fuel appeared to make up the difference. One indication of this was that there were no 'sparklers' for this test; however, there were 'sparklers' for the 100% test in the upper furnace, which indicated that the larger fuel particles burned quite well. Notably, after calibration of the flame detectors for the type of flame derived from torrefied biomass, the devices can detect the new flame characteristics. The flame detectors provide two principal functions: (1) flame quality assessment; and (2) protection against an uncontrolled re-ignition by the ignitors should a burner flame out. The former helps operators optimize plant performance, while the latter is an important operational safety feature.

12.4 Plant emissions remained normal

Even though less than one-sixth of the overall power plant infeed was torrefied biomass, the plant emissions were steady and showed no unexplainable variation.

12.5 Fire in the pulverizer

As noted, the Plant operated on all six pulverizers, but only the 'B' pulverizer was tested on biomass. The silo feeding this pulverizer was loaded with 170 t of torrefied biomass for the first trial, and an additional 100 t were loaded later to accommodate the second trial. With about 30 tons left in the silo, the testing was concluded at 5:00 pm. It was decided that the pulverizer should be allowed to gradually cool down so that it could be opened the next day in support of a detailed inspection to see where the biomass deposited and to assess for the nature of the plugging that caused the high DP (*Note:* Pulverizer temperatures were continuously monitored throughout the night.). As part of this process, the retained fuel was quite hot in the pulverizer interior; unfortunately, it ignited, causing a fire inside the pulverizer. On the following day, the fire was put out. Pulverizer fires, while undesirable, are not uncommon, and the staff at the Plant did a good job in both assessing the problem and resolving it.

12.6 Assessment and adjustment

Over a period of 8 h, the Boardman Plant operated one of six pulverizers entirely on torrefied biomass, to mostly good effect. Around 240 t of torrefied biomass was consumed in two consecutive test trials, yielding about 60 MW of renewable power. Due to the high DP concern, the plant modified pulverizer 'B' by removing certain components typically needed for coal use (the diffusers and the down-comer pipes). This was to lessen the designed 'tortuous path' that was tuned for coal pulverization, but likely not needed for torrefied biomass particles, which are both larger and less dense than coal. The test also pointed out that there is a potential need for better friability for the torrefied materials to facilitate grinding.

12.7 Second test burn–resolving the pressure differential concern

December 12, 2016

Two concerns were raised in the first test: (1) the pulverizer plugged; and (2) pulverizer material buildup was too high–as much as 25% of the delivered fuel. To deal with the former, the plant elected to remove the diffusers and the dip tube to allow more unfettered passage of the lighter and larger torrefied biomass particles from the pulverizer grinding section to the burners above. There were good prospects that this would resolve the plugging concern and perhaps also reduce the material buildup. On December 12 at 2:00 pm, the modified 'B' pulverizer was put back on line using 100% torrefied biomass and re-tested, using the previous test procedure to assess the effect of the modification. The other five pulverizers continued to burn coal. The plant was at 543 MW Net, 574 MW Gross power generation.

The actions taken during this test were twofold: remove obstructions to the flow path for the materials and increase the air flow rate to the pulverizer to allow more dynamic flow conditions for moving the torrefied materials. Elimination of the flow constraints eliminated the previous pressure drop issues and the increase in primary air flow allowed much reduced material buildup in the pulverizer. This test covered 6 h and consumed 130 t of torrefied biomass fuel, yielding around 43 MW of renewable power over the test period.

During the next few months, the plant operators continued to refine their practices and the culmination of these tests was continued progress on how to effectively use the torrefied fuel materials.

12.8 Hundred percent biomass test burn on February 23, 2017

On February 23, 2017, a 100% biomass test burn was conducted at the Boardman Power Plant.

Goals

1. Successfully transition from burning coal to burning biomass.
2. Run the plant in a safe and stable condition while operating on 100% biomass.
3. Successfully run four pulverizers on 100% biomass for approximately 4–6 h.
4. Run the unit in turbine follow mode in a manner where operations do not have to manually operate individual pulverizers.
5. Perform emission testing for particulate matter, hydrochloric acid, sulfur dioxide, carbon monoxide, and oxides of nitrogen (NO_x) for future permitting efforts.
6. Gather operating data that may be used for future biomass test burns.

12.8.1 Preliminary modifications

A series of test burns were performed from December 2016 to February 2017 on the B pulverizer to both practice safe operating conditions with biomass fuel and determine the necessary pulverizer modifications required for biomass processing. The torrefied fuel possessed less moisture than coal and ground differently and these changes required pulverizer and air flow modifications.

The following modifications were made to the pulverizers to accommodate torrefied biomass pellets in place of coal:

1. Removing the outlet section of the pulverizer. This included removing the outlet diffusers, pumpkin teeth, and the inner extension ring (dip tube).
2. Removing the reject chute gates.
3. Increasing the spring tension on the roll wheels by removing the plunger gap, therefore increasing the force that the wheels exerted on the table.

12.8.2 Conducting the trial

Four pulverizers were selected to process biomass for the February 23rd full-unit test burn, and each of these pulverizers was modified using the lessons learned from the single-mill test.

The planned use of 100% torrefied fuel in the four pulverizers was achieved and Boardman successfully burned 1653 tons of biomass and successfully operated while burning 100% biomass. In addition, the plant was able to reach average loads of 38 636 kg/h (85 KPPH) on each pulverizer. This exceeded the original target for the program. This was accomplished safely and without any plant equipment failures. In summary, the test went very well and is considered a success. The trials at the PGE Boardman plant were well orchestrated and demonstrated that torrefied biomass fuels could be routinely used at their power plant.

The results indicate that modifications can be done to accommodate the fuel used, but the results also indicate that under-torrefied materials can lead to grinding difficulties if certain actions are not taken. In addition, since the material is lighter than coal, the injection system must compensate for this difference. In the case of the Boardman facility, they could increase air flow and material feed rate to allow stable operation of the boiler.

After the base tests were completed, the Boardman plant used the remaining torrefied materials until they were gone, on a planned basis. However, the results also point out that certain properties in the torrefied fuels could be improved, including: degree of torrefaction to eliminate fibrous, hard-to-grind materials, better moisture resistance to lessen atmospheric breakdown of the products in storage, improved overall physical properties to lessen dust generation in handling, and improved consistency in overall properties from batch to batch. Work by the NRRI indicates that breakdown of the fibrous structure increases as the degree of torrefaction increases, but this also impacts the ease of densification and potential physical properties that can be attained. Therefore, continued development work in this area is needed to refine the product for large-scale use while meeting all properties desired for end users, including moisture resistance.

12.9 Use of briquette blends in stationary hearths of two one-quarter-scale steam locomotives at the Milwaukee Zoo[6,32,34]

As noted in the previous sections, various techniques were used to evaluate the creation of densified fuel materials under various production and test conditions. Usage of these materials requires an intimate knowledge of the fuel processing systems that will be employed in storing, handling, comminuting, and combusting the fuel products. During the NRRI development program, various tests have been summarized in using commercially produced pellets in PCI systems. In the previous sections, the use of up to 100% substitution of torrefied fuels was demonstrated at power facilities using PCI boiler systems.

However, other significant coal-based fuel systems do not pulverize the coal before it is combusted to provide heat for boiler or industrial use. This is often achieved by using stoker systems to feed larger pieces of coal onto a grate, where it then burns and provides hot flue gases that are used to heat pipes to generate steam for various purposes. As part of NRRI's program, investigations on the use of densified briquettes that more closely simulate this type of stationary hearth were undertaken. These tests were conducted at the Milwaukee Zoo using two one-quarter-scale steam engines that traverse a 1.1-mile oval and haul passengers around the zoo during normal operations at the facility. The scale of these engines is shown in Fig. 25. As can be seen, even at this scale, the equipment is significant. The engines provide a very good test bed for determining the suitability of using torrefied fuels as a direct substitute for the coals typically used at the zoo.

Three test periods took place using the fuels produced by NRRI. In the first case, ram briquettes were produced using the B-100 machine shown in Fig. 26 and illustrated in Fig. 27. In the second case, torrefied pellets were employed and supplied from New Biomass Energy. In the third case, two blends of materials were employed: (1) 100% torrefied woody biomass, and (2) a 50%:50% blend of sub-bituminous coal and torrefied woody biomass. For the woody torrefied biomass, briquettes were made with and without binders. The blend with coal used a binder. The binder employed was guar gum. In the third test period, the briquettes were made using one pocketed roll and one smooth roll based on previous trials using various roll configurations. The combination of one smooth and one pocketed roll gave excellent densification from the

Figure 25 Stationary hearth engines tested.

Figure 26 Ram briquetting equipment.

Figure 27 Briquettes used in phase 1.

outside to the interior of the briquette. A comparison of the fuel briquettes to the typical coal size employed is shown in Fig. 28. The materials produced using this system is shown in Fig. 29.

Various criteria were established by the zoo train operators to gauge the potential utility of the fuel, and these parameters were (1) safe operation of the locomotive without sparklers emitting from the locomotive stack; (2) full bed coverage of the grate in order to protect the grate bars; (3) improved emissions, including opacity and reduced sulfur emission from the locomotive stack; (4) equivalent or better fuel content compared to the coals typically used; (5) good flame characteristics to ensure efficient heat transfer to the boiler tubes; and (6) enough energy to allow the engines to perform their function of pulling a fully loaded zoo train around the zoo.

The results from the first trial using the B-100 ram briquetter were poor. It was found that, when the train was fully engaged in pulling its load, the draft

Figure 28 Coal and briquette comparison.

100% Torrefied Biomass **50% TB:50%PRBCoal**

Figure 29 Briquettes used in phase 3 of program.

on the combustion system caused sparklers to be emitted from the stack (Fig. 30). The emission of the sparklers is a safety hazard and was an indication of lack of full compaction throughout its volume from the ram-briquetted materials. The materials had adequate energy content and could be handled by the train operators, but this result was unsatisfactory. The heat transfer to the boiler tubes was very good during the test, and the materials provided adequate energy to generate the steam required to run the engine. The flame characteristics are illustrated in Fig. 31. The briquettes also provided good hearth coverage.

In the second test, torrefied pellets were employed. A screen had to be placed on the hearth to keep the pellets from falling through the grate. These materials provided good energy for the boilers, but their use was not deemed to be practical as materials closer in size to that of the coal typically used was desired.

Based on the results from the first trials at the zoo, it became evident that more thorough compaction was a key need to have a usable fuel product. This

Figure 30 Sparklers emitted under draft.

Figure 31 Flame character using ram briquettes.

became the focus of the development and rotary briquetting was thought to be the best option for achieving all the targets desired by the zoo personnel. In addition, based on our development experience, it was thought that in addition to total replacement of the coal with a pretreated biomass, partial replacement of the coal with a coal/torrefied fuel blend was desirable. Fuels were subsequently prepared using both a blend of coal and torrefied fuel and torrefied fuel alone. Enough material of both types was prepared to allow a full day of testing at the zoo. Various characteristics of the produced fuels were measured after production: fuel content, drop durability, Kansas State tumbling can, crush strength, moisture uptake, grindability, and volatile matter/fixed carbon.

As can be seen in Table 11, the coal blend had a higher net caloric value compared to the 100% torrefied biomass material but also had significantly higher ash and sulfur contents. The typical coals used at the zoo had both

Table 11 Analytical comparisons for the fuel products tested during phase 3

Item	Torrefied biomass	Coal/torrefied blend
Moisture (%)	4.91	4.73
Ash (%)	1.54	3.34
Volatile matter (%)	61.53	64.66
Fixed carbon (%)	33.68	27.27
Sulfur (%)	0.009	0.151
Net caloric value (kJ/kg)	20763	23166
Carbon (%)	58.22	60.76
Hydrogen (%)	5.28	4.71
Nitrogen (%)	0.27	0.53
Oxygen (%)	34.60	25.78

higher ash and significantly higher sulfur levels than that of the briquettes supplied during this test program.

The physical properties of the two types of fuels provided were excellent and the materials could be handled just like the coal typically used and were not dusty in handling or feeding to the boiler.

Both fuels were used during the day and provided adequate energy to allow the zoo to pull a full-size train around the oval for multiple laps without difficulty. In addition, when the train was under full load and was under high draft conditions, the sparkler issue encountered with the use of the ram briquettes was eliminated. The locomotives were typically pulling a load of cars that approached 23 000 kg (~50 000 lbs.). The energy content of both material types was deemed adequate for use at the zoo.

The other attributes of concern included comparisons with opacity and sulfur-related emissions as well as heat transfer to the boiler tubes. In addition, hearth coverage and hearth protection were also important concerns. Figure 32 illustrates some of the key comparisons with coal.

Figure 32 Flame characteristics for coal (left) and 100% biocoal (right).

Figure 33 Comparison of stack opacity for coal (Left) and biocoal (right).

Both the blend with coal and the 100% biomass blends worked well under low and high draft conditions. The two materials outperformed the coal in terms of more effective flame characteristics that would then enhance steam production for the boiler itself.

The stack emissions generated in using the fuels were noticeably improved when using either the 100% torrefied fuel briquettes or in using the blend, compared to the conventional coal employed. This is shown in Fig. 33.

After the full day of testing was completed, a discussion on the utility of using the new fuel products was undertaken with all involved in the test program, and the basic conclusion was that the new fuel briquettes were a good substitute for the coal they typically would use. The briquettes provided adequate power to pull the train, the emissions were greatly reduced, and the sparkler issue that had occurred with the ram briquettes was overcome using the rotary briquetting configuration employed to produce the fuel products.

12.10 Comparison of the economics of conventional pellet use versus torrefied pellet use delivered to the power plant based on European experience[29]

A significant analysis was undertaken by the Europeans to compare the potential economic consequences for using torrefied fuels in comparison with the normal conventional pellets that typically serve the power industry in Europe. This investigation was completed in 2012 and uses the projected costs at that time. It does serve as a useful comparison in the utility of using the advanced wood fuel product but does not factor in costs such as massive storage bunkers that are necessary for conventional pellets or the avoidance of capital charges in power plant modification as illustrated for the Ontario Power situation. The costs used in the study are summarized in Table 12 and are given in US dollars per GJ of delivered fuel.[29] The results show that even though it may cost more for the torrefied pellets at the point of production, by the time the savings due to logistics and power plant operation are considered,

Table 12 Economic comparison in using torrefied pellets versus conventional pellets[29]

Cost components	Wood pellets ($/GJ)	Torrefied pellets ($/GJ)	Savings ($/GJ)
Cost of biomass	4.28	4.28	0
Cost of electricity	0.6	0.74	−0.14
Cost of labor	0.47	0.47	0
Financial costs	1.01	1.49	−0.48
Other costs	0.40	0.43	−0.03
Cost price at production site	**6.76**	**7.41**	**−0.65**
Inland logistics from plant to port	1.12	0.57	0.55
Deep sea shipment	2.04	1.28	0.76
Inland logistics from port to utility	0.94	0.55	0.39
Cost price delivered at utility	**10.87**	**9.81**	**1.06**
Extra costs at the power plant	1.93	0	1.93
Total costs of coal replacement	**12.80**	**9.81**	**2.99**

the torrefied pellets outperform the conventional wood pellets in actual power generation from an economic standpoint.

12.11 Biomass as a carbon neutral substitute for coal use in power plants[39]

Some controversy exists on the use of biomass materials in power plant operations. Strauss discusses this issue in a detailed analysis and indicates that the use of the various types of biomass can be viewed as a carbon neutral or beneficial carbon reduction method relative to the use of coal in power plant situations. He states that some think 'making lumber is okay, but using wood to make energy products for energy is not'. He cites a study by the Chatham House as an example that does not consider all relevant factors relative to carbon neutrality. He notes that contrary to what many believe that 'forests are not being depleted, and that the stock of carbon held in the forests is not being reduced'. He further states that 'forests are dynamic systems that are in many stages of growth. There are mature trees that are ready to harvest, areas of new growth, and many plots that are in stages of growth between seedlings and mature trees'. His analysis in contrast to the Chatham House study indicates that 'contrary to either/or implications ... regarding harvesting trees versus carbon sequestration, a steady supply of raw materials for the forest product industries, including pellet production, does not mean reducing carbon stocks'. He notes that 'sustainable tree harvesting, i.e. continuously renewing forests, does not degrade the carbon sink function of working forests'. His model for a tree farm illustrates that, as long as the harvest rate does not exceed the growth rate of the harvested plot, no net carbon is added to the atmosphere. His argument assumes that the CO_2 released by combustion of the biomass materials is absorbed 'contemporaneously' by the new growth in the many younger plots that match or exceed the removal of biomass from mature plots. This is the essence of sustainable forest and biomass practices. In practice, many forms of biomass including waste wood and residuals, herbaceous materials, and agricultural by-products can be made into fuel products. If these products are not captured and used, they will degrade and release greenhouse gasses naturally.

13 Summary

- Biomass use in the US Power Generation is widespread, but generally at low levels of biomass if the plant is not modified to accommodate the raw biomass materials.
- Even if dried and pelletized biomass is employed, significant capital investment is necessary to reach high levels of substitution. This has been demonstrated widely in Europe.
- Co-firing with conventional coal boilers appears to be limited to around 10% if no plant modifications are made.
- Plant modifications can be very capital intensive if advanced wood fuels are not employed.
- Advanced wood fuels (including steam exploded, torrefied, and hydrothermally carbonized) can lead to 100% substitution at low capital cost for the power plant.
- Canadian experience is a useful benchmark for the capital requirements in using conventional pellet fuels versus advanced wood fuels manufactured from torrefaction, steam explosion, or HTC.
- European experience considers the net cost per GJ of delivered fuel to the power plant. Their findings indicate the net cost of the advanced wood fuels is lower than conventional pellets when transportation, logistics, and actual power plant costs are considered in the final economic assessment.
- The quality of torrefied pellets used at the major plant test noted in this chapter still needs to be significantly improved in terms of physical properties and storage considerations.
- The degree of torrefaction has a direct impact on the ability of the densification process to produce sound compacts without the use of binding agents. The greater the degree of processing employed, the more

difficult it is to form stable, moisture-resistant compacts using conventional pelleting techniques. Other agglomeration methods such as ram or rotary briquetting may be useful alternative techniques.

- Various binders have been used in development of sound physical properties for the advanced wood fuels. Some binders do not appear to impart moisture resistance and can cause the resultant fuel product to fail due to weathering. More work is needed to simultaneously meet the energy, physical properties, and storage properties for the advanced fuel products, but various naturally made products show great promise.
- Steam-exploded materials appear to produce densified fuels with improved physical properties and moisture resistance but cannot achieve the higher-energy densities of torrefied materials that have undergone more rigorous time and temperature processing. In addition, steam-exploded materials are tougher to grind than torrefied or hydrothermally carbonized products.
- Interestingly, both steam explosion and hydrothermally processed materials may be useful binding agents when combined with blends of torrefied materials and the resultant compacts can possess good energy, physical, and moisture-resistant properties.
- Hydrothermally carbonized products can possess very good fuel value and physical properties, but the state of development appears to be behind that of torrefaction or steam explosion.
- More development work for all the technologies is required, but great progress has been made especially in the last few years in using these pretreatment methods to concentrate the energy value in biomass to make it more useful in subsequent end-user applications.

14 Acknowledgements

The authors acknowledge and thank the Consortium for Advanced Wood-to-Energy Solutions (CAWES) with funding provided by the USDA Forest Service (www.fs.fed.us), the US Endowment for Forestry and Communities (www.usendowment.org) and consortia members. CAWES is a public/private partnership initially formed by the USDA Forest Service, the US Endowment for Forestry and Communities, and several innovative private companies and university partners committed to advancing sustainable, scalable, distributed wood-to-energy solutions that stimulate forest restoration and rural economic development. In addition, the research team at the Natural Resources Research Institute, University of Minnesota Duluth, is thanked for their contributions to the development program at the Institute and their thoughtful contributions to the preparation of this manuscript. Special thanks to Timothy Hagen, Mathew Young, Andriy Khotkevych, Kendall Carden, Richard Kiesel, and the Coleraine

Bioconversion Laboratory for their work in developing fuel substitutes from various biomass materials.

15 References

1. Nicholls, D. 2013. Coal plants burning wood. USDA Forest Service, Living on Earth.org.
2. Thran, D., Peetz, D. and Schaubach, K. 2017. Global wood pellet industry and trade study 2017. IEA Bioenergy Task 40, June 2017.
3. Fasina, O. and Sokhansanj, S. 1996. Storage and handling characteristics of alfalfa pellets. *Powder Handling and Processing* 8(4): 361-5.
4. Vild, M. 2014. Torrefaction: Status and expected next steps. *AEBIOM Bioenergy Conference*, Brussels.
5. Christiansen, R. 2014. The art of biomass pelletizing. *Biomass Magazine*.
6. Fosnacht, D. R. (Ed.). 2018. Final report: Demonstration of use of biomass in electric power generation. Technical Report NRRI/TR 2018/08. Natural Resources Research Institute, University of Minnesota Duluth, 102p. + Appendix.
7. DOE, EERE. *Biopower, Technical Strategy Workshop, Summary Report*, December 2010.
8. U.S. Energy Information Administration. 2012. Today in Energy, 2 March 2012.
9. infohouse.p2ric.org/ref/36/35394.pdf, 1996.
10. http://epa.gov/airmarkets/progsreg/epa-ipm/BaseCasev410.html.
11. FutureMetrics. 2015. A rational and pragmatic off-ramp to a decarbonized future, 24 June 2015.
12. Henderson, C. 2015. Cofiring of biomass in coal-fired power plants - European experience. *FCO/IEA CCC Workshops*, January 2015.
13. Sourcewatch.org/indexphp/SchillerStation.
14. energydesk.greenpeace.org/2015/06/10drax.
15. Zimmerling, S. 2013. *White Paper: Torrefied/Refined Pellets for Biomass Co-Firing*. VGB Powertech e.V.
16. Bach, Q. and Skreiberg, O. 2016. Upgrading biomass fuels via wet torrefaction: A review and comparison with dry torrefaction. *Renewable and Sustainable Energy Reviews* 54: 665–77.
17. Mani, S. 2008. *Recent Developments in Biomass Densification Technology*. University of Georgia.
18. Kiel, J. 2013. *Torrefaction–Product Quality Optimization in View of Logistics and End-use*. World Biomass Power Markets Jaap Kiel, Amsterdam, the Netherlands, 17 May 2013.
19. Wilson, T. 2010. Factors affecting wood pellet durability. MS Thesis, Penn State University.
20. Huang, J. 2013. GEMCOM Energy.
21. Stelte, W. 2012. *Steam Explosion for Biomass Treatment*. Danish Technological Institute.
22. Kaliyan, N. and Morey, R. 2010. *Bioresource Technology* 101: 1082–90.
23. Arbaflame. 2015. Arbaflame.com.
24. VGB Powertech. 2013. White paper: Torrefied/refined pellets for biomass co-firing.
25. Anca-Couce, A. and Obernberger, I. 2016. Application of detailed biomass pyrolysis kinetic scheme to hardwood and softwood torrefaction. *Fuel* 167: 158–67.

26. Metin, S. 2013. *Considerations for Grading Agricultural Residue*. Ontario Federation of Agriculture.
27. Kumar, L., Koukoulas, A., Mani, S. and Satyavolu, J. 2017. Integrating torrefaction in the wood pellet industry: A critical review. *Energy and Fuels* 31: 37–54.
28. Status overview of torrefaction technologies. IEA Bioenergy Task 32 Report, 2012, December 2012.
29. Cremers, M., Koppejan, J., Middelkamp, J., Witkamp, J., Sokhansanj, S., Melin, S. and Madrali, S. 2015. Status overview of torrefaction technologies. IEA Bioenergy Report.
30. Fosnacht, D., Hagen, T. and Khotkevych, A. 2014. Manufacture of moisture resistant torrefied biomass agglomerates. Presentation at *2014 International Biomass Symposium*, 26 March 2014.
31. Fosnacht, D. R., Hagen, T., Young, M., Carden, K. and Kiesel, R. 2018. Final report: Use of improved densification conditions for producing high fuel content products from biomass processed by torrefaction, hydrothermal carbonization, and various densification methodologies. Technical Report NRRI/TR 2018/09. Natural Resources Research Institute, University of Minnesota Duluth, 79p.
32. Fosnacht, D. R., Hagen, T., Carden, K., Young, M., Grochowski, J. and Kangas, K. 2017. Development of consolidated solid biofuels using torrefaction with subsequent densification and use of the fuel materials. *Proceedings of the 35th* Biennial *Conference of the Institute for Briquetting and Agglomeration*, St. Petersburg Beach, Florida, 12–15 November.
33. Reza,M.T., Graves, D. T.,Lynam, J. G. and Uddin, M. H. 2012. Engineering pellets of biomass blends. AIChE Annual *Conference*, University of Nevada, Reno, Pretreatment in Forest Biorefineries Session.
34. Torrefied wood field tests at Gulf Power's Plant Scholz. EPRI 2014 Technical Report, Palo Alto, CA, Report #3002003268.
35. Dutta, A. and Leon, M. 2014. *Pros and Cons of Torrefaction of Woody Biomass*. University of Guelph.
36. Boyko, B. 2014. Antikokan generating station biomass conversion project update. *International Bioenergy Conference*, 13 June 2014.
37. Boyko, B. 2015. Experience with coal to biomass conversions at Ontario power generation. *4th Industrial Wood Pellets for Coal Conversions*, 17 June 2015.
38. Li, Y. and Liu, H. 2000. High-pressure densification of wood residues to form an upgraded fuel. *Biomass and Bioenergy* 19: 177–86.
39. Strauss, W. 2017. Addressing biomass energy 'Alternative Facts'. *Biomass Magazine*, April 2017, pp. 56–61.

16 Appendix: key sources on torrefaction

Appendix A: advanced wood and energy literature reference collection

Reference categories

Recent literature

1. Simet, A. 2018. Shifting market. *Pellet Mill Magazine*, January/February 2018, pp. 12–19.

2. Transforming biomass. *Bioenergy Insight*, September/October 2017, pp. 30–2.
3. Thran, D., Witt, J., Schauback, K., Kiel, J., Carbo, M., Maier, J., Ndibe, C., Koppejan, J., Alakangas, E., Majer, S. and Schipfer, F. 2016. Moving torrefaction towards market introduction – Technical improvements and economic-environmental assessment along the overall torrefaction supply chain through the SECTOR project. *Biomass and Bioenergy*, 89: 184–200.
4. Strauss, W. and Schmidt, L. 2018. Are black pellets ready to compete with white pellets. *Biomass Magazine*, March/April 2018, pp. 48–51.
5. Gillette, M. 2017. Providing clarity of industry terminology. *Biomass Magazine*, April 2017, p. 10.
6. Una Nowling, P. E. 2018. Successful torrefied biomass test burn at a coal power plant. *Power*, March 2018, pp. 46–52.
7. Nakayama, S. Coal-fired power generation in Japan and the world. *Cornerstone Magazine*. www.cornerstonemag.net/coal-fired-power-generaon-in-Japan-and-the-World.

A. Use of fuel at power plants

2015

1. Ohm, T., Chae, J. and Kim, J. 2015. Study on the characteristics of biomass for co-combustion in coal power plant. *Journal of Material Cycles Waste Management* 17: 249–57.

2014

1. Agar, D. and Wihersaari, M. 2014. Bio-coal, torrefied lignocellulosic resources – Key properties for its use in co-firing with fossil coal – Their status. *Biomass and Bioenergy* 44: 107–11.
2. Du, S., Chen, W. and Lucas, J. 2014. Pretreatment of biomass by torrefaction and carbonization for coal blend used in pulverized coal injection. *Bioresource Technology* 161: 333–39.
3. Fryda, L., Daza, C., Pels, J., Janssen, A. and Zwart, R. 2014. Lab-scale co-firing of virgin and torrefied bamboo species *Guadua angustifolia* Kunth as a fuel substitute in coal fired power plants. *Biomass and Bioenergy* 65: 28–41.
4. Kaliyan, N., Morey R. V., Tiffany, D. G. and Lee, W. F. 2014. Life cycle assessment of a corn stover torrefaction plant integrated with a corn ethanol plant and a coal fired power plant. *Biomass and Bioenergy* 63: 92–100.
5. Li, J., Zhang, X., Pawlak-Kruczek, H., Yang, W., Kruczek, P. and Blasiak, W. 2014. Process simulation of co-firing torrefied biomass in a 220 MWe. *Energy Conversion and Management* 84: 503–11.
6. Nunes, L., Matias, J. and Catalão, J. 2014. A review on torrefied biomass pellets as a sustainable alternative to coal in power. *Renewable and Sustainable Energy Reviews* 40: 153–60.
7. Pérez-Fortes, M., Laínez-Aguirre, J. M., Bojarski, A. D. and Puigjaner, L. 2014. Optimization of pre-treatment selection for the use of woody waste in co-combustion plants. *Chemical Engineering Research and Design* 92: 1539–62.

8. Prando, D., Patuzzi, F. and Baggio, P. 2014. CHP gasification systems fed by torrefied biomass: Assessment of the energy performance. *Waste Biomass* 5: 147–55.
9. Tchapda, A. and Pisupati, S. 2014. A review of thermal co-conversion of coal and biomass/waste. *Energies* 7: 1098–148.
10. Tsalidis, G.-A., Joshi, Y., Korevaar, G. and de Jong, W. 2014. Life cycle assessment of direct co-firing of torrefied and/or pelletised woody biomass with coal in the Netherlands. *Journal of Cleaner Production* 81: 168–77.

2013

1. Chen, W., Chen, C. and Hung, C., 2013. Taguchi approach for co-gasification optimization of torrefied biomass and coals. *Bioresource Technology* 144: 615–22.
2. Goldfarb, J. and Liu, C. 2013. Impact of blend ratio on the co-firing of a commercial torrefied biomass and coal via analysis of oxidation kinetic. *Bioresource Technology* 149: 208–15.
3. Huang, Y.-F., Syu, F.-S., Chiueh, P.-T. and Lo, S.-L. 2013. Life cycle assessment of biochar cofiring with coal. *Bioresource Technology* 131: 166–71.

2012

1. Agar, D. and Wihersaari, M. 2012. Bio-coal, torrefied lignocellulosic resources e key properties for its use in co-firing with fossil coal: Their Status. *Biomass and Bioenergy* 44: 107–11.
2. Chiueh, P.-T., Lee, K.-C., Syu, F.-S. and Lo, S.-L. 2012. Implications of biomass pretreatment to cost and carbon emissions: Case study of rice straw and Pennisetum in Taiwan. *Bioresource Technology* 108: 285–94.
3. Nicholls, D. and Zerbe, J. 2012. Cofiring biomass and coal for fossil fuel reduction and other benefits – Status of North American facilities in 2010. Gen. Tech. Rep. PNW-GTR-867. U.S. Department of Agriculture, Forest Service: 1–22.

2011 and previous

1. Bridgeman, T. G., Jones, J. M., Shield, I. and Williams, P. T. 2008. Torrefaction of reed canary grass, wheat straw and willow to enhance solid fuel qualities and combustion properties. *Fuel* 87: 844–56.
2. De, S. and Assadi, M. 2009. Impact of cofiring biomass with coal in power plants-A techno-economic assessment. *Biomass and Bioenergy* 33: 283–93.
3. Savolainen, K. 2003. Co-firing of biomass in coal-fired utility boilers. *Applied Energy* 74: 369–81.

B. Process types

2015

1. Fluid, S. 2015. Co-firing of biomass in coal-fired utility boilers. *WO Patent 2015/005807 A1*.

2. Zheng, A., Zhao, Z., Chang, S., Huang, Z., Zhao, K., Wei, G., He, F. and Li, H. 2015. Comparison of the effect of wet and dry torrefaction on chemical structure and pyrolysis behavior of corncobs. *Bioresource Technology* 176: 15–22.

2014

1. Berrueco, C., Montané, D., Güell and del Alamao, G. 2014. Effect of temperature and dolomite on tar formation during gasification of torrefied biomass in a pressurized fluidized bed. *Energy* 66: 849–59.
2. Carrere, N., Muller, S. and Mitzkat, M. 2014. REVE-a new industrial technology for biomass torrefaction: Pilot studies. *Fuel Processing Technology* 126: 155–62.
3. Pérez-Fortes, M., Laínez-Aguirre, J. M., Bojarski, A. D. and Puigjaner, L. 2014. Optimization of pre-treatment selection for the use of woody waste in co-combustion plants. *Chemical Engineering Research and Design* 92: 1539–62.
4. Reza, M. T., Uddin, M. H., Lynam, J. G. and Coronella, C. J. 2014. Engineered pellets from dry torrefied and HTC biochar blends. *Biomass and Bioenergy* 63: 229–38.
5. Srinivasan, V., Adhikari, S., Chattanathan, S. A., Tu, M. and Park, S. 2014. Catalytic pyrolysis of raw and thermally treated cellulose using different acidic zeolites. *Bioenergy Research* 7: 867–75.
6. Thanapal, S. 2014. Carbon dioxide torrefaction of woody biomass. *Energy and Fuels* 28: 1147–57.
7. Wang, L., Varhegyi, G. and Skreiberg, O. 2014. CO_2 gasification of torrefied wood: A kinetic study. *Energy and Fuels* 28: 7582–90.
8. Zheng, A., Zhao, Z., Huang, Z., Zhao, K., Wei, G., Wang, X., He, F. and Li, H. 2014. Catalytic fast pyrolysis of biomass pretreated by torrefaction with varying severity. *Energy and Fuels* 28: 5804–11.

2013

1. Bach, Q.-V., Tran, K.-Q., Khalil, R. A., Skreiberg, Ø. and Seisenbaeva, G. 2013. Comparative assessment of wet torrefaction. *Energy and Fuels* 27: 6743–53.
2. Haddou, J. V. H., Ellis, N., Bi, X. and Epstein, N. 2013. Spouting characteristics of SPF wood pellets. *Canadian Journal of Chemical Engineering* 91: 808–13.

C. Fundamental studies

2015

1. Boskovic, A., Basu, P. and Amyotte, P. 2015. An exploratory study of explosion potential of dust from torrefied biomass. *Canadian Journal of Chemical Engineering* 93: 658–63.
2. Cao, L., Yuan, X., Li, H., Li, C., Xiao, Z., Jiang, L., Huang, B., Xiao, Z., Chen, X., Wang, H. and Zeng, G. 2015. Complementary effects of torrefaction and co-pelleting: Energy consumption and characteristics of pellets. *Bioscience Technology* 185: 254–62.

3. Cavagnol, S., Roesler, J. F., Sanz, E., Nastoll, W., Lu, P. and Perré, P. 2015. Exothermicity in wood torrefaction and its impact on product mass yields: From micro to pilot scale. *Canadian Journal of Chemical Engineering* 93: 331-9.
4. Chen, W., Peng, J. and Bi, X., 2015. A state-of-the-art review of biomass torrefaction, densification and applications. *Renewable and Sustainable Energy Reviews* 44: 847-66.
5. Li, T., Geier, M., Wang, L., Ku, X., Güell, B. M., Løvås and Shaddix, C. R. 2015. Effect of torrefaction on physical properties and conversion behavior of high heating rate char of forest residue. *Energy and Fuels* 29: 177-84.
6. Na, B., Ahn, B. and Lee, J. 2015. Changes in chemical and physical properties of yellow poplar *Liriodendron tulipifera* during torrefaction. *Wood Science Technology* 49: 257-72.
7. Nachenius, R. W., van de Wardt, T. A., Ronsse, F. and Prins, W. 2015. Residence time distributions of coarse biomass particles in a screw conveyor reactor. *Fuel Processing Technology* 130: 87-95.
8. Park, C., Zahid, U., Lee, S. and Han, C. 2015. Effect of process operating conditions in the biomass torrefaction: A simulation study using one-dimensional reactor and process model. *Energy* 79: 127-39.
9. Pedrazzi, S., Allesina, G., Belló, T., Rinaldini, C. A. and Tartarini, P. 2015. Digestate as bio-fuel in domestic furnaces. *Fuel Processing Technology* 130: 172-78.
10. Peng, J., Wang, J., Bi, X. T., Lim, C. J., Sokhansanj, S., Peng, H. and Jia, D. 2015. Effects of thermal treatment on energy density and hardness of torrefied wood pellets. *Fuel Processing Technology* 129: 168-73.
11. Toptas, A., Yildirim, Y., Duman, G. and Yanik, J. 2015. Combustion behavior of different kinds of torrefied biomass and their blends with lignite. *Bioresource Technology* 177: 328-36.
12. Zheng, A., Zhao, Z., Chang, S., Huang, Z., Zhao, K., Wei, G., He, F. and Li, H. 2015. Comparison of the effect of wet and dry torrefaction on chemical structure and pyrolysis behavior of corncobs. *Bioresource Technology* 176: 15-22.
13. Ziv, E. and Shonnard, D., 2015. Predicting properties of torrefied biomass by intrinsic kinetics. *Energy and Fuels* 29: 171-76.

2014

1. Bada, S., Falcon, R. and Falcon, L. 2014. Investigation of combustion and co-combustion characteristics of raw and thermal treated bamboo with thermal gravimetric analysis. *Thermochimica Acta* 589, 207-14.
2. Basu, P., Sadhukhan, A. K., Gupta, P., Rao, S., Dhungana, A. and Acharya, B. 2014. An experimental and theoretical investigation on torrefaction of a large wet wood particle. *Bioresource Technology* 159: 215-22.
3. Beruero, C. 2014. Pressurized gasification of torrefied woody biomass in a lab scale fluidized bed. *Energy* 70: 68-78.
4. Blasi, C., Branca, C., Sarnataro, F. and Gallo, A. 2014. Thermal runaway in the pyrolysis of some lignocellulosic biomasses. *Energy and Fuels* 28: 2684-96.
5. Branca, C., Di Blasi, C., Galgano, A. and Broström, M. 2014. Effects of the torrefaction conditions on the fixed-bed pyrolysis of Norway spruce. *Energy and Fuels* 28: 5882-91.

6. Clausen, M. 2014. Integrated torrefaction vs. external torrefaction - A thermodynamic analysis for the case of a thermochemical biorefinery. *Energy* 77: 597–607.
7. Fick, G., Mirgaux, P. and Patisson, F. 2014. Using biomass for pig iron production: A technical, environmental and economical assessment. *Waste and Biomass Valorization* 5: 43–55.
8. Granados, D., Velasquez, H. and Chejne, F. 2014. Energetic and exergetic evaluation of residual biomass in a torrefaction process. *Energy* 74: 181–89.
9. Grigiante, M. and Antolini, D. 2014. Experimental results of mass and energy yield referred to different torrefaction pathways. *Waste Biomass* 5: 11–17.
10. Isaksson, J., Åsblad, A. and Berntsson, T. 2014. Pretreatment methods for gasification of biomass and Fischer–Tropsch crude production integrated with a pulp and paper mill. *Clean Technology and Environmental Policy* 16: 1393–402.
11. Joshi, Y., Mangkusaputra, V., de Vries, H. and de Jong, W. 2014. Effect of mechanical fractionation on the torrefaction of grass. *Environmental Progress and Sustainable Energy* 33: 721–5.
12. Kangas, P., Koukkari, P. and Hupa, M. 2014. Modeling biomass conversion during char gasification, pyrolysis, and torrefaction by applying constrained local thermodynamic equilibrium: *Energy and Fuels* 28: 6361–70.
13. Keipi, T., Tolvanen, H., Kokko, L. and Raiko, R. 2014. The effect of torrefaction on the chlorine content and heating value of eight woody biomass samples. *Biomass and Bioenergy* 66: 232–9.
14. Kymäläinen, M., Havimo, M., Keriö, S., Kemell, M. and Solio, J. 2014. Biological degradation of torrefied wood and charcoal. *Biomass and Bioenergy* 71: 170–7.
15. Lee, S. and Lee, J. 2014. Optimization of biomass torrefaction conditions by the Gain and Loss method and regression model analysis. *Bioresource Technology* 172: 438–43.
16. Nocquet, T., Dupont, C., Commandre, J.-M., Grateau, M., Thiery, S. and Salvador, S. 2014. Volatile species release during torrefaction of wood and its macromolecular constituents: Part 1 Experimental study. *Energy* 72: 180–7.
17. Patuzzi, F. and Gasparella, A. 2014. Thermochemical and fluid dynamic model of a bench-scale torrefaction reactor. *Waste Biomass Valorization* 5: 165–73.
18. Peduzzi, E., Boissonnet, G., Haarlemmer, G., Dupont, C. and Maréchal, F. 2014. Torrefaction modelling for lignocellulosic biomass conversion processes. *Energy* 70: 58–67.
19. Saadon, S., Uemura, Y. and Mansor, N. 2014. Torrefaction in the presence of oxygen and carbon dioxide: The effect on yield of oil palm kernel shell. *Procedia Chemistry* 9: 194–201.
20. Trop, P., Anicic, B. and Goricanec, D. 2014. Production of methanol from a mixture of torrefied biomass and coal. *Energy* 77: 125–32.
21. Vincent, S., Mahinpey, M. and Aqsha Aqsha, A. 2014. Mass transfer studies during CO_2 gasification of torrefied and pyrolyzed chars. *Energy* 67: 319–27.
22. Walsem, J. 2014. Process for making chemical derivatives. U.S. Patent Application 0024769.
23. Weiland, F., Nordwaeger, M., Olofsson, I., Wiinikka, H. and Nordin, A. 2014. Entrained flow gasification of torrefied wood residues. *Fuel Processing Technology* 125: 51–8.
24. Xue, G., Kwapinska, M., Horvat, A., Zhonglai, L., Dooley, S., Kwapinski, W. and Leahy, J. J. 2014. Gasification of torrefied *Miscanthus* x *giganteus* in an air-blown bubbling fluidized bed gasifier. *Bioresource Technology* 159: 397–403.

25. Yang, Z., Sarkar, M., Kumar, A., Tumuluru, J. S. and Huhnke, R. L. 2014. Effects of torrefaction and densification on switchgrass pyrolysis products. *Bioresource Technology* 174: 266–73.

2013

1. Ampulski, R. S., Laska, T. E., Turner, J. T. and White, S. P. 2013. Optimization of torrefaction volatiles for producing liquid fuel from biomass. U.S. Patent US 2013/0247448 A1.
2. Atienza-Martinez, M., Fonts, I., Ábrego, J., Ceamanos, J. and Gea, G. 2013. Sewage sludge torrefaction in a fluidized bed reactor. *Chemical Engineering Journal* 222: 534–45.
3. Basu, P., Rao, S. and Dhungana, A. 2013. An investigation into the effect of biomass particle size on its torrefaction. *Canadian Journal of Chemical Engineering* 91: 466–74.
4. Bates, R. and Ghoniem, A. 2013. Biomass torrefaction: Modeling of reaction thermochemistry. *Bioresource Technology* 134: 331–40.
5. Brown, D., Rowe, A. and Wild, P. 2013. A techno-economic analysis of using mobile distributed pyrolysis facilities to deliver a forest residue resource. *Bioresource Technology* 150: 367–76.
6. Cavagnol, S., Sanz, E., Nastoll, W., Roesler, J. F., Zymla, V. and Perré, P. 2013. Inverse analysis of wood pyrolysis with long residence times in the temperature range 210–290°C: Selection of multi-step kinetic models based on mass loss residues. *Thermochimica Acta* 572: 1–9.
7. Chen, W. H., Lu, K. M., Liu, S. H., Tsai, C. M., Lee, W. J. and Lin, T. C. 2013. Biomass torrefaction characteristics in inert and oxidative atmospheres at various superficial velocities. *Bioresource Technologies* 146: 152–60.
8. Funke, A., Reebs, F. and Kruse, A. 2013. Experimental comparison of hydrothermal and vapothermal carbonization. *Fuel Processing Technology* 115: 261–9.
9. Khalil, R. A., Bach, Q.-V., Skreiberg, Ø. and Tran, K.-Q. 2013. Performance of a residential pellet combustor operating on raw and torrefied spruce and spruce-derived residues. *Energy and Fuels* 27: 4760–9.
10. Kokonya, S., Castro-Diaz, M., Barriocanal, C. and Snape, C. E. 2013. An investigation into the effect of fast heating on fluidity development and coke quality for blends of coal and biomass. *Biomass and Bioenergy* 56: 295–306.
11. Li, J., Biagini, E., Yang, W., Tognotti, L. and Blasiak, W. 2013. Flame characteristics of pulverized torrefied-biomass combusted with high-temperature air. *Combustion and Flame* 160: 2585–94.
12. Medina, C. H., Phylaktou, H. N., Sattar, H. andrews, G. E. and Gibbs, B. M. 2013. The development of an experimental method for the determination of the minimum explosible concentration of biomass powders. *Biomass and Bioenergy* 53: 95–104.
13. Nikolopoulos, N., et al. 2013. Modeling of wheat straw torrefaction as a preliminary tool for process design. *Waste Biomass Valorization* 4: 409–20.
14. Pirraglia, A., Gonzalez, R., Denig, J. and Saloni, D. 2013. Technical and economic modeling for the production of torrefied lignocellulosic biomass for the U.S. densified fuel industry. *Bioenergy Resources* 6: 263–75.

15. Retsina, T., Pylkkanen, V. and O'Connor, R. 2013. Processes for producing energy-dense biomass and sugars or sugar derivatives, by integrated hydrolysis and torrefaction. WO Patent WO 2013/166225 A2.
16. Rousset, P., Fernandes, K., Vale, A., Macedo, L. and Benoist, A. 2013. Change in particle size distribution of torrefied biomass during cold fluidization. *Energy* 51: 71-7.
17. Saleh, S. B., Hansen, B. B., Jensen, P. A. and Dam-Johansen, K. 2013. Efficient fuel pretreatment: Simultaneous torrefaction and grinding of biomass. *Energy and Fuels* 27: 7531-40.
18. Saleh, S. B., Hansen, B. B., Jensen, P. A. and Dam-Johansen, K. 2013. Influence of biomass chemical properties on torrefaction characteristics. *Energy and Fuels* 27: 7541-48.
19. Sarvaramini, A., Assima, G. and Larachi, F. 2013. Dry torrefaction of biomass-torrefied products and torrefaction kinetics using the distributed activation energy model. *Chemical Engineering Journal* 229: 498-507.
20. Sheikh, M. M., Kim, C. H., Park, H. J., Kim, S. H., Kim, G. C., Lee, J. Y., Sim, S. W. and Kim, J. W. 2013. Effect of torrefaction for the pretreatment of rice straw for ethanol production. *Journal of Science Food and Agriculture* 93: 3198-204.
21. Sheikh, M. M., Kim, C. H., Park, H. J., Kim, S. H., Kim, G. C., Lee, J. Y., Sim, S. W. and Kim, J. W. 2013. Influence of torrefaction pretreatment for ethanol fermentation from waste money bills. *International Union of Biochemistry and Molecular Biology* 60: 203-9.
22. Shuttleworth, P. S., Budarin, V., White, R. J., Gun'ko, V. M., Luque, R. and Clark, J. H. 2013. Molecular-level understanding of the carbonisation of polysaccharides. *Chemistry: A European Journal* 19: 9351-7.
23. Svanberg, M., Olofsson, I., Flodén, J. and Nordin, A. 2013. Analysing biomass torrefaction supply chain costs. *Bioresource Technology* 142: 287-96.
24. Tapasvi, D., Khalil, R., Várhegyi, G., Tran, K.-Q., Grønli, M. and Skreiberg, Ø. 2013. Thermal decomposition kinetics of woods with an emphasis on torrefaction. *Energy and Fuels* 27: 6134-45.
25. Tremel, A., Becherer, Dominik, Fendt, S., Gaderer, M. and Spliethoff, H. 2013. Performance of entrained flow and fluidised bed biomass gasifiers on different scales. *Energy Conversion and Management* 69: 95-106.
26. Via, B., Adhikari, S. and Taylor, S. 2013. Modeling for proximate analysis and heating value of torrefied biomass with vibration spectroscopy. *Bioresource Technology* 133: 1-8.
27. Zwart, R., Willem, R. and Remmert, J. 2013. Use of torrefaction condensate. WO Patent WO 2013/019111 A1.

2012

1. Bates, R. and Ghoniem, A. 2013. Biomass torrefaction: Modeling of volatile and solid product evolution kinetics. *Bioresource Technology* 124: 460-9.
2. Fisher, E. M., Dupont, C., Darvell, L. I., Commandré, J. M., Saddawi, A., Jones, J. M., Grateau, M., Nocquet, T. and Salvador, S. 2012. Combustion and gasification characteristics of chars from raw and torrefied biomass. *Combustion and Gasification* 119: 157-65.

3. Khazraie Shoulaifar, T., Demartini, N., Ivaska, A., Fardim, P. and Hupa, M. 2012. Measuring the concentration of carboxylic acid groups in torrefied spruce wood. *Bioresource Technology* 123: 338–43.
4. Peng, J. H., Bi, H. T., Sokhansanj, S. and Lim, J. C. 2013. A study of particle size effect on biomass torrefaction and densification. *Energy and Fuels* 26: 3826–19.
5. Pétrissans, A., Younsi, R., Chaouch, M., Gérardin, P. and Pétrissans, M. 2012. Experimental and numerical analysis of wood thermodegradation. *Journal of Thermal Analysis and Calorimetry* 109: 907–14.
6. Srinivasan, V., Adhikari, S., Chattanathan, S. A. and Park, S. 2012. Catalytic pyrolysis of torrefied biomass for hydrocarbons production. *Energy and Fuels* 26: 7347–53.
7. Tumuluru, J. S., Hess, J. R., Boardman, R. D., Wright, C. T. and Westover, T. L. 2012. Formulation, pretreatment, and densification options to improve biomass specifications for co-firing high percentages with coal. *Industrial Biotechnology* 8: 113–32.
8. Zheng, A., Zhao, Z., Chang, S., Huang, Z., He, F. and Li, H. 2012. Effect of torrefaction temperature on product distribution from two-staged pyrolysis of biomass. *Energy and Fuels* 26: 2968–74.

2011 and previous

1. Almeida, G., Brito, J. and Perré, P., 2010. Alterations in energy properties of eucalyptus wood and bark subjected to torrefaction: The potential of mass loss as a synthetic indicator. *Bioresource Technologies* 101: 9778–84.
2. Arias, B., Pevida, C., Fermoso, J., Plaza, M. G., Rubiera, F. and Pis, J. J. 2008. Influence of torrefaction on the grindability and reactivity of woody biomass. *Fuel Processing Technology* 89: 169–75.
3. Bergman, P. C. A., Boersma, A. R., Kiel, J. H. A., Prins, M. J., Ptainski, K. J. and Janssen, F. J. J. G. 2005. Torrefaction for entrained-flow gasification of biomass. *ECN Biomass* 1: 1–50.
4. Chen, W. and Kuo, P. 2010. A study on torrefaction of various biomass materials and its impact on lignocellulosic structure simulated by a thermogravimetry. *Energy* 35: 2580–6.
5. Chen, W. and Kuo, P., 2011. Torrefaction and co-torrefaction characterization of hemicellulose, cellulose and lignin as well as torrefaction of some basic constituents in biomass. *Energy* 36: 803–11.
6. FitzPatrick, M., Champagne, P., Cunningham, M. F. and Whitney, R. A. 2010. A biorefinery processing perspective: Treatment of lignocellulosic materials for the production of value-added products. *Bioresource Technologies* 101: 8915–22.
7. Phanphanich, M. and Mani, S., 2011. Impact of torrefaction on the grindability and fuel characteristics of forest biomass. *Bioresource Technology* 102: 1246–53.
8. Prins, M. J., Ptasinski, K. J. and Janssen, F. J. J. G. 2006. More efficient biomass gasification via torrefaction. *Energy* 31: 3458–70.
9. Repellin, V., Govin, A., Rolland, M. and Guyonnet, R. 2010. Energy requirement for fine grinding of torrefied wood. *Biomass and Bioenergy* 34: 923–30.
10. Rousset, P., Davrieux, F., Macedo, L. and Perré, P. 2011. Characterisation of the torrefaction of beech wood using NIRS: Combined effects of temperature and duration. *Biomass and Bioenergy* 35: 1219–26.

11. Uslu, A., Faaij, A. and Bergman, P. 2008. Pre-treatment technologies, and their effect on international bioenergy supply chain logistics. Techno-economic evaluation of torrefaction, fast pyrolysis and pelletisation. *Energy* 33: 1206–23.
12. Yan, W., Acharjee, T. C., Coronella, C. J. and Vásquez, V. R. 2009. Thermal pretreatment of lignocellulosic biomass. *Environmental Progress and Sustainable Energy* 28: 435–40.
13. Yang, Y. B., Ryu, C., Khor, A., Sharifi, V. N. and Swithenbank, J. 2005. Fuel size effect on pinewood combustion in a packed bed. *Fuel* 84: 2026–38.

D. Type of material processed

2015

1. Atienza-Martinez, M., Fonts, I., Lázaro, L., Ceamanos, J. and Gea, G. 2015. Fast pyrolysis of torrefied sewage sludge in a fluidized bed reactor. *Chemical Engineering Journal* 259: 467–80.
2. Atienza-Martinez, M., Mastral, J. F., Ábrego, J., Ceamanos, J. and Gea, G. 2015. Sewage sludge torrefaction in an auger reactor. *Energy and Fuels* 29: 160–70.
3. Benavente, V. and Fullana, A. 2015. Torrefaction of olive mill waste. *Biomass and Bioenergy* 73: 186–94.
4. Brachi, P., Miccio, F., Miccio, M. and Ruoppolo, G. 2015. Isoconversional kinetic analysis of olive pomace decomposition under torrefaction operating conditions. *Fuel Processing Technology* 130: 147–54.
5. Chen, W.-H., Huang, M.-Y., Chang, J.-S., Chen, C.-Y. and Lee, W.-J. 2015. An energy analysis of torrefaction for upgrading microalga residue as a solid fuel. *Bioresource Technology* 185: 285–93.
6. Chiou, B., Valenzuela-Medina, D., Bilbao-Sains, C., Klamczynski, A. K., Avena-Bustillos, R. J., Milczarek, R. R., Du, W. X., Glenn, G. M. and Orts, W. J. 2015. Torrefaction of pomaces and nut shells. *Bioresource Technology* 177: 58–65.
7. Mei, Y., Liu, R., Yang, Q., Haiping, Y., Shao, J. and Draper, C. 2015. Torrefaction of cedarwood in a pilot scale rotary kiln and the influence of industrial flue gas. *Bioresource Technology* 177: 355–60.
8. Poudel, J., Ohm, T. I., Lee, S. H. and Oh, S. C. 2015. A study on torrefaction of sewage sludge to enhance solid fuel qualities. *Waste Management* 40: 112–18.
9. Strandberg, M., Olofsson, I., Pommer, L., Wiklund-Lindström, S., Åberg, K. and Nordin, A. 2015. Effects of temperature and residence time on continuous torrefaction of spruce wood. *Fuel Processing Technology* 134: 387–98.
10. Volpe, R., Messineo, A., Millan, M., Volpe, M. and Kandiyoti, R. 2015. Assessment of olive wastes as energy source: Pyrolysis, torrefaction and the key role of H loss in thermal breakdown. *Energy* 82: 119–27.

2014

1. Asadullah, M., Adi, A. M., Suhada, N., Malek, N. H., Saringat, M. I. and Azdarpour, A. 2014. Optimization of palm kernel shell torrefaction to produce energy densified bio-coal. *Energy Conversion and Management* 88: 1086–93.
2. Bada, S., Falcon, R. and Falcon, L. 2014. Investigation of combustion and co-combustion characteristics of raw and thermal treated bamboo with thermal gravimetric analysis. *Thermochimica Acta* 589: 207–14.

3. Branca, C., Di Blasi, C., Galgano, A. and Broström, M. 2014. Effects of the torrefaction conditions on the fixed-bed pyrolysis of Norway spruce. *Energy and Fuels* 28: 5882–91.
4. Chen, D., Zhou, J. and Zhang, Q. 2014. Properties of rice husk by using TG-FTIR and Py-GC/MS. *Energy and Fuels* 28: 5857–63.
5. Chen, W.-H., Kuo, P. C., Liu, S.-H. and Wu, W. 2014. Thermal characterization of oil palm fiber and eucalyptus in torrefaction. *Energy* 71: 40–8.
6. Fryda, L., Daza, C., Pels, J., Janssen, A. and Zwart, R. 2014. Lab-scale co-firing of virgin and torrefied bamboo species *Guadua angustifolia* Kunth as a fuel substitute in coal fired power plants. *Biomass and Bioenergy* 65: 28–41.
7. Joshi, Y., Mangkusaputra, V., de Vries, H. and de Jong, W. 2014. Effect of mechanical fractionation on the torrefaction of grass. *Environmental Progress and Sustainable Energy* 33: 721–5.
8. Kaliyan, N., Morey R. V., Tiffany, D. G. and Lee, W. F. 2014. Life cycle assessment of a corn stover torrefaction plant integrated with a corn ethanol plant and a coal fired power plant. *Biomass and Bioenergy* 63: 92–100.
9. Kosov, V. V., Sinelschchikov, V. A., Sytchev, G. A. and Zaichenko, V. M. 2014. Effect of torrefaction on properties of solid granulated fuel of different biomass types. *High Temperature* 52: 907–12.
10. Li, R., Li, B., Yang, T., Xie, Y. and Kai, X. 2014. Production of bio-oil from rice stalk supercritical ethanol. *Energy and Fuels* 88: 1948–55.
11. Marias, F. and Casajus, C. 2014. Torrefaction of corn stover in a macro-thermobalance: Influence of operating conditions. *Waste Biomass Valorization* 5: 157–64.
12. Nicksy, D., Pollard, A., Strong, D. and Hendry, J. 2014. In-situ torrefaction and spherical pelleting of partially pre-torrefied hybrid poplar. *Biomass and Bioenergy* 70: 452–60.
13. Onwudili, J. A., Nahil, M. A., Wu, C. and Williams, P. T. 2014. High temperature pyrolysis of solid products obtained from rapid hydrothermal pre-processing of pinewood sawdust. *Royal Society of Chemistry* 4: 34784–92.
14. Poudel, J. and Oh, S. 2014. Effect of torrefaction on the properties of corn stalk to enhance solid fuel qualities. *Energies* 7: 5586–600.
15. Reckamp, J., Garrido, R. and Satrio, J. 2014. Selective pyrolysis of paper mill sludge by using pretreatment processes to enhance the quality of bio-oil and biochar products. *Biomass and Bioenergy* 71: 235–4.
16. Saadon, S., Uemura, Y. and Mansor, N. 2014. Torrefaction in the presence of oxygen and carbon dioxide: The effect on yield of oil palm kernel shell. *Procedia Chemistry* 9: 194–201.
17. Shoulaifar, T., DeMartini, N., Willför, S., Pranovich, A., Smeds, A. I., Petteri Virtanen, T. A., Maunu, S.-L., Verhoeff, F., Kiel, J. H. A. and Hupa, M. 2014. Impact of torrefaction on the chemical structure of birch wood. *Energy and Fuels* 28: 3863–72.

2013

1. Ábrego, J., Sánchez, J. L., Arauzo, J., Fonts, I., Gil-Lalaguna, N. and Atienza-Martinez, M. 2013. Technical and energetic assessment of a three-stage thermochemical treatment for sewage sludge. *Energy and Fuels* 27: 1026–34.

2. Atienza-Martinez, M., Fonts, I., Ábrego, J., Ceamanos, J. and Gea, G. 2013. Sewage sludge torrefaction in a fluidized bed reactor. *Chemical Engineering Journal* 222: 534–45.
3. Hill, S., Grigsby, W. and Hall, P. 2013. Chemical and cellulose crystallite changes in *Pinus radiata* during torrefaction. *Biomass and Bioenergy* 56: 92–8.
4. Khalil, R. A., Bach, Q.-V., Skreiberg, Ø. and Tran, K.-Q. 2013. Performance of a residential pellet combustor operating on raw and torrefied spruce and spruce-derived residues. *Energy and Fuels* 27: 4760–69.
5. Na, B., Kim, Y.-H., Lim, W.-S., Lee, S.-M., Lee, H.-W. and Lee, J.-W. 2013. Torrefaction of oil palm mesocarp fiber and their effect on pelletizing. *Biomass and Bioenergy* 52: 159–65.
6. Nikolopoulos, N., Isemin, R., Atsonios, K., Kourkoumpas, D., Kuzmin, S., Mikhalev, A., Nikolopoulos, A., Agraniotis, M., Grammelis, P. and Kakaras, Em. 2013. Modeling of wheat straw torrefaction as a preliminary tool for process design. *Waste Biomass Valorization* 4: 409–20.
7. Pirraglia, A., Gonzalez, R., Saloni, D. and Denig, J. 2013. Technical and economic assessment for the production of torrefied ligno-cellulosic biomass pellets in the U.S. *Energy Conversion and Management* 66: 153–64.
8. Sabil, K. M., Aziz, M. A., Lal, B. and Uemura, Y. 2013. Effects of torrefaction on the physiochemical properties of oil palm empty fruit bunches, mesocarp fiber and kernel shell. *Biomass and Bioenergy* 56: 351–60.
9. Shoulaifar, T. K., Demartini, N., Zevenhoven, M., Verhoeff, F., Kiel, J. and Hupa, M. M. 2013. Ash-forming matter in torrefied birch wood: Changes in chemical association. *Energy and Fuels* 27: 5684–90.
10. Stelte, W., Nielsen, N. P. K., Hansen, H. O., Dahl, J., Shang, L. and Sanadi, A. R. 2013. Pelletizing properties of torrefied wheat straw. *Biomass and Bioenergy* 49: 214–21.
11. Wang, C., Peng, J., Li, H., Bi, X. T., Legros, R., Lim, C. J. and Sokhansanj, S. 2013. Oxidative torrefaction of biomass residues and densification of torrefied sawdust to pellets. *Bioresource Technology* 127: 318–25.
12. Zheng, A., Zhao, Z., Chang, S., Huang, Z., Wang, X., He, F. and Li, H. 2013. Effect of torrefaction on structure and fast pyrolysis behavior of corncobs. *Bioresource Technology* 128: 370–7.

2012

1. Aziz, M. A., Sabil, K. M., Uemura, Y. and Ismail, L. 2012. A study on torrefaction of oil palm biomass. *Journal of Applied Sciences* 12: 1130–5.
2. Chiueh, P.-T., Lee, K.-C., Syu, F.-S. and Lo, S.-L. 2012. Implications of biomass pretreatment to cost and carbon emissions: Case study of rice straw and Pennisetum in Taiwan. *Bioresource Technology* 108: 285–94.
3. Medic, D., Darr, M. J., Shah, A. and Rahn, S. J. 2012. Effect of torrefaction on water vapor adsorption properties and resistance to microbial degradation of corn stover. *Energy and Fuels* 26: 2386–93.
4. Shang, L., Ahrenfeldt, J., Holm, J. K., Sanadi, A. R., Barsberg, S., Thomsen, T., Stelte, W. and Henriksen, U. B. 2012. Changes of chemical and mechanical behavior of torrefied wheat straw. *Biomass and Bioenergy* 40: 63–70.

2011 and previous

1. Almeida, G., Brito, J. and Perré, P. 2010. Alterations in energy properties of eucalyptus wood and bark subjected to torrefaction: The potential of mass loss as a synthetic indicator. *Bioresource Technologies* 101: 9778–84.
2. Bridgeman, T. G., Jones, J. M., Shield, I. and Williams, P. T. 2008. Torrefaction of reed canary grass, wheat straw and willow to enhance solid fuel qualities and combustion properties. *Fuel* 87: 844–56.
3. Rousset, P., Davrieux, F., Macedo, L. and Perré, P. 2011. Characterisation of the torrefaction of beech wood using NIRS: Combined effects of temperature and duration. *Biomass and Bioenergy* 35: 1219–26.

E. Densification methods

2015

1. Chen, W., Peng, J. and Bi, X., 2015. A state-of-the-art review of biomass torrefaction, densification and applications. *Renewable and Sustainable Energy Reviews* 44: 847–66.

2014

1. Asadullah, M., Adi, A. M., Suhada, N., Malek, N. H., Saringat, M. I. and Azdarpour, A. 2014. Optimization of palm kernel shell torrefaction to produce energy densified bio-coal. *Energy Conversion and Management* 88: 1086–93.
2. Bringas, C. and Skreiberg, Ø. 2014. Torrefaction influence on pelletability and pellet quality of Norwegian forest residues. *Energy and Fuels* 28: 2554–61.
3. Kosov, V. V., Sinelschchikov, V. A., Sytchev, G. A. and Zaichenko, V. M. 2014. Effect of torrefaction on properties of solid granulated fuel of different biomass types. *High Temperature* 52: 907–12.
4. Nicksy, D., Pollard, A., Strong, D. and Hendry, J. 2014. In-situ torrefaction and spherical pelleting of partially pre-torrefied hybrid poplar. *Biomass and Bioenergy* 70: 452–60.
5. Reza, M. T., Uddin, M. H., Lynam, J. G. and Coronella, C. J. 2014. Engineered pellets from dry torrefied and HTC biochar blends. *Biomass and Bioenergy* 63: 229–38.
6. Yang, Z., Sarkar, M., Kumar, A., Tumuluru, J. S. and Huhnke, R. L. 2014. Effects of torrefaction and densification on switchgrass pyrolysis products. *Bioresource Technology* 174: 266–73.

2013

1. Na, B., Kim, Y.-H., Lim, W.-S., Lee, S.-M., Lee, H.-W. and Lee, J.-W. 2013. Torrefaction of oil palm mesocarp fiber and their effect on pelletizing. *Biomass and Bioenergy* 52: 159–65.
2. Peng, J. H., Bi, H. T., Lim, C. J. and Sokhansanj, S. 2013. Study on density, hardness, and moisture uptake of torrefied wood pellet. *Energy and Fuels* 27: 967–74.

3. Stelte, W., Nielsen, N. P. K., Hansen, H. O., Dahl, J., Shang, L. and Sanadi, A. R. 2013. Pelletizing properties of torrefied wheat straw. *Biomass and Bioenergy* 49: 214–21.
4. Wang, C., Peng, J., Li, H., Bi, X. T., Legros, R., Lim, C. J. and Sokhansanj, S. 2013. Oxidative torrefaction of biomass residues and densification of torrefied sawdust to pellets. *Bioresource Technology* 127: 318–25.

2012

1. Peng, J. H., Bi, H. T., Sokhansanj, S. and Lim, J. C. 2013. A study of particle size effect on biomass torrefaction and densification. *Energy and Fuels* 26: 3826–39.

F. General reviews

2015

1. Dudyński, M., van Dyk, J. C., Kwiatkowski, K. and Sosnowski, M. 2015. Biomass gasification: Influence of torrefaction on syngas production and tar formation. *Fuel Processing Technology* 131: 203–12.
2. Liaw, S.-S., Frear, C., Lei, W., Zhang, S. and Garcia-Perez, M. 2015. Anaerobic digestion of C1–C4 light oxygenated organic compounds derived from the torrefaction of lignocellulosic materials. *Fuel Processing Technology* 131: 150–8.
3. Tapasvi, D., Kempegowda, R. S., Tran, K.-Q., Skreiberg, Ø. and Grønli, M. 2015. A simulation study on the torrefied biomass gasification. *Energy Conversion and Management* 90: 446–57.
4. Vold, J., Ulven, C. and Chisholm, B. 2015. Torrefied biomass filled polyamide biocomposites: Mechanical and physical property analysis. *Journal of Material Science* 50: 725–32.

2014

1. Batidzirai, B., van der Hilst, F., Meerman, H., Junginger, M. H. and Faaij, A. P. C. 2014. Optimization potential of biomass supply chains with torrefaction technology. *Biofuels, Bioproducts and Biorefining* 8: 253–82.
2. Gerssen-Gondelach, S. J., Saygin, D., Wicke, B., Patel, M. K. and Faaij, A. P. C. 2014. Competing uses of biomass: Assessment and comparison of the performance of bio-based heat, power, fuels and materials. *Renewable and Sustainable Energy Reviews* 40: 964–98.
3. Peduzzi, E., Boissonnet, G., Haarlemmer, G., Dupont, C. and Maréchal, F. 2014. Torrefaction modelling for lignocellulosic biomass conversion processes. *Energy* 70: 58–67.
4. Sandeep, K. and Dasappa, S. 2014. First and second law thermodynamic analysis of air and oxy-steam biomass gasification. *International Journal of Hydrogen Energy* 39: 19474–84.
5. Sarvaramini, A. and Larachi, F. 2014. Fe/Mg silicate mining residues as solid oxygen carriers for chemical looping combustion of torrefaction volatiles. *Energy and Fuels* 28: 1983–91.

6. Wang, L., Lurina, M., Hyytiäinen, J. and Mikkonen, E. 2014. Bio-coal market study: Macro and microenvironment of the bio-coal business in Finland. *Biomass and Bioenergy* 63: 198–209.

2013

1. Batidzirai, B., Mignot, A. P. R., Schakel, W. B., Junginger, H. M. and Faaij, A. P. C. 2013. Biomass torrefaction technology: Techno-economic status and future prospects. *Energy* 62: 196–214.
2. Brown, D., Rowe, A. and Wild, P. 2013. A techno-economic analysis of using mobile distributed pyrolysis facilities to deliver a forest residue resource. *Bioresource Technology* 150: 367–76.
3. Pirraglia, A., Gonzalez, R., Denig, J. and Saloni, D. 2013. Technical and economic modeling for the production of torrefied lignocellulosic biomass for the U.S. densified fuel industry. *Bioenergy Resources* 6: 263–75.
4. Svanberg, M., Olofsson, I., Flodén, J. and Nordin, A. 2013. Analysing biomass torrefaction supply chain costs. *Bioresource Technology* 142: 287–96.

2012

1. Acharya, B., Sule, I. and Dutta, A. 2012. A review on advances of torrefaction technologies for biomass processing. *Biomass Conversion and Biorefining* 2: 349–69.
2. Tumuluru, J. S., Hess, J. R., Boardman, R. D., Wright, C. T. and Westover, T. L. 2012. Formulation, pretreatment, and densification options to improve biomass specifications for co-firing high percentages with coal. *Industrial Biotechnology* 8: 113–32.

Appendix B: powerpoint and PDF presentations on biomass and torrefaction

Powerpoint and PDF Presentations

1. Bertrand, S. 2015. Biomass to biocoal technology. *Airex Energie: 4th Industrial Wood Pellets for Coal Plant Co-firing/Conversions Summit*, Minneapolis.
2. Boyko, B. 2015. Experience with coal to biomass conversions at Ontario Power Generation. *Ontario Power Generation: 4th Industrial Wood Pellets for Coal Plant Co-firing/Conversions Summit*, Minneapolis.
3. Calderon, C. 2013. IBTC: Setting the standards for torrified biomass. *Bioenergy Insight*.
4. Chauvin, H. 2015. Torrefaction: TORSPYD® - Fast continuous biomass depolymerisation system: Thermya.
5. Hazel, D. 2015. Forest landowner perspectives on the wood pellet industry - A North Carolina view. *North Carolina State University: 4th Industrial Wood Pellets for Coal Plant Co-firing/Conversions Summit*, Minneapolis.
6. Henderson, C. 2015. Cofiring of biomass in coal-fired power plants - *European Experience: FCO/IEA Clean Coal Workshops*, Hebei and Shandong Provinces, China.

7. IEA-ETSAP and IRENA. 2013. Biomass co-firing. Technology Brief: E21, pp. 1–8.
8. Jostrom, M. 2015. Wood for energy? A forest owner's perspective. *Plumb Creek Timber Company: 4th Industrial Wood Pellets for Coal Plant Co-firing/Conversions Summit*, Minneapolis.
9. Kiesel, R., et al. 2015. Renewable energy biofuel: Natural Resources Research Institute. *UMD: 4th Industrial Wood Pellets for Coal Plant Co-firing/Conversions Summit*, Minneapolis.
10. Kingsley, E. 2015. How to Compensate Generators for the Higher Cost of Generation. *Innovative Natural Resource Solutions: 4th Industrial Wood Pellets for Coal Plant Co-firing/Conversions Summit*, Minneapolis.
11. Koppejan, J., Sokhansanj, S., Melin, S. and Madrali, S. 2012. Status overview of torrefaction technologies. IEA Bioenergy Task 32 Report, 54p.
12. Kristiansen, A. 2015. Advanced wood pellets – a game changer for the biomass industry. *Arbaflame: 4th Industrial Wood Pellets for Coal Plant Co-firing/Conversions Summit*, Minneapolis.
13. McCormick, M. 2015. Experimental investigation of continuous torrefaction conditions of biomass residues for the subsequent use of torrefied pellets in domestic and district heating systems. *10th European Conference on Industrial Furnaces and Boilers*, Portugal.
14. McGlynn, E. 2015. U.S. policy support for biomass co-firing, the earth partners. *4th Industrial Wood Pellets for Coal Plant Co-firing/Conversions Summit*, Minneapolis.
15. McRae, T. 2015. Wood pellet supply, security and opportunity. *Pinnacle Renewable Energy: 4th Industrial Wood Pellets for Coal Plant Co-firing/Conversions Summit*, Minneapolis.
16. Michel, J.-B. 2010. Energy crisis or energy opportunities – from utopia to reality. *International Conference on Frontiers in Mechanical Engineering*.
17. Michel, J.-B., Mahmed, C., Ropp, J., Richard, J., Sattler, M. and Schmid, M. 2011. Combustion and life-cycle evaluation of torrefied wood for decentralized heat and power production. *9th European Conference on Industrial Furnaces and Boilers*, Portugal.
18. Michel, J. B., Mahmed, C., Ropp, J., Richard, J., Sattler, M. and Schmid, M. 2010. Combustion and evaluation of torrefied wood pellets on a 50KW boiler. *18th European Biomass Conference and Exhibition*, Lyon.
19. *Ontario Power Generation, Pembina*. 2011. Biomass Sustainability Analysis Summary Report.
20. Rheuban, J. 2015. Evolution challenge for torrefaction plants. *Solvay Biomass Energy: 4th Industrial Wood Pellets for Coal Plant Co-firing/Conversions Summit*, Minneapolis.
21. Sabo, J. 2015. Properties of torrified biomass. *TCI: 4th Industrial Wood Pellets for Coal Plant Co-firing/Conversions Summit*, Minneapolis.
22. Saines, R. 2015. State-by-state RPS and RPS variants summary: implications and opportunities for coal plant conversions. *4th Industrial Wood Pellets for Coal Plant Co-firing/Conversions Summit*, Minneapolis.
23. Skov, M. 2015. Bioconversion projects. *Ramboll Energi: 4th Industrial Wood Pellets for Coal Plant Co-firing/Sonversions Summit*, Minneapolis.
24. Stearns, C. 2015. Industrial "Black"® pellets for coal plant co-firing/conversions. *Zilkha Biomass Energy: 4th Industrial Wood Pellets for Coal Plant Co-firing/Conversions Summit*, Minneapolis.

25. Strauss, W., 2015. A rational and pragmatic off-ramp to a decarbonized future. *FutureMetrics*, LLC, pp. 1–10.
26. Sussman, R. 2015. EPA clean power plan: Opportunities for Wood Pellets, Sussman and Associates. *4th Industrial Wood Pellets for Coal Plant Co-firing/Conversions Summit*, Minneapolis.
27. Svaan, J. 2015. Industrial wood type analysis. *Futurmetrics: 4th Industrial Wood Pellets for Coal Plant Co-firing/Conversions Summit*, Minneapolis.
28. Thomsen, P. 2015 Biomass conversions. *Dong Energy: 4th Industrial Wood Pellets for Coal Plant Co-firing/Conversions Summit*, Minneapolis.
29. Todd, J. 2011. Thermal repowering program. *Ontario Power Generation: Canadian Bioenergy Association Annual Conference*.
30. Wiberg, C. 2015. Industrial wood pellets for coal plant co-firing/conversions, biomass energy labs. *4th Industrial Wood Pellets for Coal Plant Co-firing/conversions Summit*, Minneapolis.
31. Wild, M. 2015. Torrefied biomass: The perfect CO_2 neutral coal substitute is maturing. *VGB PowerTech*, July 2015.

Part 2

Materials

Chapter 4

Production of biodiesel from renewable sources

Dan Zeng, Daidi Fan, Le Wu and Yuqi Wang, Northwest University, China

1 Introduction

Continued use of petroleum-based fuels is now widely recognized as unsustainable, because of diminishing supplies and the contribution of these fuels to the accumulation of carbon dioxide in the environment (Chisti, 2007). Biodiesel, produced from renewable resources, is becoming more attractive due to its positive environmental benefits (Fukuda et al., 2001; Ma and Hanna, 1999). This fuel is sustainable, biodegradable and nontoxic and generates lower emissions in comparison to petroleum-based diesel. Consumption of biodiesel will help promote the balance sought between agriculture, economic development and the environment.

Feedstock represents more than 75% of the overall cost of biodiesel production, making it important to select the appropriate feedstock to control costs (Atabani et al., 2012). The four main categories of raw materials for biodiesel are (Yusuf et al., 2011):

- edible vegetable oil;
- non-edible vegetable oil;
- waste or recycled oil; and
- animal fats.

http://dx.doi.org/10.19103/AS.2019.0027.10

Over the last decade, the majority of biodiesel production over the world has used edible vegetable oils, including countries such as USA, Germany, Malaysia and Brazil (Balat, 2009; Biermann et al., 2011; Mekhilef et al., 2011). However, the dramatic increases in prices of these oils over this period have both undermined the economics of biodiesel production and raised questions about the diversion of land from food to fuel feedstock production.

This has led to interest in non-edible oils as a potentially more efficient and economic alternative. Above all, the usage of non-edible oils eliminates the competition between food and fuels, revitalizing the long-term development of the biodiesel production industry. Non-edible oil sources include waste or recycled oil and animal fats (Canoira et al., 2008; Dufour and Iribarren, 2012; Encinar et al., 2011; Kulkarni and Dalai, 2006; Ma et al., 1998; Yaakob et al., 2013). Industrial application of these feedstocks faces challenges like efficient collection of waste oils from diverse sources and the need for pre-treatment of the high amount of saturated fatty acids.

The four main methods to produce biodiesel are:

- direct use and blending;
- micro-emulsion;
- pyrolysis; and
- transesterification.

Among these, the transesterification is the most commonly used method to convert triglycerides (TGs) from different types of lipids with an alcohol (Demirbas, 2009b). Conventionally, the transesterification reaction often proceeds using an acid, alkali or enzyme catalyst. However, the reaction with an acid catalyst is time consuming and requires acid-resistant equipment (Ong et al., 2013). In the alkali-catalyzed transesterification method, water and free fatty acids (FFAs) reduce yields of methyl esters since the presence of water and FFAs consumes the catalyst, reduces the effectiveness of the catalyst and causes soap formation (Demirbas, 2009a). In the enzymatic production of biodiesel, the primary obstacles are the cost of lipases, the slower reaction rate and lipase inactivation caused by methanol and the by-product glycerol (Gog et al., 2012).

In seeking to eliminate the negative effects of catalytic processes in biodiesel production, supercritical fluid (SCF) technology has attracted significant attention (Savage et al., 1995; Tan and Lee, 2011; Hoang et al., 2013). SCFs have physical and chemical properties between those of a fluid and a gas and a unique dissolving capacity (Anikeev and Yermakova, 2011). At supercritical conditions, chemical reactions can be manipulated by changing pressure in order to increase the solubility of reactants and products, to dispel transport limitations between the phases in reaction rates and to integrate reaction and separation processes (Savage et al., 1995). Triglyceride transesterification using supercritical lower alcohols like methanol, ethanol and propanol has proven to be the most promising biodiesel production process (Demirbas, 2008). The temperature and pressure of reactants need to reach a critical temperature 239°C and pressure of 8.1 MPa. In these conditions, a single phase solution will be formed as the solubility parameter of alcohol is reduced to a value near to TGs (Tan et al., 2010). This chapter discusses the production of biodiesel from renewable sources using supercritical technology with a focus on supercritical methanol.

2 Types of feedstocks for biodiesel production

Biodiesel consisting of fatty acid methyl esters (FAMEs) is made using TGs though transesterification with an alcohol (mostly methanol). There is a range of sources of TG, including edible or non-edible vegetable oil, waste or recycled oil, animal fats and microalgae (Chisti, 2007; Singh and Singh, 2010; Maddikeri et al., 2012; Yusuf et al., 2011). As has been noted, it is not economically feasible to use edible vegetable oils as feedstocks due to the high prices of these feedstocks, which makes biodiesel uneconomic compared with conventional diesel fuel. Alternative sources are low-grade feedstocks with high FFAs such as waste fat from restaurants, animal fats, waste cooking oils and soapstocks left over from vegetable oil processing (Canakci and Sanli, 2008). Recycling these waste products avoids potential pollution issues from disposal (Ong et al., 2013).

In 1900 a small diesel engine running on peanut oil was exhibited by the Otto Company at the Paris Exhibition. In the same period Rudolph Diesel also demonstrated the use of vegetable oil to run an unmodified diesel engine (Chalkley, 1912). Since then, a range of edible vegetable oils have provided first-generation biodiesel feedstocks, using crops such as soybeans, rapeseed, safflower, sunflower, palm oil, coconut and peanut (Atabani et al., 2012). Although high yield could be achieved by converting vegetable oils into FAMEs is obtained, high material costs and competition with food production have made these feedstocks much less attractive. This has led to interest in so-called low-grade feedstocks such as non-edible oils, waste cooking oils, soapstocks and animal fats. Each of these has different fatty acid compositions. Table 1 summarizes the fatty acid distributions of commonly used low-grade feedstocks.

Non-edible oils, such as cotton, castor, karanja, jatropha and mahua, can grow in marginal lands and have been investigated, particularly jatropha, which can be grown easily on marginal land at low cost (Pinzi et al., 2014; Samniang et al., 2014). Research suggests that biodiesel produced from waste cooking oils has the same engine performance and exhaust emissions as other types of biodiesel (Lapuerta et al., 2008). Waste cooking oil is also 2.5–3.5 times cheaper than vegetable oils (Pinzi et al., 2014; Talebian-Kiakalaieh et al., 2013).

Soapstock is the by-product of edible oil refining and less expensive than edible refined oils. It consists of an alkaline aqueous emulsion of lipids including

Table 1 Comparison of the fatty acid distributions of various biodiesel feedstocks

	Myristic	Palmitic	Palmitoleic	Stearic	Oleic	Linoleic	Linolenic	
Fatty acid (% w/w)	14:0	16:0	16:1	18:0	18:1	18:2	18:3	Reference
Soybean soapstock	-	17.2	-	4.4	15.7	55.6	7.1	Haas (2005)
Pork lard	-	26.4	-	12.1	44.7	12.7	1.0	Jeong et al. (2009)
Waste vegetable oil	0.65	27.79		4.57	38.82	24.92	0.96	Ghoreishi and Moein (2013)
Refined waste lard	1.8	24.7	2.50	12.1	44.4	11.9	1.50	Shin et al. (2012)
Waste cooking oil from sunflower oil	-	6.8	0.4	3.7	22.8	65.2	0.1	Demirbas (2009a)
Sunflower seed oil	-	5.4	0.1	2.9	18.7	72.9	0	Demirbas (2009a)
Chicken fat	-	7.7	21.0	5.5	48.5	17.3	0	Marulanda et al. (2010)
Yellow grease	2.43	23.24	3.79	12.96	44.32	6.97	0.67	Canakci and Van Gerpen (2001)
Brown grease	1.66	22.83	3.13	12.54	42.36	12.09	0.82	Canakci and Van Gerpen (2001)
Rapeseed oil	-	3.49	-	0.85	64.4	22.3	8.23	Goering et al. (1982)
Soybean oil	-	10.58	-	4.76	22.52	52.34	7.19	Canakci and Van Gerpen (2001)
Croton megalocarpus oil	0.1	6.5	0.1	3.8	11.6	72.7	3.9	Kafuku et al. (2011)

about 50% water and a small percentage of FFAs, triacylglycerols, pigments, phosphoacylglycerols and other nonpolar components. As half the content of soapstock is water, which forms an emulsion with the lipids, it is hard to remove and significantly inhibits the transesterification reaction, especially when using catalytic methods, thus reducing its value as a feedstock (Haas, 2005).

Animal fats, including lard, tallow and chicken fat, are commonly used for biodiesel production. They are derived from animal by-products, which have little or no value as potential food ingredients (to avoid competing with food production). Their low cost makes them an attractive feedstock. About 20% of the total feedstock used in global biodiesel production is animal fats, compared to soybean oil (28%) and palm oil (22%) (Banković-Ilić et al., 2014). As shown in Table 1, animal fats have a high concentration of saturated fatty acids, which, while affecting some chemical and physical properties of biodiesel, results in a higher calorific value, better oxidation stability and a high cetane number (Gürü et al., 2009).

All of these low-grade feedstocks have the same limitation of a high FFA content. In the case of alkali-catalyzed transesterification, tolerances are lower than 1–to 0.06 wt.% (Canakci, 2007). Supercritical transesterification can tolerate water and FFA content up to 36 and 30 wt.%, respectively (Warabi et al., 2004). In addition, the process requires a relatively short reaction time with a relatively high ester conversion yield (Anikeev et al., 2012; Demirbas, 2009b; He et al., 2007). This makes supercritical transesterification of low-grade feedstocks with high FFAs an attractive method of biodiesel production.

3 Understanding transesterification for biodiesel production

As noted earlier, TG transesterification using supercritical lower alcohols, such as methanol, ethanol and propanol, has proven to be a promising process for biodiesel production (Demirbas, 2008). Among these alcohols, methanol is most popular commercially, due in part to its low cost and its physical and chemical attributes (highly polar molecule with shortest aliphatic moiety) (De Boer and Bahri, 2011; Demirbas, 2005; Demirbas, 2007). Ethanol is also popular because it is more miscible than methanol (Santana et al., 2012).

Previous research on the application of SCFs in biodiesel production focused predominantly on the optimization of process variables and understanding the kinetics and phase equilibrium of oil transesterification in batch or flow reactors (Fang et al., 2008; Kusdiana and Saka, 2001; Maçaira et al., 2011; Meher et al., 2006; Negi et al., 2006; Saka and Kusdiana, 2001; Stepanov et al., 2010). There have been a few investigations of chemical equilibria in the transesterification reaction using supercritical lower alcohols (Anikeev et al., 2012; Anikeev and Ermakova, 2012; Anikeev and Ermakova,

2011; Anikeev et al., 2011b). Anikeev et al. have provided the thermodynamic data for compounds in the transesterification reaction and constructed phase diagrams for these materials (Anikeev and Ermakova, 2011; Anikeev et al., 2011b). They have also studied the chemical equilibria during the stepwise transesterification reactions between mixed TGs and methanol, allowing the prediction of the transesterification behavior of mixed TGs in the supercritical state (Anikeev and Ermakova, 2012).

Thermodynamic studies have proven particularly useful for predicting the reaction equilibria and evaluating the thermodynamic feasibility of a given process. However, to date, there have been virtually no thermodynamic investigations of the transesterification of TGs with supercritical ethanol or isopropanol. Very few reports are available to thermodynamically justify the selection of methanol as the most suitable alcohol for transesterification with TGs. The intention of this chapter is to investigate the differences between the main three transesterification reaction systems using thermodynamic analysis:

- triolein and supercritical methanol (T&SCM);
- triolein and supercritical ethanol (T&SCE); and
- triolein and supercritical isopropanol (T&SCI).

The chapter explains why methanol has been selected as an appropriate alcohol for biodiesel production and how the equilibrium constant for the transesterification reaction varies with the number of carbon atoms in alcohols.

Triolein, which is the major constituent of TGs containing similar fatty acid residues, has been chosen for the analysis of the three stepwise transesterification reactions with three alcohols (methanol, ethanol and isopropanol) in the supercritical state. In this condition, di- and monoolein are formed as intermediates (Freedman et al., 1986). The three reaction systems can be generally represented by Eqn. (1):

$$\begin{aligned} &\text{Step 1: T} + \text{A} \rightleftharpoons \text{D} + \text{E} \\ &\text{Step 2: D} + \text{A} \rightleftharpoons \text{M} + \text{E} \\ &\text{Step 3: M} + \text{A} \rightleftharpoons \text{G} + \text{E} \end{aligned} \tag{1}$$

Where A denotes methanol, ethanol or isopropanol, respectively, and E denotes MeOL, EtOL or IpOL, respectively.

3.1 Calculation of thermodynamic parameters for individual compounds

To conduct the thermodynamic analysis of the reactions in the process, the basic thermodynamic parameters of each component involved in transesterification

Table 2 Calculated critical temperatures and pressures for triolein by various methods

			Methods		
Parameter	Literature data		Lydersen	Joback	GANI
T_c, K	947.1 (Weber et al., 1999)	Cal.	3,439.8	2,663.3	926.6
		% error	263.19	181.21	−2.16
P_c, MPa	0.4682 (Weber et al., 1999)	Cal.	0.4700	0.2359	0.3259
		% error	0.38	−49.62	−30.39

must be estimated using an appropriate method. The thermodynamic data for methanol, ethanol, isopropanol and glycerol are available (Ma, 2006). However, the parameters for the other six components are absent from the literature and need to be calculated.

To select a suitable method, the critical temperatures and pressures for triolein were calculated and compared using different group methods, including the Lydersen, Joback and GANI methods (Constantinou and Gani, 1994). The results are listed in Table 2. The critical temperature, T_c, was calculated by the GANI method, which provides both a first-order approximation and a more accurate second-order prediction with the smallest average relative error (0.85%). The critical pressure, P_c, was calculated by the Lydersen method (one of the earliest successful methods to estimate critical parameters with an average error lower than 2%). These two methods obtained the closest values to those reported in other studies. The reported data were calculated using the Ambrose method with an average relative error of approximately 4.6%, which is higher than for the Lydersen and Joback methods. Therefore, the GANI method for T_c and the Lydersen method for P_c were chosen to estimate the critical values for all of the system's reactants. The normal boiling point and the acentric factor of the reaction materials were also predicted using the GANI method and the CSGC-AF method, respectively. The data obtained (shown in Table 3) are consistent with other studies, indicating that the calculation methods are accurate. Those critical parameters contribute to the thermodynamic database for biodiesel production and will help to estimate conditions for transesterification experiments with supercritical alcohols in future research.

3.2 Calculation of standard formation enthalpy, entropy and the heat capacity values based on temperature

Most published studies have shown that the reaction temperature, rather than pressure, is the main parameter influencing the extent of the transesterification reaction for biodiesel production, especially in the region above the critical temperature of methanol (514.0 K) (Sawangkeaw et al., 2010). Fang et al. (2008) have reported the phase equilibrium data for mixtures of supercritical

Table 3 Critical parameters, boiling temperature and acentric factor of compounds used for calculations

Compound	Formula	Formula weight	T_c, K	P_c, MPa	T_b, K	ω
T	$C_{57}H_{104}O_6$	885.4	926.6, 946.0 (Anikeev et al., 2011a)	0.47, 0.46 (Anikeev et al., 2011a)	819.6, 854 (Anikeev et al., 2011a)	0.982, 0.83 (Anikeev et al., 2011b)
D	$C_{39}H_{72}O_5$	621.0	870.4, 870.0 (Anikeev et al., 2011a)	0.68, 0.87 (Anikeev et al., 2011a)	759.4, 757 (Anikeev et al., 2011a)	1.499, 1.68 (Anikeev et al., 2011a)
M	$C_{21}H_{40}O_4$	356.5	804.8, 800.0 (Anikeev et al., 2011a)	1.24, 1.29 (Anikeev et al., 2011a)	676.0, 660 (Anikeev et al., 2011a)	1.962
MeOL	$C_{19}H_{36}O_2$	296.5	734.5	1.15, 1.13 (Anikeev et al., 2011a)	595.9	1.840
EtOL	$C_{20}H_{38}O_2$	310.5	732.5	1.23, 1.10 (Stepanov et al., 2010)	605.9	1.864
IpOL	$C_{21}H_{40}O_2$	324.5	737.7	1.18, 1.01 (Stepanov et al., 2010)	610.6	1.841
G	$C_3H_8O_3$	92.1	850.0 (Ma, 2006)	7.5 (Ma, 2006)	563.2 (Ma, 2006)	1.320 (Ma, 2006)
MeOH	CH_4O	32.0	512.5 (Ma, 2006)	8.1 (Ma, 2006)	337.9 (Ma, 2006)	0.566 (Ma, 2006)
EtOH	C_2H_6O	46.1	514.0 (Ma, 2006)	6.1 (Ma, 2006)	351.4 (Ma, 2006)	0.637 (Ma, 2006)
IPA	C_3H_8O	60.1	508.3 (Ma, 2006)	5.2 (Ma, 2006)	355.4 (Ma, 2006)	0.669 (Ma, 2006)

methanol + C18 methyl esters and observed that the binary system became homogeneous at pressures a little higher than the critical pressure of methanol (8.09 MPa). A similar conclusion has been obtained from the vapor-liquid equilibria data for methanol + methyl laurate and methanol + methyl myristate systems (Shimoyama et al., 2007). The particular phase behavior indicates that, at the supercritical state, pressure has little influence on the improvement of FAME yield. Thermodynamic properties are therefore calculated on the basis of temperature.

3.2.1 Calculation of standard values of the formation enthalpy and entropy in the gas state

To obtain the chemical equilibrium constant for any chemical reaction, it is necessary to calculate the values of the three required thermodynamic properties: formation enthalpy, entropy and Gibbs free energy. As few researchers have shown interest in the subject reaction systems, the related thermodynamic parameters have not been reported in the literature, with the exception of alcohols, glycerol and methyl oleate. Benson's Group Method (Benson, 1976) is employed here to estimate the standard values of formation enthalpy and entropy at 298.15 K:

$$\Delta_f H^{\theta}_{298.15}(\mathrm{g}) = \sum_i n_i \Delta_f H^{\theta}_{298.15,i} \tag{2}$$

$$S^{\theta}_{298.15}(\mathrm{g}) = \sum_i n_i S^{\theta}_{298.15,i} - R\ln\sigma + R\ln\eta \tag{3}$$

The calculated values of standard enthalpy and entropy of the formation for the ten substances of interest are listed in Table 4.

3.2.2 Calculation of standard values of the formation enthalpy and entropy in the liquid state

To calculate the standard values for formation enthalpy and entropy in the liquid state, the standard values for vaporization entropy were firstly calculated using Zhao's empirical formula (Zhao and Jin, 1983), as shown in Eqn. (4), where the empirical constants a and b are 125.4 and −222.376 J mol^{-1} K^{-1}, respectively. The standard values for vaporization enthalpy were calculated using the GANI method. The standard values for formation enthalpy and entropy in the liquid state were obtained using Eqs. (5) and (6). The final results are shown in Table 4:

$$\Delta S^{\theta}_{v} = a\lg T_b + b(\mathrm{J \cdot mol^{-1} \cdot K^{-1}}) \tag{4}$$

Table 4 Calculated standard enthalpy and entropy of the formation for all compounds

Component	$\Delta H^{\theta}_{f298.15}(g)$ (kJ mol^{-1})	ΔH^{θ}_{v} (kJ mol^{-1})	$\Delta H^{\theta}_{f298.15}(l)$ (kJ mol^{-1})	$S^{\theta}_{298.15}(g)$ (J mol^{-1} K^{-1})	S^{θ}_{v} (J mol^{-1} K^{-1})	$S^{\theta}_{298.15}(l)$ (J mol^{-1} K^{-1})
T	−1,843.30	285.77	−2,129.07	2,548.80	142.99	2,405.81
D	−1,420.00	222.69	−1,651.79	1,832.80	138.84	1,693.96
M	−995.45	151.86	−1,147.31	1,122.60	132.50	990.10
MeOL	−649.90 (Ma, 2006)	84.6 (Ma, 2006)	−734.50 (Ma, 2006)	955.67	125.63	830.04
EtOL	−658.10	65.44	−721.68	996.67	126.54	870.13
IpOL	−697.10	66.83	−761.30	1027.80	126.96	900.84
G	−582.80 (Ma, 2006)	85.80 (Ma, 2006)	−688.52 (Ma, 2006)	400.90	196.43	204.47 (Ma, 2006)
MeOH	−201.50 (Ma, 2006)	37.60 (Ma, 2006)	−239.10 (Ma, 2006)	239.88 (Ma, 2006)	112.64	127.24 (Ma, 2006)
EtOH	−234.00 (Ma, 2006)	42.70 (Ma, 2006)	−276.98 (Ma, 2006)	280.64 (Ma, 2006)	119.60	161.04 (Ma, 2006)
IPA	−272.40 (Ma, 2006)	45.30 (Ma, 2006)	−317.86 (Ma, 2006)	309.20 (Ma, 2006)	128.62	180.58 (Ma, 2006)

Table 5 Calculated coefficients in the temperature dependence of the constant-pressure heat capacity: $C_{pL} = A + BT + CT^2/(\text{J mol}^{-1}\ \text{K}^{-1})$

Component	A	B	$C \times 10^3$
T	1,475.0791	−0.7072	5.8483
D	1,105.3142	−1.2734	6.1244
M	735.5490	−1.8397	6.4005
MeOL	509.4171	−0.3055	2.0693
EtOL	521.5520	−0.1836	1.8436
IpOL	549.5995	−0.3010	2.2179
G	365.7840	−2.4060	6.6766
MeOH	7.2703	0.1328	−0.0610
EtOH	6.3610	0.2278	−0.1154
IPA	3.7702	0.3287	−0.1730

$$\Delta H_f^\theta(l) = \Delta H_f^\theta(g) - \Delta H_v^\theta \tag{5}$$

$$S_{298.15}^\theta(l) = S_{298.15}^\theta(g) - S_v^\theta \tag{6}$$

3.2.3 Calculation coefficients of the constant-pressure heat capacity in the liquid state

In these systems, because the process pressure is constant and the main reaction phase is liquid, the constant-pressure heat capacity C_{pL} in relation to temperature was calculated using the Rozicka-Domalski method (Ma, 2003), with the results shown in Table 5.

3.3 Calculation of Gibbs free energy and chemical equilibrium constant

3.3.1 Calculation of standard values of Gibbs free energy and chemical equilibrium constant for each reaction step at 298.15 K

The enthalpy and entropy change for each reaction step can be evaluated using the standard values for formation enthalpy and entropy at 298.15 K in Eqs. (7) and (8):

$$\Delta_r H_{298.15}^\theta = \sum v_B \Delta_f H_{298.15}^\theta \tag{7}$$

$$\Delta_r S_{298.15}^\theta = \sum v_B S_{298.15}^\theta \tag{8}$$

The standard Gibbs free energy change in every chemical reaction is defined by Eqn. (9), and the equilibrium constant for each reaction step is calculated directly by Eqn. (10):

$$\Delta_r G^{\theta}_{298.15} = \Delta_r H^{\theta}_{298.15} - T\Delta_r S^{\theta}_{298.15} \tag{9}$$

$$K^{\theta}(T) = \exp[-\Delta_r G^{\theta}(T)/RT] \tag{10}$$

The changes in enthalpy, entropy, Gibbs free energy and the corresponding values of the equilibrium constant for each step in triolein transesterification with three different alcohols are shown in Table 6.

As shown in Table 6, for the triolein and methanol reaction system, the enthalpy changes for the first and third steps of transesterification are −18.12 and −36.61 kJ mol^{-1}, respectively, indicating that these reactions are slightly exothermic. The enthalpy change of the second step of transesterification is low and positive, indicating a slightly endothermic reaction. The changes in Gibbs free energy for the three reaction steps are similar to the enthalpy changes. Although in the first and third reaction steps $\Delta G < 0$, the equilibrium position of the reaction corresponds to the presence of more products than reactants at equilibrium. The chemical equilibrium constant for each reaction step is so low that more energetic conditions are required to promote the reaction, as illustrated by the spontaneous nature of triolein transesterification with methanol at 298.15 K.

For the other two reaction systems, the enthalpy change of each reaction step is greater than 0 but <60 kJ mol^{-1}, indicating that each reaction step is slightly endothermic. The results verify the fact that the reaction heat of

Table 6 Changes in enthalpy, entropy, Gibbs free energy and the corresponding values of the equilibrium constant for each individual step of triolein transesterification reaction at 298.15 K

Reaction	$\Delta_r H^{\theta}_{298.15}$ (kJ mol^{-1})	$\Delta_r S^{\theta}_{298.15}$ (J mol^{-1} K^{-1})	$\Delta_r G^{\theta}_{298.15}$ (kJ mol^{-1})	K	Composition of equilibrium mixture
M1	−18.12	−9.05	−15.42	5.03×10^{2}	Products dominate
M2	9.08	−1.06	9.40	2.26×10^{-2}	Reactants dominate
M3	−36.61	−82.83	−11.91	1.22×10^{2}	Products dominate
E1	32.58	−2.76	33.40	1.41×10^{-6}	Negligible products
E2	59.78	5.23	58.22	6.30×10^{-11}	
E3	14.09	−76.54	36.91	3.41×10^{-7}	
I1	33.84	8.41	31.33	3.24×10^{-6}	
I2	61.04	16.40	56.15	1.45×10^{-10}	
I3	15.35	−65.37	34.84	7.87×10^{-7}	

transesterification is small at room temperature. The Gibbs free energy of each reaction step is positive, and the chemical equilibrium constant is so small that it can be discounted.

3.3.2 Calculating standard values of Gibbs free energy and chemical equilibrium constant for the overall reaction at 298.15 K

To verify the validity of and compare transesterification results with different alcohols, the standard values for Gibbs free energy and chemical equilibrium constants for the overall reaction were calculated at 298.15 K. The three-step transesterification reactions expressed by Eqn. (1) can be integrated into a one-step reaction. The three reaction systems with different alcohols can be represented in Eqn. (11) as:

$$\begin{aligned} &\text{Overall}-1{:}\text{T}+3\text{MeOH} \rightleftharpoons 3\text{E}+\text{G} \\ &\text{Overall}-2{:}\text{T}+3\text{EtOH} \rightleftharpoons 3\text{E}+\text{G} \\ &\text{Overall}-3{:}\text{T}+3\text{IPA} \rightleftharpoons 3\text{E}+\text{G} \end{aligned} \quad (11)$$

The changes in enthalpy, entropy, Gibbs free energy and the equilibrium constant for the three overall reaction systems under standard conditions are shown in Table 7.

From the results of the three overall reaction systems, it is obvious that the changes in enthalpy and Gibbs free energy for triolein transesterification with methanol are negative, whereas the other two reaction systems are both positive. The results indicate that only the transesterification between triolein and methanol is exothermic and can proceed spontaneously at 298.15 K. The equilibrium constant for each overall reaction system shows the similar trend as the stepwise reaction calculation. As the overall reaction calculation results are consistent with the stepwise calculations, they show that ethanol and isopropanol are not commonly used in making biodiesel due to the lower reactivity in comparison to methanol. Methanol is clearly demonstrated as the most suitable alcohol for biodiesel production via transesterification.

Table 7 Changes in enthalpy, entropy, Gibbs free energy and the corresponding values of the equilibrium constant for each overall triolein transesterification system at 298.15 K

Reaction system	$\Delta_r H^{\theta}_{298.15}$ (kJ mol^{-1})	$\Delta_r S^{\theta}_{298.15}$ (J mol^{-1} K^{-1})	$\Delta_r G^{\theta}_{298.15}$ (kJ mol^{-1})	K
Overall−1	−45.65	−92.94	−17.94	1.39×10^{3}
Overall−2	106.45	−74.07	128.53	3.03×10^{-23}
Overall−3	110.23	−40.56	122.32	3.71×10^{-22}

3.3.3 Calculating the standard values of Gibbs free energy and chemical equilibrium constant dependence on temperature

According to the data in Table 5, the enthalpy of each reaction step at temperature T can be related to the values at 298.15 K by integration using Eqn. (12) to obtain:

$$\Delta_r H^\theta(T) = \left[\int_{298.15K}^{T} \sum v_B C_{\mathrm{pL,B}}\right] dT + \Delta_r H^\theta_{298.15K} \tag{12}$$

The Gibbs free energy at temperature T can be derived from the Gibbs-Helmholtz equation (Eqn. 13). Similarly, ln$K$ at temperature T can be derived from the Van't Hoff Isochore equation (Eqn. 14):

$$\left[\frac{\partial\left(\Delta G^\theta / T\right)}{\partial T}\right]_P = -\frac{\Delta H^\theta}{T^2} \tag{13}$$

$$\left[\frac{\partial \ln K}{\partial T}\right]_P = \frac{\Delta H^\theta}{RT^2} \tag{14}$$

3.4 Results and discussion

The relationships of ΔH-T, ΔG-T and ln$K$-T for each reaction step were calculated and are depicted in Figs. 1–3 (Zeng et al., 2014b), respectively.

From 300 to 700 K, the enthalpy appears to increase. For T&SCE and T&SCI, the enthalpy of each transesterification step is positive. However, for T&SCM, the enthalpy for each transesterification step became positive for $T > 545.31$ K (in the supercritical methanol region). At high temperatures the reaction equilibrium is displaced in the direction of a higher concentration of esters. Thus, an increase in temperature over the supercritical alcohol region leads to a rise in the ester yield.

It can be seen from Fig. 2 that, for $T > 563.37$ K, the Gibbs free energy of each transesterification step for T&SCM is negative, so each reaction step can proceed spontaneously. Unfortunately, for the other two reaction systems, that is, T&SCE and T&SCI, the Gibbs free energy is positive in each reaction step. Consequently, it is necessary to provide additional external energy for these systems to make them function. The comparison based on thermodynamic analysis illustrates that only supercritical methanol can react with triolein to form methyl esters under favorable reaction conditions, making it more attractive for industrial scale-up.

In the integrated form of Eqn. (14) (assuming that ΔH is independent of temperature) we get Eqn. (15):

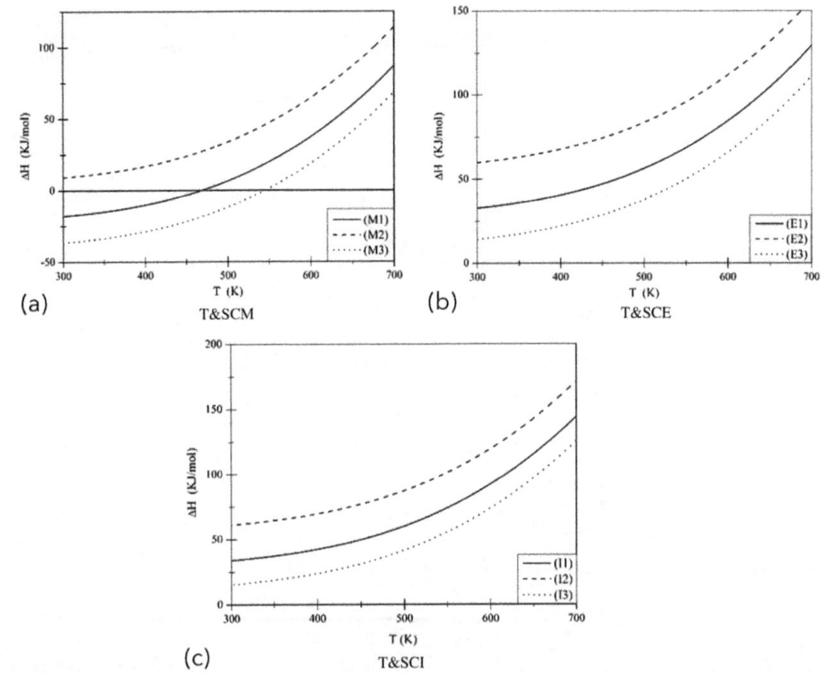

Figure 1 Temperature dependences of the enthalpy for each transesterification step corresponding to the number of the reactions from set (1).

$$\ln\frac{K_2}{K_1} = \frac{\Delta_r H_m^{\theta}}{R} \cdot \frac{(T_2 - T_1)}{T_1 T_2} \tag{15}$$

K decreases inversely with temperature and vice versa for an exothermic reaction. For $T > 563.37$ K, the chemical equilibrium constant of the transesterification steps for all three systems increases directly with increasing temperature, thus promoting the forward rate of reaction.

Hanh et al. (2009) have investigated triolein transesterification with various alcohols in an ultrasonic field. The results confirmed that the rate of ester formation depends upon the type of alcohol used. The greater the number of carbon atoms in alcohol, the slower the rate of reaction. In addition, it was suggested that the steric hindrance of secondary alcohols, such as isopropanol, isobutanol and isohexanol, strongly affects transesterification of triolein, leading to minimal ester conversion.

Chen et al. (2011) have studied the reaction processes and the mechanism of supercritical transesterification using in-situ ATR infrared spectroscopy. The results showed that for $T > 498.15$ K, the vibrating peaks of the hydroxyl methanol group split, which was not the case with ethanol and propanol. This forms

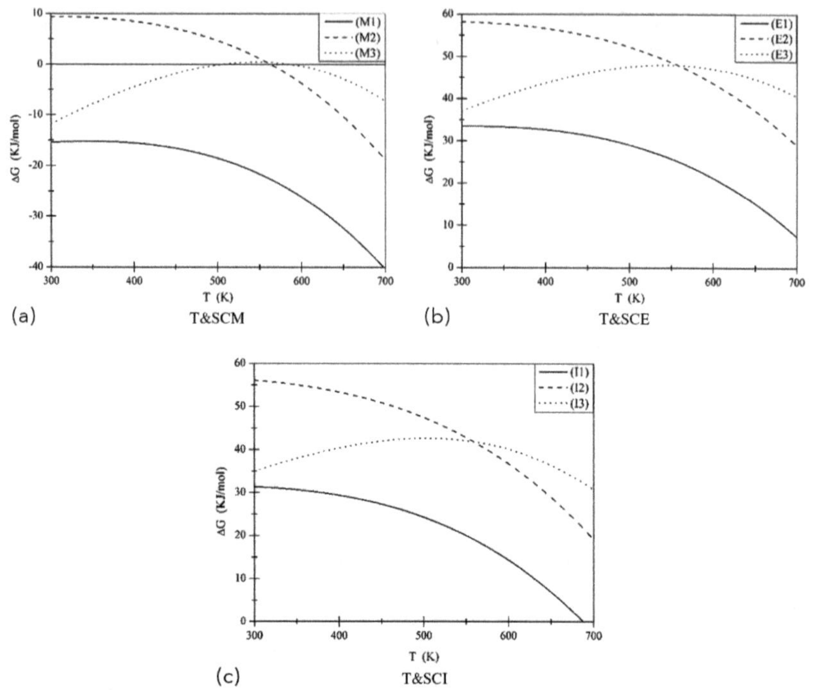

Figure 2 Temperature dependences on the Gibbs free energy of each transesterification step corresponding to the number of the reactions from set (1).

C^+ ... O^- ... H^+, which strengthens the electrophilicity and the nucleophilicity of methanol, accelerating the rate of T&SCM supercritical transesterification. The results in this study fit with the conclusions of other studies, showing that the group-contribution analysis method is feasible for analyzing the chemical equilibrium and feasibility of triolein transesterification with supercritical lower alcohols (Anikeev et al., 2011a). As noted earlier, the results show that only triolein transesterification using supercritical methanol can proceed under the right reaction conditions and that the transesterification reaction rarely occurs when the alcohol reactant is changed from methanol to either ethanol or isopropanol.

4 Optimizing operating conditions for biodiesel synthesis

The four main approaches to synthesize biodiesel are (Sharma et al., 2008):

- direct use and blending of vegetable oils;
- micro-emulsions;
- thermal cracking (pyrolysis); and
- transesterification.

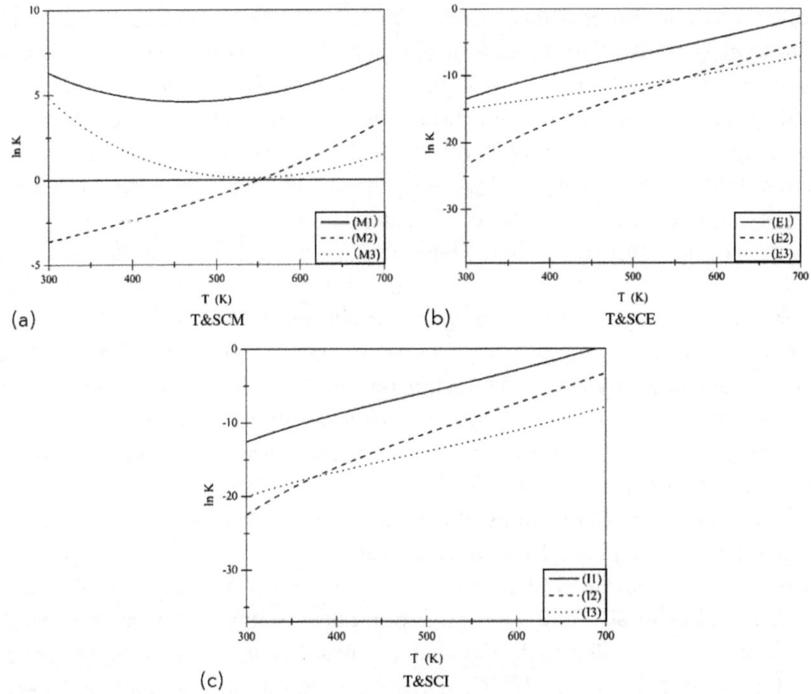

Figure 3 Temperature dependences on ln*K* of each transesterification step corresponding to the numbers of the reactions from set (1).

Among these, transesterification is the most common method used for the production of biodiesel catalyzed by acids, alkalis and lipases (Guldhe et al., 2017; Kansedo and Lee, 2013; Ma et al., 2012; Saydut et al., 2016; Sendzikiene et al., 2016; Veny et al., 2014; Wu et al., 2016a). Methanol is the most frequently used alcohol in transesterification because of its low cost and its physical and chemical advantages as the shortest chain alcohol.

The relatively slow reaction rate of acid-catalyzed transesterification and the high cost of lipases limit the use of these methods in the commercial production of biodiesel (Aguieiras et al., 2015; Pinnarat and Savage, 2008). In contrast, alkali-catalyzed transesterification has been widely adopted, given the ability to catalyze the reaction at low temperature and atmospheric pressure with high conversion within a practical time frame (Leung et al., 2010). This has been shown in a number of studies with a range of feedstocks.

Rashid et al. (2009) synthesized biodiesel using base-catalyzed transesterification of rice bran oil, achieving a biodiesel yield of 83.3% with the following operating parameters: 7.5:1 methanol-to-oil molar ratio, 0.88% catalyst concentration, 55°C and 60 min. Amini-Niaki and Ghazanfari (2013) obtained a similar yield of 83.4% with sunflower oil in the following conditions: a molar ratio of oil to methanol of 1:5.5, 1.0% KOH catalyst at 51.7°C for 65.5 min. Vicente et al. (2004) compared four primary alkali catalysts (CH_3ONa, CH_3OK, NaOH, KOH) for methanolysis of vegetable oil. This study demonstrated nearly 100% biodiesel yields with methoxide catalysts because the methoxide anion is strongly basic and has a high catalytic activity in transesterification (Ma and Hanna, 1999; Vicente et al., 2004). Among these, CH_3ONa is the most commonly used catalyst for chemical interesterification (Liu, 2004).

Borugadda and Goud 2012) have reviewed biodiesel production using various feedstocks under different conditions with homogeneous alkali catalysts. The catalyst amount required is usually over 1.0%, and can be as high as 6.0%, while the reaction takes several hours using conventional catalytic transesterification methods. When the raw material has a high percentage of FFAs or water, the alkali catalyst is consumed by saponification reactions, and the process needs complicated downstream separation and purification processes to work (Leung et al., 2010).

To overcome the limitations of conventional catalytic processes, Saka and Kusdiana (2001) proposed a new supercritical methanol (SC-MeOH) process for biodiesel production with a shorter reaction time and a simpler purification procedure. Lee et al. (2012) studied supercritical methanol transesterification from waste canola oil and found that biodiesel yield close to 100% under the following conditions: 270°C, a methanol-to-oil weight ratio of 2, and a processing period of 45 min. Román-Figueroa et al. (2016) investigated a non-catalytic transesterification of crude castor oil into FAMEs and obtained 96.5% of biodiesel yield at optimized conditions of a methanol-to-oil molar ratio of 43:1 within 90 min at 300°C. García-Martínez et al. (2017) reviewed biodiesel produced from tobacco seed oil by supercritical methanolysis and achieved a maximum yield of 91.1% at 303.4°C and 90 min with a molar ratio of methanol-to-oil 43:1. These studies show significantly enhanced biodiesel yields using supercritical technology. However, a relatively high temperature around 300°C and reaction time of 45–90 min are needed for optimum reaction conditions. This requires high equipment costs and high energy consumption, which limit the application of supercritical technology in the biodiesel industry. This highlights the need to better understand and improve the efficiency of the process.

5 Optimizing supercritical transesterification technology for biodiesel production: a case study

Response surface methodology (RSM), first developed by Box and Wilson (1951), has attracted attention in recent decades as an important branch of experimental design. It is an efficient way to find the relationship between reaction and response variables (Myers et al., 2009). Many studies have shown that RSM is a suitable technique for optimizing transesterification of biodiesel production (Chakraborty and Sahu, 2014; Mohamad et al., 2017; Noshadi et al., 2012; Rashid et al., 2009).

This following case study explores a novel integrated process for supercritical transesterification of methanol using a small amount of CH_3ONa catalyst to produce biodiesel from soybean oil. This proposed method combines the advantages of supercritical technology and base-catalyzed transesterification. It ensures a simple reaction procedure with a relatively low reaction temperature in contrast to most supercritical technology. It also requires less catalyst, a much shorter reaction time and no saponification problems compared with conventional base-catalyzed transesterification processes. The case study reviews process parameters including reaction temperature, molar ratio of methanol to soybean oil and catalyst amount (mass percent of CH_3ONa to soybean oil). It also includes kinetic and thermodynamic analysis of the process.

5.1 Materials and methods

Soybean oil was purchased from Xi'an Jiali Oil Industry Co., Ltd. (Xi'an, Shaanxi Province, China). Methanol (analytical grade purity) was obtained from Tianjin Fuyu Fine Chemical Co., Ltd. CH_3ONa (analytical grade purity) was purchased from Tianjin Fuchen Chemical Reagents Factory, and methyl oleate (CAS number: 112-62-9) from Yake Chemical Reagent Co., Ltd. (Suzhou) with a purity of ≥ 99.0%. The experiment apparatus for biodiesel production (using equipment from Huaxing Petroleum Instrument Co., Ltd., Nantong, Jiangsu Province, China) is presented in Fig. 4 (Zeng et al., 2017). It consists of a

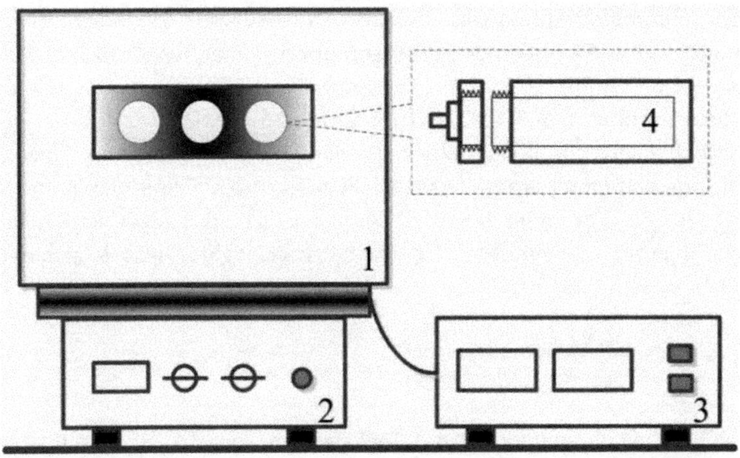

Figure 4 Schematic diagram of the experimental apparatus (1 – high-temperature furnace; 2 – oscillator; 3 – temperature controller; 4 – batch reaction vessel).

batch reaction vessel (with an inner volume of ~10 ml, and with the maximum pressure and temperature of 50 MPa and 500°C, respectively) and a heating furnace.

Soybean oil and CH_3ONa-dissolved methanol solution (6 ml) were charged into the three batch reaction vessels, and then the reactor was placed in the high-temperature furnace that had been preheated to the set temperature. After the transesterification (50–60 reactor oscillations/min), the reactors were taken out of the furnace and quenched with cold water for 5 min. The product from each reactor was collected, and the oil phase was completely separated by washing with hot water (85–90°C) to prepare for the analysis. The yield data were obtained by calculating the average yield of three reactors with a deviation lower than 5%.

5.2 Properties of the raw material

The saponification value (SV) and the acid value (AV) of soybean oil were measured according to the methods specified by the American Oil Chemists' Society (AOCS) in our previous study (Zeng et al., 2012). They were 193.5 mg KOH g^{-1} oil and 6.4 mg KOH g^{-1} oil, respectively. The average molecular weight (M_r) was 841.8, calculated using the following Eqn. (16):

$$M_r = \frac{56.1 \times 1000 \times 3}{(SV - AV)} \qquad (16)$$

5.3 FAME analysis

The contents of FAME in the samples diluted by *n*-hexane were analyzed by the gas chromatography-mass spectrometry (GC–MS) 5973-6890 (Agilent Technologies Inc., Palo Alto, CA, USA) equipped with an HP-5 capillary column (30 m × ∅ 0.25 mm × 0.25 μm). The analysis conditions were as follows: a detector temperature of 300°C, injector temperature of 250°C, a carrier gas of helium with a flow rate of 1.0 mL/min, a split ratio of 30 and a sampling volume of 1 μL. The oven temperature was initially held at 200°C for 10 min and then elevated to 230°C at a ramp rate of 5°C/min, finally held at 230°C for 2 min. The external standard was methyl oleate. The biodiesel yield (*Y*) was determined by Eqn. (17):

$$Y = \frac{C_S \times A_{FAME}}{A_S \times C_B} \qquad (17)$$

where C_S is the concentration of external standard (g/g), A_{FAME} is the total peak area of FAME, A_S is the peak area of external standard and C_B is the concentration of the biodiesel product (g/g).

5.4 Experimental design

The reaction time was constant at 20 min. The experiment analyzed key transesterification process parameters, including reaction temperature from 150°C to 250°C, methanol-to-oil molar ratio from 23:1 to 69:1 and CH_3ONa amount from 0.5 to 1.5 wt.%. Analysis used RSM coupled with a Box–Behnken design (BBD). As this model uses points outside the design space (−1.00, 1.00 levels), providing a reliable estimation over the entire design space, it is more efficient and economical than the corresponding full three-level factorial design (Bezerra et al., 2008). The central point was performed three times to assess reproducibility. The results of the experimental design matrix and the biodiesel yield (response value) are shown in Table 8. The experiments were performed in a randomized order and in triplicate to achieve repeatable results.

5.5 Statistical analysis

A quadratic polynomial model was developed to predict the biodiesel yield as a function of the independent variables. The general form of this second-order polynomial can be expressed as (Eqn. 18):

$$X = \beta_0 + \sum_{i=1}^{3} \beta_i x_i + \sum_{i=1}^{3} \beta_{ii} x_i^2 + \sum_{i=1}^{2} \sum_{j=i+1}^{3} \beta_{ij} x_i x_j \quad (18)$$

Table 8 Experimental design matrix and the corresponding biodiesel yield in transesterification

Std. order	Run order	Temperature (°C)	Methanol-to-oil molar ratio	Catalyst amount (wt. %)	Observed yield (%)	Predicted yield (%)
1	1	150	23	1.0	63.47	62.81
4	2	250	69	1.0	90.01	90.67
13	3	200	46	1.0	94.09	94.90
12	4	200	69	1.5	78.91	79.39
7	5	150	46	1.5	64.78	65.14
8	6	250	46	1.5	90.16	89.03
3	7	150	69	1.0	74.79	73.96
15	8	200	46	1.0	94.98	94.90
10	9	200	69	0.5	92.89	92.59
9	10	200	23	0.5	78.84	78.36
14	11	200	46	1.0	95.62	94.90
6	12	250	46	0.5	92.32	91.96
5	13	150	46	0.5	62.57	63.70
2	14	250	23	1.0	97.42	98.25
11	15	200	23	1.5	89.76	90.06

where X is the predicted response; x_i and x_j are the experimental factor levels; β_0 is the offset term; β_i, β_{ii} and β_{ij} are the linear, quadratic and interactive coefficients, respectively.

A complete analysis of variance (ANOVA) was performed using Design Expert Software Version 8.0.6 (Stat-Ease, Inc. 2021 East Hennepin Ave., Suite 480 Minneapolis, MN 55413) at a significance level of 95%. Subsequently, the quality of the proposed model was evaluated using a correlation coefficient (R^2).

5.6 Process optimization: analysis of FAME with GC-MS

A detailed analysis of the composition of the biodiesel was undertaken using GC-MS. The GC chromatogram is shown in Fig. 5 (Zeng et al., 2017). The FAME profile was conducted at optimum conditions (250°C, methanol-to-oil molar ratio of 23:1, 1.0 wt.% CH_3ONa and 20 min). The profile is summarized in Table 9. 9-Octadecenoic acid, methyl ester (9*E*) (C18:1) and 9,12-octadecadienoic acid, methyl ester (C18:2) were the two major constituents present. The presence of these monounsaturated compounds (46.636%) gave a high cetane number and decreased the viscosity of the biodiesel (Ramos et al., 2009).

5.7 Process optimization: development of the mathematical model

Table 10 shows the ANOVA results to explore model fitness. The *F* value of 174.84 indicates the model is significant, and the *P* value is less than 0.0001,

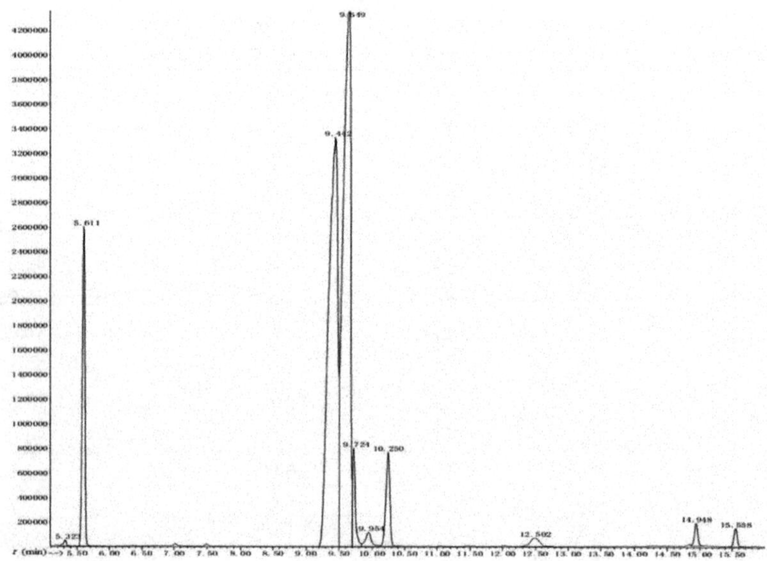

Figure 5 A typical GC chromatogram of biodiesel.

Table 9 Fatty acid profile of biodiesel (as% of methyl esters)

Residence time	Type of FAME	% Area
5.323	9-Hexadecenoic acid, methyl ester (C16:1)	0.198
5.611	Hexadecanoic acid, methyl ester (C16:0)	8.691
9.442	9,12-Octadecadienoic acid, methyl ester (C18:2)	38.452
9.649	9-Octadecenoic acid, methyl ester (9*E*)-(C18:1)	42.686
9.724	11-Octadecenoic acid, methyl ester (C18:1)	2.915
9.954	9,12,15-Octadecatrienoic acid, methyl ester, (Z,Z,Z)-(C18:3)	0.861
10.250	Octadecanoic acid, methyl ester (C18:0)	3.898
12.502	Docosanoic acid, methyl ester (C22:0)	0.836
14.948	cis-11-Eicosenoic acid, methyl ester (C20:1)	0.837
15.558	Eicosanoic acid, methyl ester (C20:0)	0.625
Saturated fatty acid		14.050
Monounsaturated fatty acid		46.636
Polyunsaturated fatty acid		39.313

Table 10 Result of ANOVA for response surface quadratic model

Source	Sum of square	Degree of freedom	Mean square	*F* value	Prob. > *F*	Significant
Model	0.22	9	0.024	174.84	< 0.0001	Yes
A-A	0.14	1	0.14	984.29	< 0.0001	Yes
B-B	6.301E−004	1	6.301E−004	4.56	0.0858	No
C-C	1.133E−004	1	1.133E−004	0.82	0.4067	No
AB	8.780E−003	1	8.780E−003	63.56	0.0005	Yes
AC	4.774E−004	1	4.774E−004	3.46	0.1221	No
BC	0.016	1	0.016	112.22	0.0001	Yes
A^2	0.041	1	0.041	298.07	< 0.0001	Yes
B^2	3.142E−003	1	3.142E−003	22.75	0.0050	Yes
C^2	0.017	1	0.017	126.52	< 0.0001	Yes
Residual	6.906E−004	5	1.381E−004			
lack of fit	5.725E−004	3	1.908E−004	3.23	0.2452	No
Pure error	1.181E−004	2	5.904E−005			
cor. Total	0.22	14				

$R^2 = 0.9968$, R^2 (adj) = 0.9911, R^2 (pred) = 0.9568.

implying only a 0.01% chance that a model with such large *F* value could occur due to noise. For individual terms, only those with Prob. > $F < 0.05$ are deemed as significant. In this case, *A*, *AB*, *BC*, A^2, B^2, C^2 were found to be significant. The 'lack-of-fit *F* value' of 3.23 implies the lack of fit is not significant relative to pure error. This suggests the proposed model is valid and statistically significant. Subsequently, the final equation regarding coded factors after eliminating the insignificant parameters is given below as (Eqn. 19):

$$X = 0.95 + 0.13 \times A - 0.047 \times A \times B - 0.062 \times B \times C - 0.11 \times A^2 - 0.029 \times B^2 - 0.069 \times C^2 \quad (19)$$

A correlation coefficient $R^2 = 0.9968$, which was found to be very close to unity, shows the excellent correlation between the experimental and predicted data. It reveals that the regression model can successfully predict biodiesel yield by transesterification. The predicted R^2 of 0.9568 was in reasonable agreement with the adjusted R^2 of 0.9911 in the refined model. The signal-to-noise ratio is measured with sufficient precision, and a ratio greater than 4 is desirable. In this case, the ratio of 36.935 indicated an adequate signal. Thus, the model can be used to navigate the design space.

An additional experiment was performed under the optimum reaction conditions to evaluate the accuracy of the model. The biodiesel yield obtained was 97.83%, which only had a percentage error of 4.27%, with the predicted yield of 98.25%. The verification value agreed well with the estimated value of the model equation, which shows a reasonable accuracy of the regression model to predict the biodiesel yield.

5.8 Effect of experimental factors

The surface plots of the biodiesel yield versus interactions of two variables are presented in Figs. 6–8 (Zeng et al., 2017). In each graph, the remaining variables are set constant at the center point.

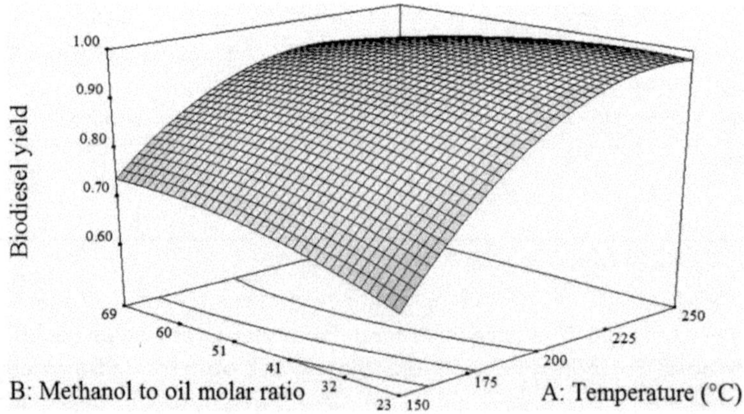

Figure 6 Effects of temperature and methanol-to-oil molar ratio on biodiesel yield.

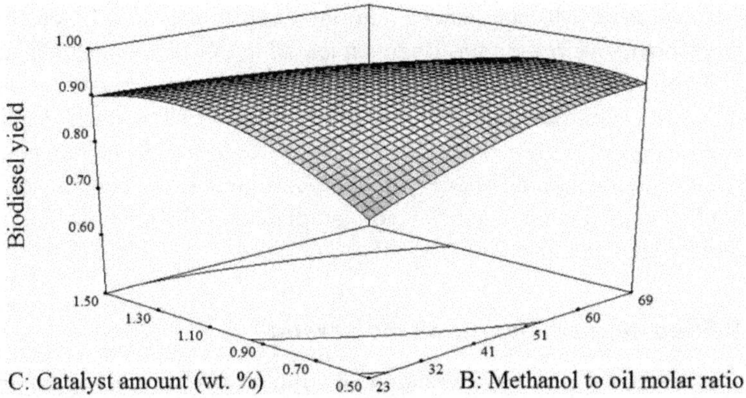

Figure 7 Effects of methanol-to-oil molar ratio and catalyst amount on biodiesel yield.

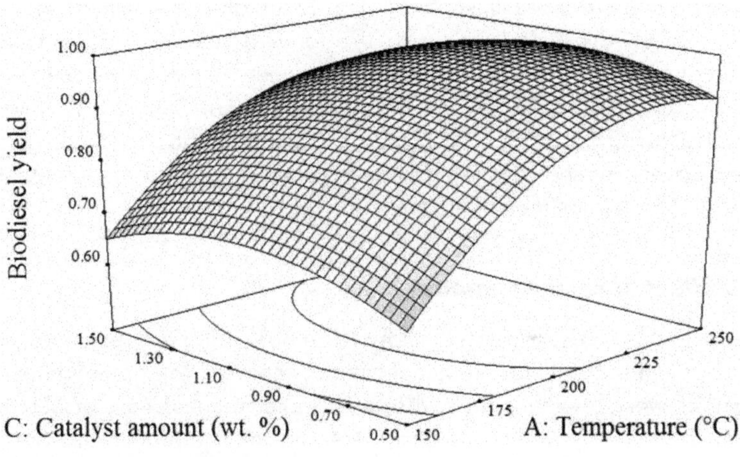

Figure 8 Effects of temperature and catalyst amount on biodiesel yield.

5.8.1 Effect of temperature

Figure 6 depicts the 3D response curve of the biodiesel yield against reaction temperature and methanol-to-oil molar ratio. The higher *F* value of the model term represents a greater effect on biodiesel yield. The relative importance of the parameters was reaction temperature (*A*) > methanol-to-oil molar ratio (*B*) > catalyst amount (*C*).

The temperature significantly affects the conversion of soybean oil to FAME compared with the other two parameters. As temperature increases from 150°C to 250°C, the yield increased from 63.47% to 97.42%. The variation tendency suggests that a higher temperature could lead to a higher conversion yield. It is consistent with the conclusion by Demirbas (2002) that increasing reaction temperature, especially supercritical temperatures, improved biodiesel yield. The thermodynamic reason has been explained in our previous article (Zeng et al., 2014b). Since the enthalpy for transesterification has become positive in the SC–MeOH temperature domain, the reaction equilibrium is displaced in the direction of higher FAME concentration. Considering higher reaction conditions require higher energy consumption, and to avoid the negative effects of thermal decomposition at high temperatures, the biodiesel yield was found to be satisfactory at 250°C.

5.8.2 Effect of methanol-to-oil molar ratio

As mentioned earlier, the methanol-to-oil molar ratio is the second-most influential parameter in transesterification. The experiment results indicated that the methanol-to-oil molar ratio has a gradually increasing impact on the conversion into FAME, as illustrated in Fig. 7. A high molar ratio of methanol pushes the chemical equilibrium to the forward reaction side. An excess amount of methanol increases the contact between methanol and soybean oil, improving the transesterification reaction (Kusdiana and Saka, 2001). However, increasing the methanol ratio increases the solubility of the by-product glycerol in the biodiesel. As shown in Table 8, the highest value of 97.42% was obtained with a methanol-to-oil molar ratio at 23:1.

5.8.3 Effect of catalyst amount

In our previous study, biodiesel yield was 94.80% using transesterification of soybean oil in supercritical methanol without catalysts. The optimum reaction conditions were: methanol-to-soybean oil molar ratio of 46:1 at 300°C for 45 min (Zeng et al., 2012). This study assessed the use of a small amount of CH_3ONa (0.5–1.5 wt.%) on reaction conditions. Figure 8 shows the interaction between temperature and catalyst amount on biodiesel yield. Biodiesel yield initially increased with an increase of CH_3ONa. CH_3ONa dissociation in methanol produces the necessary methoxide anion to begin the reaction. Methoxide anion (CH_3O^-) is strongly basic and has a high catalytic activity in transesterification reactions (Ma and Hanna, 1999).

As shown in Fig. 9, transesterification is a nucleophilic substitution reaction in the presence of methanol, with methoxide functioning as a nucleophile (Liu, 2004). The methoxide anion first attaches to the carbonyl carbon atom of the TGs to form a tetrahedral intermediate. The tetrahedral intermediate reacts with methanol to generate methoxide anion. Diglycerides (DGs) and alkyl ester are then formed by the rearrangement of the tetrahedral intermediate. DGs and monoglycerides (MGs) are converted by the same mechanism to produce alkyl ester and the by-product glycerol, where R_1, R_2 and R_3 represent long-chain

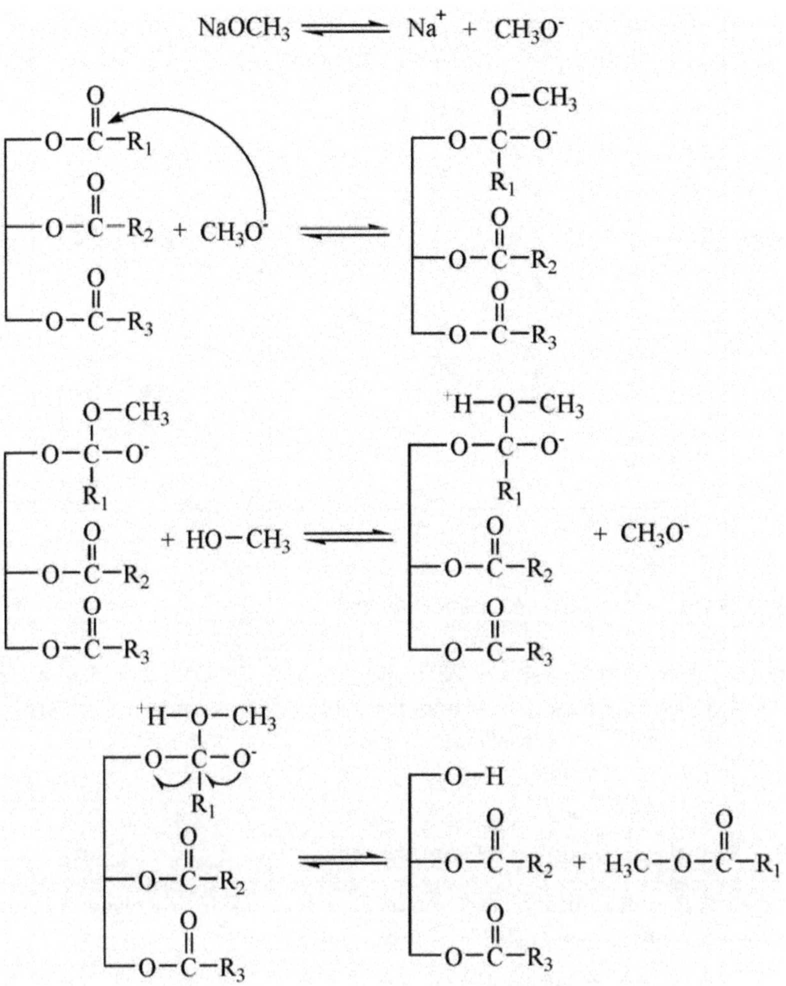

Figure 9 Mechanism of methoxide-catalyzed transesterification of soybean oil.

alkyl groups, respectively. The maximum yield of 97.42% shown in Table 8 was obtained with a CH_3ONa amount of 1.0 wt.%.

5.9 Kinetic study of the transesterification process

5.9.1 Experiment

The kinetics of the transesterification process were also investigated. The variation curves in biodiesel yield from 150°C to 250°C with respect to time are shown in Fig. 10 (Zeng et al., 2017). The other conditions were fixed at

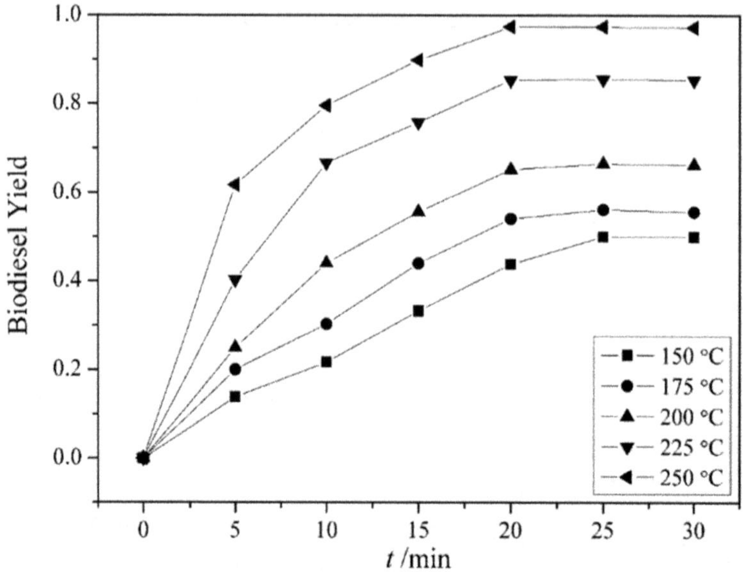

Figure 10 Effect of temperature on biodiesel yield.

a methanol-to-oil molar ratio of 23:1 and 1.0 wt.% CH_3ONa with the reactor oscillating velocity of 50–60 per min. From Fig. 10, it is evident that, with an increase in time, the biodiesel yield increases gradually and becomes almost constant from 25 to 30 min.

5.9.2 Kinetic parameters of the process

The overall transesterification reaction of TG to form FAME and glycerol (GL) in the presence of methanol (MeOH) is shown in Eqn. (20):

$$TG + 3MeOH \rightarrow 3FAME + GL \tag{20}$$

The rate of formation of the product has been found to correlate well using the following rate equation (Eqn. 21):

$$-r_A = k_0 c_A^n c_B^m \tag{21}$$

We may assume that the concentration of methanol is constant. In addition: $k = k_0 c_B^m$. This produces (Eqn. 22):

$$-r_A = -\frac{dc_A}{dt} = kc_A^n \tag{22}$$

Because the concentration of TG at any moment of the reaction can be expressed as $c_A = c_0(1-x)$, so (Eqn. 23):

$$-r_A = -\frac{dc_A}{dt} = -\frac{d[c_0(1-x)]}{dt} = c_0\frac{dx}{dt} = k[c_0(1-x)]^n \tag{23}$$

From Eqn. (23), the rate equation can be expressed as Eqn. (24):

$$\frac{dx}{dt} = \frac{k}{c_0}[c_0(1-x)]^n \tag{24}$$

By taking the natural logarithm, Eqn. (24) can be written as Eqn. (25):

$$\ln\frac{dx}{dt} = n\ln[c_0(1-x)] + \ln k' \tag{25}$$

In Eqn. (21), $k' = k/c_0$.

As shown in Eqn. (25), the order of the reaction and the rate constant are determined from the plot of ln(dx/dt) and ln[c_0(1 − x)]. Here, we use the differential method to identify the order of the reaction using the exponential function in origin software. The exponential function expressed as (Eqn. 26):

$$y = y_0 + Ae^{R_0 x} \tag{26}$$

Taking the obtained plot at 150°C as an example, the fitting function can be expressed as (Eqn. 27):

$$x(t) = 0.75861 - 0.76360\,e^{(-0.03964t)} \tag{27}$$

The differential equation governing the transesterification reaction of TG is expressed as (Eqn. 28):

$$\frac{dx}{dt} = 0.03027e^{(-0.03964t)} \tag{28}$$

According to the experiment results, a well-fitted differential plot of ln(dx/dt) and ln [c_0(1 −x)] can be obtained in Fig. 11 (Zeng et al., 2017). A straight line is observed as (Eqn. 29):

$$\ln\frac{dx}{dt} = 1.3965\ln[c_0(1-x)] - 2.5979 \tag{29}$$

with the R^2 value of 0.9884.

From Eqs. (25) and (29), the reaction order is obtained as 1.3965, and the reaction rate constant is 0.03871. Similarly, the linear fitting plots at 175°C, 200°C, 225°C and 250°C can be obtained, and the results are listed in Table 11.

5.9.3 Activation energy determination

The Arrhenius equation can be written as Eqn. (30). By taking the natural logarithm, Eqn. (30) can be written as Eqn. (31):

$$k = Ae^{-\frac{E_a}{RT}} \tag{30}$$

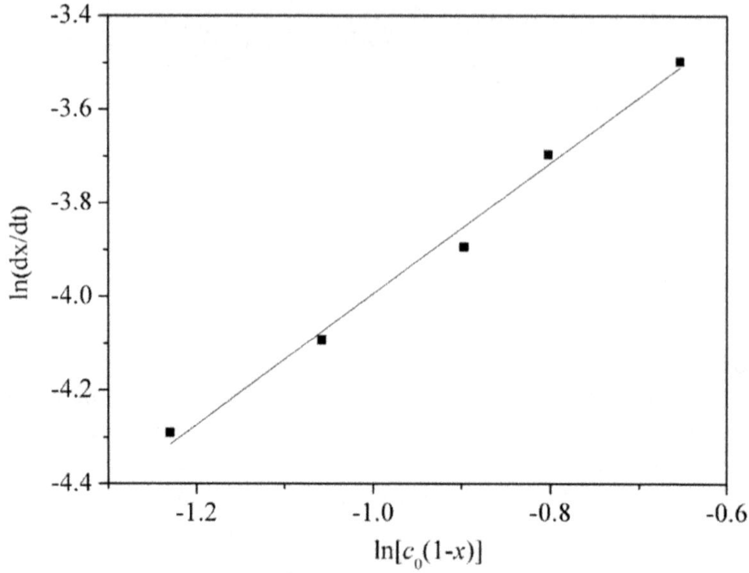

Figure 11 Plot of $\ln[c_0(1-x)]$和 $\ln\frac{dx}{dt}$ at 150°C.

Table 11 Reaction order, reaction rate constant and relative value at different temperatures

T/°C	Rate constant (k/min^{-1})	Order n	R^2
150	0.03871	1.3965	0.9884
175	0.08577	1.8785	0.9946
200	0.12067	1.7861	0.9964
225	0.15782	1.3745	0.9905
250	0.17298	1.0481	0.9722

$$\ln k = -\frac{E_a}{RT} + \ln A \tag{31}$$

where A is the pre-exponential factor and E_a is the activation energy (kJ·mol^{-1}). R is the molar universal gas constant (8.314 J mol^{-1} K^{-1}), and T represents the absolute temperature (K).

A linear correlation between ln$k$ and T^{-1} can be found in Eqn. (31). By using the rate constants listed in Table 11 and the corresponding temperature from 150°C to 250°C, the straight line with the R^2 value of 0.9016 can be obtained in Fig. 12 (Zeng et al., 2017).

The activation energy and pre-exponential factor can be calculated from the intercept and the slope as 27.06 kJ mol^{-1} and 102.71, respectively. The reaction order of SC-MeOH and CH_3ONa-catalyzed transesterification is

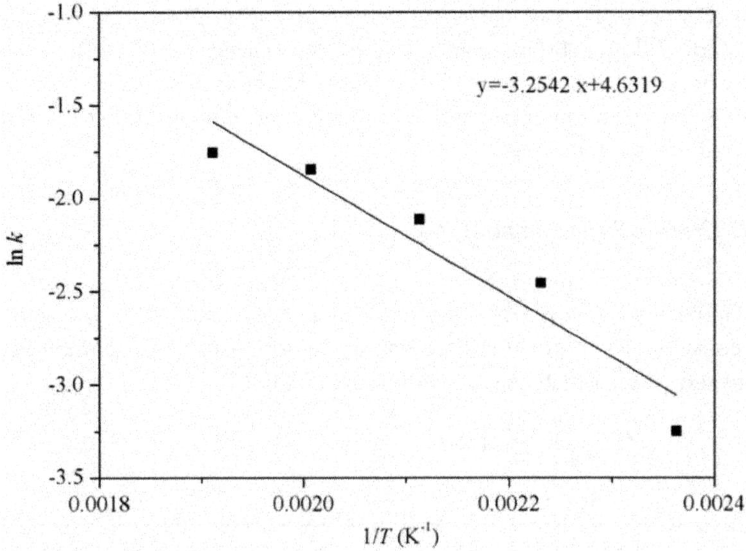

Figure 12 Plot of $1/T$ and ln k.

Table 12 Comparison of the kinetic parameters calculated by different methods

Method	Feedstock	Reaction order (N)	Pre-exponential factor (A)	Activation energy (E_a/kJ·mol^{-1})	Reference
Integrated process	Soybean oil	1.4967	102.71	27.06	This study
SC-MeOH transesterification	Tannery waste	1	1,176.67	36.01	Ong et al. (2013b)
	Linseed oil		4.68	46.50	Varma and Madras (2007)
	Castor oil		32.40	35.00	Varma and Madras (2007)
	Rapeseed oil		3.21×10^5	67.76	Kusdiana and Saka (2001)

1.4967, which is averaged by the order value at each temperature in Table 11. Using the parameters in Eqs. (22) and (30), the kinetic model of this integrated process can be expressed as Eqn. (32):

$$-\frac{dc_A}{dt} = 102.71e^{-\frac{27.06 \times 10^3}{RT}} c_A^{1.4967} \quad (32)$$

Table 12 shows the comparison of kinetic parameters between this integrated process and other research on this process using other feedstocks: tannery waste (Ong et al., 2013), linseed oil (Varma and Madras, 2007), castor oil (Varma and Madras, 2007) and rapeseed oil (Kusdiana and Saka, 2001). These results are higher than the value obtained in this study. This is because of the different fatty acid composition of oil used and the small amount of catalyst used, which reduced activation energy.

5.10 Thermodynamic analysis

Thermodynamic parameters, particularly enthalpy (ΔH), entropy (ΔS), and the Gibbs free energy (ΔG), are important in evaluating transesterification reactions. The value of Gibbs free energy can be calculated from the Eyring-Polanyi equation, which can be written as Eqn. (33):

$$k = \kappa \frac{k_B T}{h} \exp\left(-\frac{\Delta G}{RT}\right) \tag{33}$$

where κ is the transmission coefficient and usually taken as unity, k_B is the Boltzmann constant (1.38×10^{-23} J K^{-1}), and h is the Planck's constant (6.63×10^{-34} J s).

Taking the natural logarithm of Eqn. (33) and substituting $\Delta G = \Delta H - T\Delta S$ gives (Eqn. 34):

$$\ln\left(\frac{k}{T}\right) = \frac{-\Delta H}{R}\left(\frac{1}{T}\right) + \left[\ln\kappa + \ln\left(\frac{k_B}{h}\right) + \frac{\Delta S}{R}\right] \tag{34}$$

Equation (34) describes the relationship between enthalpy and entropy. The values of these two thermodynamic parameters can be obtained from the slope and the intercept of the linear plot of $1/T$ vs. $\ln(k/T)$. From Table 13, values of enthalpy and entropy were found to be 23.15 kJ mol^{-1} and −0.2185 kJ mol^{-1} K^{-1}, respectively. Thus, the value of Gibbs free energy can be calculated as 137.43 kJ mol^{-1} at 523 K. The positive value of enthalpy indicates that the external heat input is required to raise the energy level for the reactants to transform to their transition state. The negative value of entropy means the reactants joined to form transition state, which is more structured in the ground

Table 13 Thermodynamic parameters in the integrated transesterification process

Feedstock	ΔH (kJ mol^{-1})	ΔS (kJ mol^{-1} K^{-1})	ΔG (kJ mol^{-1})	Reference
Soybean oil	23.15	−0.22	137.43	This study
Tannery waste	31.37	−0.23	153.64	Ong et al. (2013b)
Rapeseed oil	19.59	−0.19	75.26-79.06	Encinar et al. (2016)
Soybean oil	28.33	−0.18	83.30-87.69	Wu et al. (2016b)

state. The positive Gibbs free energy shows that the SC–MeOH and CH_3ONa-catalyzed transesterification is unspontaneous and endergonic. Similar trends for the thermodynamic parameters of the transesterification reaction have been reported in other studies. Some examples are presented in Table 13. Different feedstocks and transesterification methods lead to differences in values of thermodynamic parameters for biodiesel production.

The process combining supercritical technology and alkali-catalyzed transesterification has the advantages of significantly decreased temperature, relatively small catalyst consumption and enhanced reaction rate compared with either single method. It was successfully used to synthesize biodiesel under optimum conditions of 250°C, methanol-to-oil molar ratio of 23:1, 1.0 wt.% CH_3ONa and 20 min process time, with a biodiesel yield exceeding 97%. The temperature was found to be the most significant parameter based on RSM analysis. The second-order polynomial regression model was validated to ensure it is appropriate in designing a similar reaction system. The kinetic model of this integrated process was found out to be (Eqn. 35):

$$-\frac{dc_A}{dt} = 102.71e^{-\frac{27.06\times10^3}{RT}} c_A^{1.4967} \quad (35)$$

The study also evaluated the thermodynamic parameters of the transesterification process. The values of ΔH, ΔS and ΔG were found to be 23.15 kJ mol^{-1} and −0.2185 kJ mol^{-1} K^{-1} and 137.43 kJ mol^{-1} at 523 K, respectively, which show that the reaction process was endothermic, endergonic and unspontaneous.

6 Improving biodiesel quality

Biodiesel produced using these catalytic processes has some quality problems, which limit acceptance by the market. The downstream separation and purification processes are complicated. The contamination of the final product by minor components (such as FFA, residual alcohol, unsaponifiable matter - plant sterols, tocopherols and hydrocarbons - and soaps) can cause problems such as engine deposits, filter clogging or fuel deterioration (Wang et al., 2008).

Various methods for testing biodiesel quality have been established, including the United States (ASTM D 664) and Europe (EN 14104) tests. In both standards, AV is one of the important quality parameters to evaluate proper ageing properties of biodiesel. Specifically, the maximum AV is restricted to 0.8 mg KOH g^{-1} for biodiesel (B100) (Barabas and Todorut, 2011). It reflects the presence of FFA or acids used in the manufacture of biodiesel and also the degradation of biodiesel due to thermal effects (Meher et al., 2006). Reducing the high content of FFA in the feedstock by pretreatment, or adding a refining step to decrease the AV of the crude biodiesel produced, can improve the quality of biodiesel.

The authors used an integrated process, with continuous esterification using supercritical methanol and a sodium methoxide catalyst, to reduce the FFA content in crude biodiesel. Crude biodiesel feedstock with an AV of 18.66 mg KOH g^{-1} was obtained from Xi'an Blue Sky Bio-engineering Co., Ltd (Xi'an city, Shaanxi Province, China). Crude biodiesel and CH_3ONa-dissolved methanol solution were continuously pumped into the preheater of the continuous reactor (Huaxing Petroleum Instrument Co., Ltd., Nantong, Jiangsu Province, China), which is shown in Fig. 13 (Zeng et al., 2014a). The ratio of these two materials was controlled by the flow rate of pumps 1 and 2, respectively. After

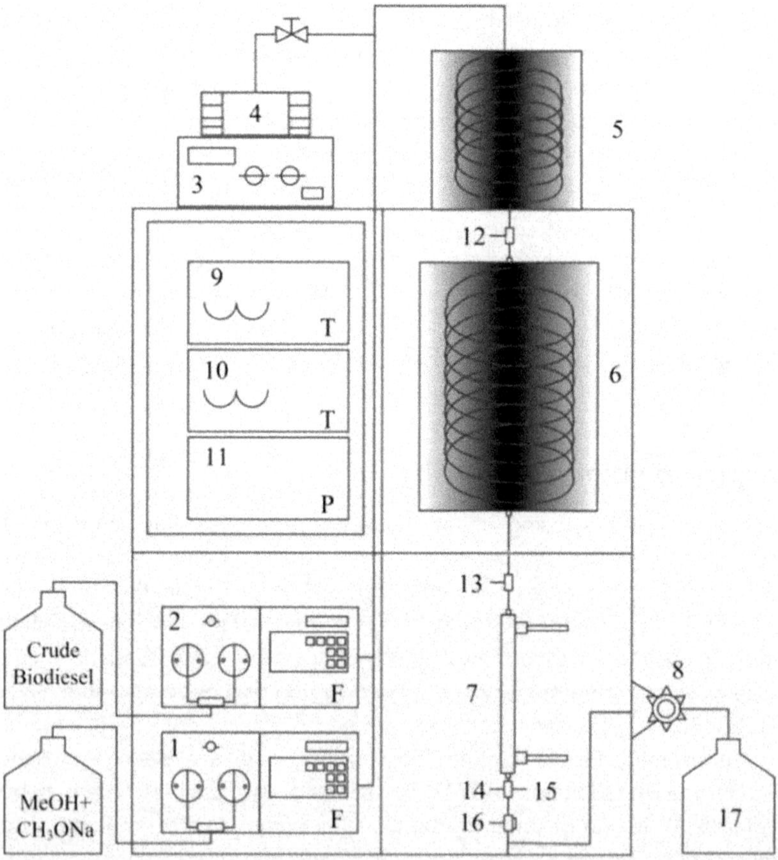

Figure 13 Continuous operation apparatus diagram (1 - solvent pump; 2 - feed pump; 3 - temperature and stirrer controller; 4 - mixer with stirrer; 5 - preheater; 6 - reactor; 7 - cooler; 8 - back pressure regulator; 9, 10 - heater controller; 11, 15 - to pressure indicator; 12, 13, 14 - to temperature indicator; 16 - filter; 17 - product)

esterification in supercritical reaction conditions, the product was collected and then the oil phase was completely separated by washing with hot water (85–90°C) to prepare for the following AV analysis. The crude biodiesel mainly consists of 89–90% FAMEs and 9.29% FFA (chiefly oleic acid). The esterification of FFA is shown by Eqn. (36):

$$\underset{\text{FFA}}{RCOOH} + \underset{\text{MeOH}}{CH_3OH} \xrightleftharpoons{CH_3ONa} \underset{\text{FAMEs}}{RCOOCH_3} + H_2O \tag{36}$$

The refined biodiesel samples were analyzed by measuring AV using standard test method ASTM D 974 by color-indicator titration. The experimental AV was determined using Eqn. (37):

$$\text{Acid Value}(\text{mg KOH} \times \text{g}^{-1}) = \frac{(A - B) \times M \times 56.1}{W} \tag{37}$$

where A and B are the KOH solution required for titration of the sample and the blank (mL), respectively, M is the molarity of the KOH solution (g), 56.1 represents the molecular weight of KOH and W is the mass of a sample (g).

Using this integrated process of supercritical methanol and sodium methoxide, the AV of the crude biodiesel was reduced from 18.66 to 0.55 mg KOH g^{-1} with reaction conditions of 350°C, 0.5% of sodium methoxide catalyst and 10 MPa pressure. This met the standard set by ASTM D 664 and EN 14104. This integrated process also reduces energy consumption by using a relatively low reaction pressure and decreases the amount of catalyst required.

7 Conclusions

As world energy consumption continues to increase with rapid economic growth, reducing reliance on the use of fossil fuels is becoming ever more important. Biodiesel produced from renewable sources, including non-edible oils, waste cooking oils, soapstocks and animal fats, has become more attractive given its range of environmental benefits. Since conventional catalytic methods are not suitable for low-grade feedstocks with high FFA and water content, transesterification with SCFs has emerged as a promising process for commercial application. Combining supercritical technology and alkali-catalyzed transesterification has the advantages of significantly decreased temperature, relatively small catalyst consumption and an enhanced reaction rate compared with either method on its own. Developing these technologies will allow the development of biodiesel as a renewable fuel source.

8 Where to look for further information

Commercial production of biodiesel, simultaneously associated with CO_2 emission reduction, waste water treatment and high-valuable products' extraction, will become competitive in comparison with other conventional sources of energy in the foreseeable future. Microalgae are also promoted as the potential biodiesel feedstock with the advantages of high lipid content and productivity.

Books and reports of expertise include:

- A thorough introduction for biodiesel production-The Biodiesel Bible is published by Keith Addison, Journey to Forever, 2013, 342 pages. (http://journeytoforever.org/books/biodiesel-bible.cgi).
- Biodiesel handling and use guide(Fifth edition) published by National Renewable Energy Laboratory of U.S. department of energy. (https://afdc.energy.gov/files/u/publication/biodiesel_handling_use_guide.pdf).

Centres of expertise include:

- The Journey to Forever Project is organised by Handmade Projects. Biodiesel is one of the biofuel sections (http://journeytoforever.org/biodiesel.html). The site includes information resources and key contacts.
- The National Biodiesel Research Institute in China (http://www.chnbd.org/).

The European biodiesel board collected the information about the associations all over the world (https://www.ebb-eu.org/links.php).

9 References

Aguieiras, E. C. G., Cavalcanti-Oliveira, E. D. and Freire, D. M. G. 2015. Current status and new developments of biodiesel production using fungal lipases. *Fuel*, 159, 52–67.

Amini-Niaki, S. R. and Ghazanfari, A. 2013. Comparison of fuel and emission properties of petro diesel and sunflower biodiesel prepared by optimized production variables. *Fuel*, 109, 384–8.

Anikeev, V. I. and Ermakova, A. 2011. Chemical equilibrium of transesterification reactions between mixed triglycerides and methanol. *Russian Journal of Physical Chemistry A*, 86(1), 19–25.

Anikeev, V. I. and Ermakova, A. 2012. Calculating chemical equilibrium in the stepwise transesterification reactions of mixed triglycerides of fatty acids. *Russian Journal of Physical Chemistry A*, 86(2), 196–9.

Anikeev, V. I. and Yermakova, A. 2011. Technique for complete oxidation of organic compounds in supercritical water. *Russian Journal of Applied Chemistry*, 84(1), 88–94.

Anikeev, V. I., Stepanov, D. A. and Ermakova, A. 2011a. Calculating the thermodynamic characteristics of the stepwise transesterification of simple triglycerides. *Russian Journal of Physical Chemistry A*, 85(12), 2082–7.

Anikeev, V. I., Stepanov, D. A. and Ermakova, A. 2011b. Phase and chemical equilibria in the transesterification reaction of vegetable oils with supercritical lower alcohols. *Russian Journal of Physical Chemistry A*, 85(8), 1336–46.

Anikeev, V., Stepanov, D. and Yermakova, A. 2012. Thermodynamics of phase and chemical equilibrium in the processes of biodiesel fuel synthesis in subcritical and supercritical methanol. *Industrial and Engineering Chemistry Research*, 51(13), 4783–96.

Atabani, A. E., Silitonga, A. S., Badruddin, I. A., Mahlia, T. M. I., Masjuki, H. H. and Mekhilef, S. 2012. A comprehensive review on biodiesel as an alternative energy resource and its characteristics. *Renewable and Sustainable Energy Reviews*, 16(4), 2070–93.

Balat, M. 2009. Prospects for worldwide biodiesel market development. *Energy Sources Part B-Economics Planning and Policy*, 4(1), 48–58.

Banković-Ilić, I. B., Stojković, I. J., Stamenković, O. S., Veljkovic, V. B. and Hung, Y.-T. 2014. Waste animal fats as feedstocks for biodiesel production. *Renewable and Sustainable Energy Reviews*, 32, 238–54.

Barabas, I. and Todorut, I.-A. 2011. Biodiesel quality, standards and properties. In: Montero, G. and Stoytcheva, M. (Eds), *Biodiesel-quality, Emissions and By-products*. InTech, Rijeka, Croatia, pp. 3–28.

Benson, S. W. 1976. *Thermochemical Kinetics: Methods for the Estimation of Thermochemical Data and Rate Parameters* (2nd edn.). John Wiley, New York.

Bezerra, M. A., Santelli, R. E., Oliveira, E. P., Villar, L. S. and Escaleira, L. A. 2008. Response surface methodology (RSM) as a tool for optimization in analytical chemistry. *Talanta*, 76(5), 965–77.

Biermann, U., Bornscheuer, U., Meier, M. A. R., Metzger, J. O. and Schafer, H. J. 2011. Oils and fats as renewable raw materials in chemistry. *Angewandte Chemie-International Edition*, 50(17), 3854–71.

Borugadda, V. B. and Goud, V. V. 2012. Biodiesel production from renewable feedstocks: status and opportunities. *Renewable and Sustainable Energy Reviews*, 16(7), 4763–84.

Box, G. E. P. and Wilson, K. B. 1951. On the experimental attainment of optimum conditions. *Journal of the Royal Statistical Society: Series B (Statistical Methodology)*, 13(1), 1–45.

Canakci, M. 2007. The potential of restaurant waste lipids as biodiesel feedstocks. *Bioresource Technology*, 98(1), 183–90.

Canakci, M. and Sanli, H. 2008. Biodiesel production from various feedstocks and their effects on the fuel properties. *Journal of Industrial Microbiology and Biotechnology*, 35(5), 431–41.

Canakci, M. and Van Gerpen, J. 2001. Biodiesel production from oils and fats with high free fatty acids. *Transactions of the ASAE*, 6(44), 1429–36.

Canoira, L., Rodriguez-Gamero, M., Querol, E., Alcantara, R., Lapuerta, M. and Oliva, F. 2008. Biodiesel from low-grade animal fat: production process assessment and biodiesel properties characterization. *Industrial and Engineering Chemistry Research*, 47(21), 7997–8004.

Chakraborty, R. and Sahu, H. 2014. Intensification of biodiesel production from waste goat tallow using infrared radiation: process evaluation through response surface methodology and artificial neural network. *Applied Energy*, 114, 827–36.

Chalkley, A. P. 1912. *Diesel Engines for Land and Marine Work* (2nd edn.). D. Van Nostrand, New York.

Chen, W., Wang, C. W., Liu, S. W., Wu, Y. X. and Tang, Z. J. 2011. Reaction processes and mechanism of supercritical transesterification in situ ATR infrared spectrum. *Journal of Fuel Chemistry and Technology*, 39(11), 817–22.

Chisti, Y. 2007. Biodiesel from microalgae. *Biotechnology Advances*, 25(3), 294–306.

Constantinou, L. and Gani, R. 1994. New group contribution method for estimating properties of pure compounds. *AIChE Journal*, 40, 1697–710.

De Boer, K. and Bahri, P. A. 2011. Supercritical methanol for fatty acid methyl ester production: a review. *Biomass and Bioenergy*, 35(3), 983–91.

Demirbas, A. 2002. Biodiesel from vegetable oils via transesterification in supercritical methanol. *Energy Conversion and Management*, 43(17), 2349–56.

Demirbas, A. 2005. Biodiesel production from vegetable oils via catalytic and non-catalytic supercritical methanol transesterification methods. *Progress Energy and Combustion Science*, 31(5–6), 466–87.

Demirbas, A. 2007. Recent developments in biodiesel fuels. *International Journal of Green Energy*, 4(1), 15–26.

Demirbas, A. 2008. Comparison of transesterification methods for production of biodiesel from vegetable oils and fats. *Energy Conversion and Management*, 49(1), 125–30.

Demirbas, A. 2009a. Biodiesel from waste cooking oil via base-catalytic and supercritical methanol transesterification. *Energy Conversion and Management*, 50(4), 923–7.

Demirbas, A. 2009b. Progress and recent trends in biodiesel fuels. *Energy Conversion and Management*, 50(1), 14–34.

Dufour, J. and Iribarren, D. 2012. Life cycle assessment of biodiesel production from free fatty acid-rich wastes. *Renewable Energy*, 38(1), 155–62.

Encinar, J. M., Sanchez, N., Martinez, G. and Garcia, L. 2011. Study of biodiesel production from animal fats with high free fatty acid content. *Bioresource Technology*, 102(23), 10907–14.

Encinar, J. M., Pardal, A. and Sánchez, N. 2016. An improvement to the transesterification process by the use of co-solvents to produce biodiesel. *Fuel*, 166, 51–8.

Fang, T., Shimoyama, Y., Abeta, T., Iwai, Y., Sasaki, M. and Goto, M. 2008. Phase equilibria for the mixtures of supercritical methanol+C18 methyl esters and supercritical methanol+α-tocopherol. *The Journal of Supercritical Fluids*, 47, 140–6.

Freedman, B., Butterfield, R. O. and Pryde, E. H. 1986. Transesterification kinetics of soybean oil. *Journal of the American Oil Chemists' Society*, 63, 1375–80.

Fukuda, H., Kondo, A. and Noda, H. 2001. Biodiesel fuel production by transesterification of oils. *Journal of Bioscience and Bioengineering*, 92, 405–16.

García-Martínez, N., Andreo-Martinez, P., Quesada-Medina, J., de los Rios, A. P., Chica, A., Beneito-Ruiz, R. and Carratala-Abril, J. 2017. Optimization of non-catalytic transesterification of tobacco (*Nicotiana tabacum*) seed oil using supercritical methanol to biodiesel production. *Energy Conversion And Management*, 131, 99–108.

Ghoreishi, S. M. and Moein, P. 2013. Biodiesel synthesis from waste vegetable oil via transesterification reaction in supercritical methanol. *The Journal of Supercritical Fluids*, 76, 24–31.

Goering, C. E., Schwab, A. W., Dangherty, M. J., Pryde, E. H. and Heakin, A. J. 1982. Fuel properties of eleven vegetable oils. *Transactions of the ASAE*, 25(6), 1472–7.

Gog, A., Roman, M., Toşa, M., Paizs, C. and Irimie, F. D. 2012. Biodiesel production using enzymatic transesterification - current state and perspectives. *Renewable Energy*, 39(1), 10–16.

Guldhe, A., Singh, P., Ansari, F. A., Singh, B. and Bux, F. 2017. Biodiesel synthesis from microalgal lipids using tungstated zirconia as a heterogeneous acid catalyst and its comparison with homogeneous acid and enzyme catalysts. *Fuel*, 187, 180–8.

Gürü, M., Artukoğlu, B. D., Keskin, A. and Koca, A. 2009. Biodiesel production from waste animal fat and improvement of its characteristics by synthesized nickel and magnesium additive. *Energy Conversion and Management*, 50(3), 498–502.

Haas, M. J. 2005. Improving the economics of biodiesel production through the use of low value lipids as feedstocks: vegetable oil soapstock. *Fuel Processing Technology*, 86(10), 1087–96.

Hanh, H. D., Dong, N. T., Okitsu, K., Nishimura, R. and Maeda, Y. 2009. Biodiesel production through transesterification of triolein with various alcohols in an ultrasonic field. *Renewable Energy* 34(3), 766–8.

He, H., Wang, T. and Zhu, S. 2007. Continuous production of biodiesel fuel from vegetable oil using supercritical methanol process. *Fuel.*, 86(3), 442–7.

Hoang, D., Bensaid, S. and Saracco, G. 2013. Supercritical fluid technology in biodiesel production. *Green Process Synth* 2: 407–25.

Jeong, G. T., Yang, H. S. and Park, D. H. 2009. Optimization of transesterification of animal fat ester using response surface methodology. *Bioresource Technology*, 100(1), 25–30.

Kafuku, G., Tan, K. T., Lee, K. T. and Mbarawa, M. 2011. Noncatalytic biodiesel fuel production from *Croton megalocarpus* oil. *Chemical Engineering and Technology*, 34(11), 1827–34.

Kansedo, J. and Lee, K. T. 2013. Process optimization and kinetic study for biodiesel production from non-edible sea mango (*Cerbera odollam*) oil using response surface methodology. *Chemical Engineering Journal*, 214, 157–64.

Kulkarni, M. G. and Dalai, A. K. 2006. Waste cooking oil-an economical source for biodiesel: a review. *Industrial and Engineering Chemistry Research*, 45(9), 2901–13.

Kusdiana, D. and Saka, S. 2001. Kinetics of transesterification in rapeseed oil to biodiesel fuel as treated in supercritical methanol. *Fuel*, 80(5), 693–8.

Lapuerta, M., Rodriguez-Fernandez, J. and Agudelo, J. R. 2008. Diesel particulate emissions from used cooking oil biodiesel. *Bioresource Technology*, 99(4), 731–40.

Lee, S., Posarac, D. and Ellis, N. 2012. An experimental investigation of biodiesel synthesis from waste canola oil using supercritical methanol. *Fuel*, 91(1), 229–37.

Leung, D. Y. C., Wu, X. and Leung, M. K. H. 2010. A review on biodiesel production using catalyzed transesterification. *Applied Energy*, 87(4), 1083–95.

Liu, L. 2004. How is chemical interesterification initiated: nucleophilic substitution or α-proton abstraction? *Journal of the American Oil Chemists' Society*, 81(4), 331–7.

Ma, P. S. 2003. *Chemical Engineering Data*. China Petrochemical Press, Beijing.

Ma, P. S. 2006. *Experimental Property Data Manual of Organic Compounds-Carbon, Hydrogen, Oxygen and Halogen Part*. Chemical Industry Press, Beijing.

Ma, F. R. and Hanna, M. A. 1999. Biodiesel production: a review. *Bioresource Technology*, 70(1), 1–15.

Ma, F. R., Clements, L. D. and Hanna, M. A. 1998. Biodiesel fuel from animal fat. Ancillary studies on transesterification of beef tallow. *Industrial and Engineering Chemistry Research*, 37(9), 3768–71.

Ma, Z., Shang, Z. Y., Wang, E. J., Xu, J. C., Xu, Q. Q. and Yin, J. Z. 2012. Biodiesel production via transesterification of soybean oil using acid catalyst in CO_2 expanded methanol liquids. *Industrial and Engineering Chemistry Research*, 51(38), 12199–204.

Maçaira, J., Santana, A., Recasens, F. and Angeles Larrayoz, M. 2011. Biodiesel production using supercritical methanol/carbon dioxide mixtures in a continuous reactor. *Fuel*, 90(6), 2280–8.

Maddikeri, G. L., Pandit, A. B. and Gogate, P. R. 2012. Intensification approaches for biodiesel synthesis from waste cooking oil: a review. *Industrial and Engineering Chemistry Research*, 51(45), 14610–28.

Marulanda, V. F., Anitescu, G. and Tavlarides, L. L. 2010. Investigations on supercritical transesterification of chicken fat for biodiesel production from low-cost lipid feedstocks. *The Journal of Supercritical Fluids*, 54(1), 53–60.

Meher, L., Vidyasagar, D. and Naik, S. 2006. Technical aspects of biodiesel production by transesterification-a review. *Renewable and Sustainable Energy Reviews*, 10(3), 248–68.

Mekhilef, S., Siga, S. and Saidur, R. 2011. A review on palm oil biodiesel as a source of renewable fuel. *Renewable and Sustainable Energy Reviews*, 15(4), 1937–49.

Mohamad, M., Ngadi, N., Wong, S. L., Jusoh, M. and Yahya, N. Y. 2017. Prediction of biodiesel yield during transesterification process using response surface methodology. *Fuel*, 190, 104–12.

Myers, R. H., Montgomery, D. C. and Anderson-Cook, C. M. 2009. *Response Surface Methodology: Process and Product Optimization Using Designed Experiments* (3rd edn.). John Wiley and Sons.

Negi, D. S., Sobotka, F., Kimmel, T., Wozny, G. and Schomacker, R. 2006. Liquid-liquid phase equilibrium in glycerol-methanol-methyl oleate and glycerol-monoolein-methyl oleate ternary systems. *Industrial and Engineering Chemistry Research*, 45, 3693–6.

Noshadi, I., Amin, N. A. S. and Parnas, R. S. 2012. Continuous production of biodiesel from waste cooking oil in a reactive distillation column catalyzed by solid heteropolyacid: optimization using response surface methodology (RSM). *Fuel*, 94(1), 156–64.

Ong, L. K., Kurniawan, A., Suwandi, A. C., Lin, C. X., Zhao, X. S. and Ismadji, S. 2013. Transesterification of leather tanning waste to biodiesel at supercritical condition: kinetics and thermodynamics studies. *The Journal of Supercritical Fluids*, 75, 11–20.

Pinnarat, T. and Savage, P. E. 2008. Assessment of noncatalytic biodiesel synthesis using supercritical reaction conditions. *Industrial and Engineering Chemistry Research*, 47(18), 6801–8.

Pinzi, S., Leiva, D., López-García, I., Redel-Macías, M. D. and Dorado, M. P. 2014. Latest trends in feedstocks for biodiesel production. *Biofuels, Bioproducts and Biorefining*, 8(1), 126–43.

Ramos, M. J., Fernandez, C. M., Casas, A., Rodriguez, L. and Perez, A. 2009. Influence of fatty acid composition of raw materials on biodiesel properties. *Bioresource Technology*, 100(1), 261–8.

Rashid, U., Anwar, F., Ansari, T. M., Arif, M. and Ahmad, M. 2009. Optimization of alkaline transesterification of rice bran oil for biodiesel production using response surface methodology. *Journal of Chemical Technology and Biotechnology*, 84(9), 1364–70.

Román-Figueroa, C., Olivares-Carrillo, P., Paneque, M., Palacios-Nereo, F. J. and Quesada-Medina, J. 2016. High-yield production of biodiesel by non-catalytic supercritical methanol transesterification of crude castor oil (*Ricinus communis*). *Energy*, 107, 165–71.

Saka, S. and Kusdiana, D. 2001. Biodiesel fuel from rapeseed oil as prepared in supercritical methanol. *Fuel*, 80, 225-31.

Samniang, A., Tipachan, C. and Kajorncheappun-ngam, S. 2014. Comparison of biodiesel production from crude Jatropha oil and Krating oil by supercritical methanol transesterification. *Renewable Energy*, 68, 351-5.

Santana, A., Maçaira, J. and Larrayoz, M. A. 2012. Continuous production of biodiesel using supercritical fluids: a comparative study between methanol and ethanol. *Fuel Processing Technology*, 102, 110-15.

Savage, P. E., Gopalan, S., Mizan, T. I., Martino, C. J. and Brock, E. E. 1995. Reactions at supercritical conditions: applications and fundamentals. *AIChE Journal*, 41(7), 1723-78.

Sawangkeaw, R., Bunyakiat, K. and Ngamprasertsith, S. 2010. A review of laboratory-scale research on lipid conversion to biodiesel with supercritical methanol (2001-2009). *The Journal of Supercritical Fluids*, 55(1), 1-13.

Saydut, A., Erdogan, S., Kafadar, A. B., Kaya, C., Aydin, F. and Hamamci, C. 2016. Process optimization for production of biodiesel from hazelnut oil, sunflower oil and their hybrid feedstock. *Fuel*, 183, 512-17.

Sendzikiene, E., Sinkuniene, D., Kazanceva, I. and Kazancev, K. 2016. Optimization of low quality rapeseed oil transesterification with butanol by applying the response surface methodology. *Renewable Energy*, 87, 266-72.

Sharma, Y. C., Singh, B. and Upadhyay, S. N. 2008. Advancements in development and characterization of biodiesel: a review. *Fuel*, 87(12), 2355-73.

Shimoyama, Y., Iwai, Y., Jin, B. S., Hirayama, T. and Arai, Y. 2007. Measurement and correlation of vapor-liquid equilibria for methanol+methyl laurate and methanol+methyl myristate systems near critical temperature of methanol. *Fluid Phase Equilibria*, 257(2), 217-22.

Shin, H. Y., Lee, S. H., Ryu, J. H. and Bae, S. Y. 2012. Biodiesel production from waste lard using supercritical methanol. *The Journal of Supercritical Fluids*, 61, 134-8.

Singh, S. P. and Singh, D. 2010. Biodiesel production through the use of different sources and characterization of oils and their esters as the substitute of diesel: a review. *Renewable and Sustainable Energy Reviews*, 14(1), 200-16.

Stepanov, D. A., Ermakova, A. and Anikeev, V. I. 2010. Calculations of phase equilibria for mixtures of triglycerides, fatty acids, and their esters in lower alcohols. *Russian Journal of Physical Chemistry A*, 86, 196-9.

Talebian-Kiakalaieh, A., Amin, N. A. S. and Mazaheri, H. 2013. A review on novel processes of biodiesel production from waste cooking oil. *Applied Energy*, 104, 683-710.

Tan, K. and Lee, K. 2011. A review of supercritical fluid technology in sustainable biodiesel production: potential and challenges. *Renewable and Sustainable Energy Reviews* 15(5): 2452-6..

Tan, K. T., Gui, M. M., Lee, K. T. and Mohamed, A. R. 2010. An optimized study of methanol and ethanol in supercritical alcohol technology for biodiesel production. *Journal of Supercritical Fluids*, 53, 82-7.

Varma, M. N. and Madras, G. 2007. Synthesis of biodiesel from castor oil and linseed oil in supercritical fluids. *Industrial and Engineering Chemistry Research*, 46(1), 1-6.

Veny, H., Aroua, M. K. and Sulaiman, N. M. N. 2014. Kinetic study of lipase catalyzed transesterification of jatropha oil in circulated batch packed bed reactor. *Chemical Engineering Journal*, 237, 123-30.

Vicente, G., Martínez, M. and Aracil, J. 2004. Integrated biodiesel production: a comparison of different homogeneous catalysts systems. *Bioresource Technology*, 92(3), 297–305.

Wang, H., Tang, H., Wilson, J., Salley, S. O. and Ng, K. Y. S. 2008. Total acid number determination of biodiesel and biodiesel blends. *Journal of the American Oil Chemists' Society*, 85(11), 1083–6.

Warabi, Y., Kusdiana, D. and Saka, S. 2004. Reactivity of triglycerides and fatty acids of rapeseed oil in supercritical alcohols. *Bioresource Technology*, 91, 283–7.

Weber, W., Petkov, S. and Brunner, G. 1999. Vapour-liquid-equilibria and calculations using the Redlich-Kwong-Aspen-equation of state for tristearin, tripalmitin, and triolein in CO_2 and propane. *Fluid Phase Equilibria*, 158, 695–706.

Wu, L., Huang, K. L., Wei, T. Y., Lin, Z. J., Zou, Y. and Tong, Z. F. 2016a. Process intensification of NaOH-catalyzed transesterification for biodiesel production by the use of bentonite and co-solvent (diethyl ether). *Fuel*, 186, 597–604.

Wu, L., Wei, T., Lin, Z., Zou, Y., Tong, Z. and Sun, J. 2016b. Bentonite-enhanced biodiesel production by NaOH-catalyzed transesterification: process optimization and kinetics and thermodynamic analysis. *Fuel*, 182, 920–7.

Yaakob, Z., Mohammad, M., Alherbawi, M., Alam, Z. and Sopian, K. 2013. Overview of the production of biodiesel from Waste cooking oil. *Renewable and Sustainable Energy Reviews*, 18, 184–93.

Yusuf, N. N.a.N., Kamarudin, S. K. and Yaakub, Z. 2011. Overview on the current trends in biodiesel production. *Energy Conversion and Management*, 52, 2741-51.

Zeng, D., Li, Y. and Fang, T. 2012. Production of biodiesel in supercritical methanol and related correlation on phase equilibrium. *Advanced Materials Research*, 560–1, 555–62.

Zeng, D., Li, R., Feng, M. and Fang, T. 2014a. Continuous esterification of free fatty acids in crude biodiesel by an integrated process of supercritical methanol and sodium methoxide catalyst. *Biotechnology and Applied Biochemistry*, 174(4), 1484–95.

Zeng, D., Li, R., Jin, T. and Fang, T. 2014b. Calculating the thermodynamic characteristics and chemical equilibrium of the stepwise transesterification of triolein using supercritical lower alcohols. *Industrial and Engineering Chemistry Research*, 53(17), 7209–16.

Zeng, D., Yang, L. and Fang, T. 2017. Process optimization, kinetic and thermodynamic studies on biodiesel production by supercritical methanol transesterification with CH3ONa catalyst. *Fuel*, 203, 739–48.

Zhao, G. L. and Jin, C. D. 1983. *Estimation of Thermodynamic Data on Organic Compounds*. Higher Education Press, Beijing.

Chapter 5

Production of biodiesel from oilseeds: *Jatropha curcas*

Rahmath Abdulla, Universiti Malaysia Sabah, Malaysia

1 Introduction

Biodiesel is mainly produced by transesterification of vegetable oils and this feedstock plays an important role in its production (Lin et al., 2011). Biodiesel generally causes a decrease in unburned hydrocarbons (HC), carbon monoxide (CO), and particulate (PM) emissions, along with an increase in nitrous oxide (NOx) emissions. Using biodiesel also reduces other environmental pollutants compared to fossil fuels (Pandey et al., 2012). Presently, more than 95% of the world's biodiesel is produced from edible vegetable oils, which can be easily obtained on a large scale from processing of agricultural products. However, there is concern that continuous production of biodiesel with edible oils as the main feedstock may compete with food production (Gui et al., 2008). This is due to the rapid increase in coarse grain production (such as corn, barley, and sorghum) as compared to primary food grains (such as wheat and rice) and the increased use of these coarse grains as feedstock for fuels. According to the UN Food and Agricultural Organization, spikes in food prices in recent years are partly due to demand for first-generation biofuels derived from crops (Srinivasan, 2009; Kumar and Sharma, 2011). The haste to develop first-generation biofuels has also led to tensions over the land use. For example, Malaysian policies promoting the production and use of palm oil-based biodiesel have accelerated deforestation of the Borneo rain forest (Rowe, 2011).

http://dx.doi.org/10.19103/AS.2019.0027.12

Table 1 Potential non-edible sources for biodiesel production

	Oil content	
Non-edible vegetable source	Seed (wt%)	Kernel (wt%)
Azadirachta indica (neem)	20-30	25-45
Aphanamixis piolystachya (wall)		35
Annona muricate		20-30
Annona squamosa	15-20	
Asclepias syriaca (milkweed)	20-25	0.019
Brassica carinata (ethiopian mustard)	42	2.2-10.8
Balanites aegyptiaca (desert date)		36-47
Bombax malabaricum	18-26	
Calophyllum inophyllum L.	65	22
Crambe abyssinica	30-38	
Ceiba pentandra	24-40	
Cerbera odollam (sea mango)	54	6.4
Croton tiglium	30-45	50-60
Cuphea	20-38	
Crotalaria retusa L.	15	
Eruca sativa gars	35	
Garcinia indica	45.5	
Guizotia abyssinica L. (niger)	50-60	
Hevea brasiliensis (rubber)	40-60	40-50
Idesia polycarpa var. vestita fruit oil	26.15-26.26	
Jatropha curcas L.	20-60	40-60
Linum usitatissimum (linseed)	35-45	
Madhuca indica	35-50	50
Melia azedarach	10-45	2.8
Michela chaampaca	45	
Mesua ferrea	35-50	
Nicotiana tabacum	36-41	17
Pongamia pinnata	25-50	30-50
Putranjiva roxburghii	41-42	
Pongamia glabra	33.6	
Raphanus sativus	40-45	
Ricinus communis	45-50	
Simmondsia chinensis	45-55	
Salvadora oleoides	45	
Sapium sebifeum L. Roxb	13-32	53-64
Sleichera triguga		55-70

(Continued)

Table 1 (Continued)

Non-edible vegetable source	Oil content	
	Seed (wt%)	Kernel (wt%)
Samadera indica		35
Sapindus mukorossi		51.8
Tomato seed		32–37
Tung		35–40
Terminala catappa		49
Ximenia americana		49–61

Source: Atabani et al. (2013).

As a solution to the food versus fuel debate, biodiesel feedstock markets are transitioning from expensive first-generation feedstocks (such as soy, rapeseed, and palm oil) to cheaper, non-edible oils (Thurmond, 2008). These form the second-generation feedstock for biofuels. As a result, biodiesel production using non-edible oils have been extensively investigated during the past few years. The main advantages of non-edible oils as biodiesel feedstocks are (Ahmad et al., 2011) as follows:

1 They do not compete with food production.
2 They are potentially more efficient and environmentally friendly than first-generation feedstocks.
3 They are comparable to edible oils in terms of quality.
4 They can potentially be grown on marginal land that is unsuitable for crop production.
5 They can produce formation of useful byproducts.

Table 1 gives an overview of the different types of non-edible plant oils that can be employed for production of biodiesel. One of these, *J. curcas*, is gaining much importance as a promising feedstock for the future because of the characteristics of the oil it produces (Gui et al., 2008; Shah et al., 2004).

2 Characteristics of *Jatropha curcas* L.

Jatropha curcas L. (known as the physic nut) is a perennial plant from the family *Euphorbiaceae* with 175 known species. The genus name *Jatropha* is derived from Greek words *jatr'os* (doctor) and *troph'e* (food), which refers to its medicinal properties. The oil-yielding varieties are *J. pohliana*, *J. gossypiifolia*, *J. multifida*, and *J. curcas*. Out of these, *J. curcas* has attracted particular attention because of its combination of oil content and hardiness as a plant. Linnaeus was the first one to name the variety as *J. curcas* Linnaeus (*J. curcas*

L.). The plant is thought to be native to Central America as well as Mexico, where it is found naturally in forests in coastal areas. Fossil evidence suggests a possible origin in Peru (Becker and Makkar, 2008). However, the 'true' center of origin is yet to be found. Jatropha was believed to be distributed by Portuguese ships through the Cape Verde islands and Guinea Bissau from where it reached Africa and Asia (Burkill, 1966). Dutch colonists are thought to have brought the plant to Malaysia. In many countries, it is known as the 'castor oil plant' or 'hedge castor oil plant', indicating its introduction as a plant cultivated for its oil or its use in hedges. The first commercial application of *J. curcas* was reported from Portugal, where the oil was used for making soaps and for lighting lamps (Rahmath et al., 2011; Nelson et al., 1996; Heller, 1996; Linnaeus, 1753).

J. curcas grows well in tropical and sub-tropical climates across the developing world in particular, with cultivation limits at 30°N and 35°S. It can grow well at lower altitudes of 0–500 m above sea level. It thrives in countries like Africa, India, Myanmar, Malaysia, Thailand, Mexico, Australia, China, and Mauritius. Figure 1 shows the areas suitable for *J. curcas* cultivation around the world.

The plant has a number of advantages (Kumar and Sharma, 2008). The seeds of *J. curcas* yield around 40–60% oil, which is an excellent feedstock for biodiesel. Jatropha oil has approximately 24.60% of crude protein, 47.25% crude fat, and has a moisture content of 5.54% (Akintayo, 2004). In addition, the plant grows well on poorer-quality soils in arid and semi-arid conditions. It can even grow in degraded soils and wasteland. The poisonous latex in the plant limits attack by pests. The plant can flower at any time of the year (Heller, 1996). The hardiness of the plant means it does not compete for land and resources

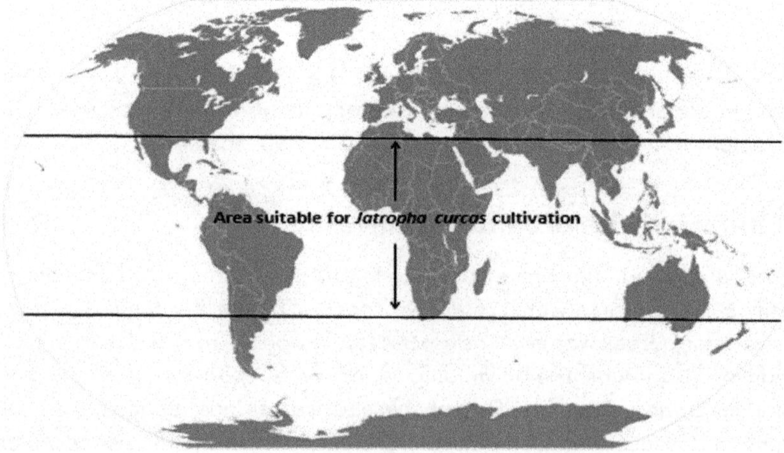

Figure 1 Cultivation limits of *Jatropha curcas*.

with food crops, making it a good substitute for major first-generation biofuel feedstocks such as palm, soya, sugarcane, and corn, as well as a way of diversifying rural incomes (Romijn, 2011).

The area of Jatropha cultivation is still small, covering approximately 900 000 ha. Asia accounts for more than 85% of cultivation, most notably in Myanmar, India, China, and Indonesia. Approximately 120 000 ha is grown in Africa and 20 000 ha in Latin America, especially Brazil. The area for Jatropha cultivation is projected to grow to 12.8 million ha by 2020. In Asia, Indonesia is expected to be the largest producer with 5.2 million ha. Figure 2 shows projections for *J. curcas* cultivation in key countries.

J. curcas is a small tree or large shrub that normally attains a height of 3–5 m, though it can reach a height of 8 m or 10 m. It has a short gestation period. *J. curcas* has soft wood, articulated growth, a straight trunk with thick branchlets. The branches contain latex. The leaves are smooth with 4–6 lobes and measure approximately 10–15 cm by length and width (Achten et al., 2007). The plant develops a main taproot with initially four narrow lateral roots. This root structure can help prevent soil erosion. The lateral roots also help the plant to grow on the surface of rocks with a little amount of water. Plants can live 50 years (Francis et al., 2005). The rainy season is the preferred period of flowering. However, in humid areas, the plant can flower throughout the year. Jatropha can flourish well on marginal soils with low nutrient content (Jones and Miller, 1992).

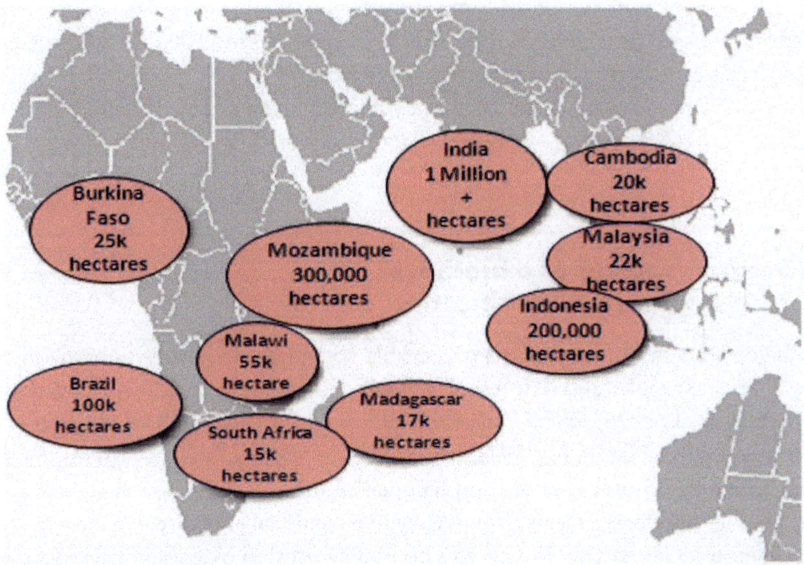

Figure 2 Major *Jatropha curcas* projects around the world. Source: Biodiesel (2020).

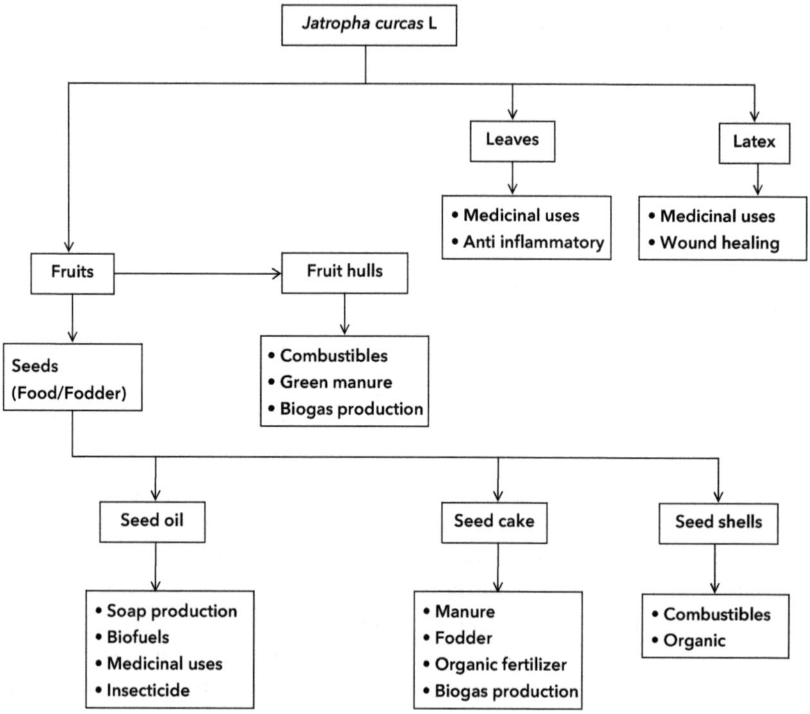

Figure 3 Potential benefits of *Jatropha curcas* L. Source: Silitonga et al. (2011).

The plant can be easily propagated and maintained. The number of Jatropha trees per hectare of planting ranges from 1600 to 2200. Wider spacing is reported to give a larger yield of fruit: 794 kg ha^{-1} and 318 g $shrub^{-1}$ (Abou Kheira and Atta, 2009). Presently *J. curcas* can produce 2000 L ha^{-1} oil per annum (Mofijur et al., 2012). The potential benefits of *J. curcas* are shown in Fig. 3.

3 *Jatropha curcas* as a biodiesel feedstock: transesterification

The importance of *J. curcas* as a non-edible source of biodiesel was highlighted by Shah and Gupta (2007). The seed kernels of *J. curcas* L. contain 40–60% (w/w) oil (Foidl et al., 1996), which indicates that it is a good raw material for biodiesel production (Makkar et al., 1997). The oil of *J. curcas* has been rendered unsafe for cooking purposes due to the presence of toxic curcin, phorbol esters, and some anti-nutritional factors. This toxicity makes the oil attractive as a non-edible vegetable oil feedstock. The oil can be mainly used in oleochemical industries such as production of biodiesel, soap, surfactants, and detergents. The fatty

Table 2 Fatty acid composition of crude *Jatropha curcas* oil

Fatty acid	Formula	Systemic name	Structure	wt%
Myristic	$C_{14}H_{28}O_2$	Tetradecanoic	14:0	0-0.1
Palmitic	$C_{16}H_{32}O_2$	Hexadecanoic	16:0	14.1-15.3
Palmitoleic	$C_{16}H_{30}O_2$	cis-9-Hexadecenoic	16:1	0-1.3
Stearic	$C_{18}H_{36}O_2$	Octadecanoic	18:0	3.7-9.8
Oleic	$C_{18}H_{34}O_2$	cis-9-Octadecenoic	18:1	34.3-45.8
Linoleic	$C_{18}H_{32}O_2$	cis-9,cis-12-Octadecedianoic	18:2	29.0-44.2
Linolenic	$C_{18}H_{30}O_2$	cis-6,cis-9,cis-12-Octadecatrienoic	18:3	0-0.3
Arachidic	$C_{20}H_{40}O_2$	Eicosanoic	20:0	0-0.3
Behenic	$C_{22}H_{44}O_2$	Docosanoic	22:0	0-0.2

Source: Berchmans and Hirata (2008).

acid composition of Jatropha oil varies depending on the geographical area. *Jatropha* oil contains approximately 24.60% of crude protein, 47.25% of crude fat, and 5.54% of moisture content (Akintayo, 2004). Table 2 shows the fatty acid composition of crude Jatropha oil. Jatropha oil has a higher amount of unsaturated fatty acids (oleic acid and linoleic acid) than that of saturated fatty acids (palmitic acid and stearic acid). Since Jatropha oil is abundant with monounsaturated oleic and polyunsaturated linoleic acid, it has semidrying properties and therefore has been found to be an efficient substitute for diesel fuel (Augustus et al., 2002).

Transesterification of triglycerides is normally carried out by reaction of an excess of alcohol with vegetable oils in the presence of a catalyst and under different reaction conditions. Since the reaction is reversible in nature (Fig. 4), an excess of alcohol is employed to shift the equilibrium of the reaction. The overall process of transesterification is normally a sequence of three consecutive steps. In the first step, triglycerides in the oil are converted to diglyceride; secondly, they are converted from diglyceride to monoglyceride; and, finally, they are converted from monoglycerides to glycerol (Marchetti et al., 2007).

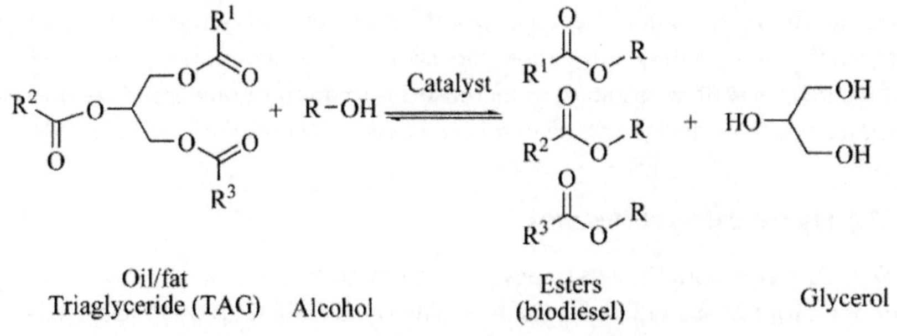

Figure 4 Transesterification reaction. Source: Ghaly et al. (2010).

Each step-in conversion produces one FAAE (Fatty acid alkyl ester) molecule. A total of three FAAE molecules are formed for one triglyceride (Murugesan et al., 2009). Alcohols that are used in transesterification reactions are methanol, ethanol, propanol, butanol, and amyl alcohol (Ibeto et al., 2011). To improve the reaction rate and yield of biodiesel, transesterification is normally catalyzed by an acid, a base, or an enzyme. Each catalytic process is different from one another in its parameters.

3.1 Alkali catalyst

Alkalis are the most common type of catalyst used for biodiesel production on a commercial scale (Ranganathan et al., 2008). The advantage of an alkali catalyst is a higher yield in a relatively shorter reaction time (Ma and Hanna, 1999). In addition, alkaline catalysts are less corrosive to industrial equipment (Agarwal, 2007). The most commonly used alkaline catalysts for biodiesel production are sodium hydroxide and potassium hydroxide (Marchetti et al., 2007). Being cheaper, sodium hydroxide is widely used in large-scale processing (Fukuda et al., 2001). Using alkaline as catalysts include the following disadvantages: a higher energy requirement; difficulty in recovery of glycerol; the need to remove the alkaline catalyst from the final product; the need to treat the alkaline waste water formed during the reaction before discharge into the environment; and the presence of free fatty acids and water, which interfere with the reaction and form soap (Barnwal and Sharma, 2005).

3.2 Acid catalyst

Acid-catalyzed transesterification is generally slower compared to alkali catalysts (Mittelbach and Tratnigg, 1990). Moreover, acids require a higher temperature and pressure along with large amounts of alcohol for the complete reaction to take place. The most commonly used acids in transesterification reactions are sulfuric acid, sulfonic acid, and hydrochloric acid (Freedman et al., 1984). Acid catalysts are preferred over alkali catalysts when the oil has a higher content of free fatty acids and water. The major disadvantage of acid catalysts for biodiesel production is that they can cause corrosion to equipment, and the reaction rate is low, as well as resulting in increased formation of unwanted secondary products such as dialkyl and glycerol derivatives (Noureddini and Zhu, 1997).

3.3 Enzyme catalyst (lipase)

Enzymes, particularly lipases (triacylglycerol hydrolase, EC 3.1.1.3), are gaining attention for transesterification as these can produce high purity biodiesel and can be separated easily from the byproduct, glycerol (Gog et al., 2012). Lipases can catalyze hydrolysis of the carboxylic ester link in the triacylglycerol moiety to form free fatty acids, di-monoglycerides and glycerol (Bajaj et al., 2010). Generally, these lipases are extracted from various microorganisms such as *Mucor miehei*, *Rhizopus oryzae*, *Candida antarctica, Pseudomonas fluorescens*, and *Burkholderia cepacia* (*Pseudomonas cepacia)*. Among these, lipase from *B. cepacia* has been found to give higher biodiesel yield (Marchetti et al., 2007). Biodiesel can be produced using both intracellular and extracellular lipases. Byproducts are biodegradable, thus making the process more environmentally friendly (Devanesan et al., 2007).

For biodiesel production, lipases can be employed as soluble enzymes (Kaieda et al., 2001) or in their whole-cell (Fukuda et al., 2008) or immobilized forms (Bernardes et al., 2007). Due to the high cost of enzymes, immobilized lipase is preferred for enzymatic transesterification (Shimada et al., 2002). After immobilization, the lipase can be reused several times for the production of biodiesel. Moreover, the operating temperature of an enzymatic process is low as compared to the use of acid and alkaline catalysts. On the other hand, the high cost of lipase remains an obstacle to industrialization of enzymatic transesterification (Nelson et al., 1996). In order to make the process economically feasible, the lipase (both intracellular and extracellular) can be reused by immobilizing in a suitable biomass support particle (Ranganathan et al., 2008, Nelson et al., 1996). In this way, biodiesel production using biocatalysts such as lipases results in a high purity product with fewer or no downstream operations required (Meher et al., 2006; Fukuda et al., 2001).

4 Factors affecting enzymatic transesterification using immobilized lipase

There are several factors that affect the transesterification rate, which determines maximum biodiesel yield. The main factors are as follows:

1 selection of alcohol,
2 lipase enzyme,
3 immobilized lipase pretreatments, and
4 water activity/moisture content of the system.

These are discussed below.

4.1 Selection of alcohol

The alcohols used for the production of biodiesel are methanol, ethanol, propanol, isopropanol, 2-propanol, n-butanol, and isobutanol (Antczak et al., 2009). Among these, methanol and ethanol are the cheapest, are produced on a large scale, and thus are most commonly used for industrial production of biodiesel. It has been found that these two alcohols are strong denaturing agents compared to the longer aliphatic alcohols and can also inactivate the enzymes in the process (Nelson et al., 1996).

When used in biodiesel production, immobilized lipase systems were found to be deactivated with lower linear alcohols. This happens because excess alcohol added forms droplets that coat the surface of immobilized lipase (Chen and Wu, 2003). It is important to find out the optimum level of oil-to-alcohol ratio for a typical reaction (Jeong and Park, 2008). The overall yield of biodiesel is dependent on the equilibrium between reaction velocity and rate of enzyme denaturation (Chen et al., 2006). For example, the presence of methanol and water in an immobilized lipase system can speed up the denaturation of the enzyme, whereas in the case of systems containing ethanol, propanol, isopropanol, butanol, and isobutanol, the presence of small amounts of water is necessary for complete transesterification (Haas et al., 2003). Several solutions have been used to overcome this problem of lipase deactivation. Controlled addition of methanol is the most common approach to avoid lipase inactivation (Lu et al., 2007; Soumanou and Bornscheuer, 2003). The second approach is to use different acyl acceptors such as methyl acetate or ethyl acetate (Modi et al., 2007; Du et al., 2004; Xu et al., 2003). The third strategy to solve lipase deactivation is to use organic solvents, but these later create problems in solvent recovery, which could make the process less efficient (Orçaire et al., 2006). On economic grounds, ethanol is the preferred source of alcohol over methanol for transesterification. This is because methanol is more toxic, while ethanol can also be regenerated from biomass, making it a renewable source (Morin et al., 2007).

4.2 Lipase enzyme

Lipases comprise the most important group of biocatalysts for biotechnological applications. Lipases (EC 3.1.1.3) can catalyze both the hydrolysis and synthesis of esters from glycerol and long-chain fatty acids (Jaeger and Eggert, 2002). Unlike other enzymes, lipases exhibit remarkable levels of stability and activity in non-aqueous environments. This can facilitate the catalysis of reactions with high selectivity such as esterification and transesterification (Jegannathan et al., 2008). A variety of lipases that are produced mainly from bacteria or fungi are available commercially. The main advantages of lipase in biotechnological reactions are due to the fact that they (Jaeger and Eggert, 2002):

1. are stable in organic solvents,
2. do not require cofactors,
3. possess broad substrate specificity, and
4. exhibit a high enantioselectivity.

Immobilized lipases from various sources have been employed for biodiesel production. The majority of researchers have chosen lipase from *Candida antarctica*. Other commonly used lipases include *Burkholderia cepacia* (formerly known as *Psuedomonas cepacia), Psuedomonas fluorescence*, *Mucor miehei*, *Thermomyces lanuginosus*, *Chromobacterium viscosum*, *Enterobacter aerogenes*, and porcine pancreas (Akoh et al., 2007). Out of these, immobilized lipase from *Burkholderia cepacia* is gaining attention because of its higher biodiesel yield (Shah and Gupta, 2007; Otero et al., 2005).

4.3 Immobilized lipase pretreatments

Pretreatment often involves soaking the immobilized lipase in a medium prior to its use in biodiesel production (Ghaly et al., 2010). The deactivation of lipase due to lower-chain alcohols can be partially prevented by this type of pretreatment (Ranganathan et al., 2008). Pretreatment in a polar organic solvent is thought to transform the hydrophobic closed active site of lipase to a hydrophobic open active site, which, in turn, enhances its activation (Jegannathan et al., 2008). Various pretreatment mediums that have been employed on a small scale include isopropanol, methyl oleate, and tert-butanol (Fjerbaek et al., 2009).

The importance of lipase pretreatment has been illustrated by pre-incubating the lipase in methyl oleate for 0.5 h and later in soybean oil for 12 h. As a result, pre-incubation, inhibitory effects were reduced, and the yield of biodiesel increased. A methyl ester yield of 97% was obtained within 3.5 h together with controlled addition of methanol at an interval of 0.25–0.40 h, which was maintained even after 20 cycles of methanolysis (Samukawa et al., 2000). In another study by Lu et al. (2007), it was shown that pretreatment of immobilized lipase in solutions of $CaCl_2$ and $MgCl_2$ improved lipase activity, tolerance to methanol, stability, and biodiesel yield. On the other hand, Chen and Wu (2003) suggested the use of tert-butanol and 2-butanol as the pretreatment medium for immobilized lipase. The use of pretreatment adds to the production cost of biodiesel.

4.4 Water content

A minimum amount of water is essential to maintaining the specific three-dimensional structure of lipase (Lu et al., 2008). The water content of a transesterification reaction is expressed as water activity or as percentage concentration (Antczak et al., 2009). Biocatalysts often require a minimum amount of water to maintain their activity (Bommarius and Riebel, 2005). Unmasking and restructuring of the active site of lipase is possible in the presence of the oil-water interface. This is because the activity of lipase is dependent on the available interfacial area. With the addition of water, the interfacial area between oil-water droplets increases (Al-Zuhair et al., 2006). However, with higher water content, lipases favor hydrolysis instead of transesterification (Noureddini et al., 2005). The optimum amount of water added should therefore be a compromise between hydrolysis and maximum lipase activity for the transesterification reaction, even when employed with lower-chain alcohols (Akoh et al., 2007).

Lipases from *Candida rugosa*, *Pseudomona cepacia*, and *Pseudomonas fluorescens* show no reaction in the absence of water (Akoh et al., 2007). These lipases exhibit an increased rate of reaction with an increase in water content from 1% to 20% (Fjerbaek et al., 2009). On the other hand, a negative response was shown with *Candida antarctica* lipase (Deng et al., 2005; Fjerbaek et al., 2009). The reason for this is that reaction rate of transesterification decreases when over 0.1 g of water is present per gram of dry enzyme. This is because the excess water content blocks the pores of immobilized lipase, thus decreasing the enzyme's exposure to the substrates (Robles et al., 2009). The optimal water content for biodiesel production using immobilized lipase is ultimately dependent on the reaction system, feedstock (since water content varies with the type of oils used), type of lipase (some commercial lipases are in powder form), immobilization technique (generally the method of immobilization involves the use of water), and the stability and type of alcohol (analytical grade or reagent grade) used for transesterification (Antczak et al., 2009).

5 Life cycle assessment of *Jatropha curcas* for biodiesel production

Life cycle assessment (LCA) is a computational tool used to analyze the sustainability of a product over its life cycle. Prior to its release into the market, the environmental superiority of the related product or technology should be examined thoroughly based on a life cycle approach (Consoli et al., 1993). Through LCA, the overall environmental impact of a product (i.e. from cradle to grave) can be assessed (Quirrin et al., 2006). Every product has a life cycle starting from extraction or production of raw materials to final disposal, with intermediate stages such as manufacturing, transport, storage, and use of the product (Pehnt, 2006).

The first step in an LCA is to define the goal and scope of the study. This step provides the description of the product system in terms of system boundaries and a functional unit. The functional unit is the most important basis that enables alternative goods, or services to be compared and analyzed. A process flow chart for each system should be put together and unit processes such as input, output, and waste disposal or recycling should be clearly identified. The life cycle inventory consists of collecting data on inputs associated with the system from databases or literature. The third part of an LCA, the life cycle impact assessment, provides the basis for analyzing the potential impact of processes on the environment, for example emissions affecting climate change, pollution, and land use (Pennington et al., 2004). The final step is life cycle interpretation, which assesses implications, for example in reducing environmental impact (Rebitzer et al., 2004; Guinée, 2002).

A comparison of the life cycle impacts of Jatropha biodiesel on 11 environmental categories from the production of 1 tonne, 5 tonnes, and 10 tonnes of biodiesel is shown in Figs. 5–7. In all these cases, biodiesel production using alkali catalysts showed a higher environmental impact especially in

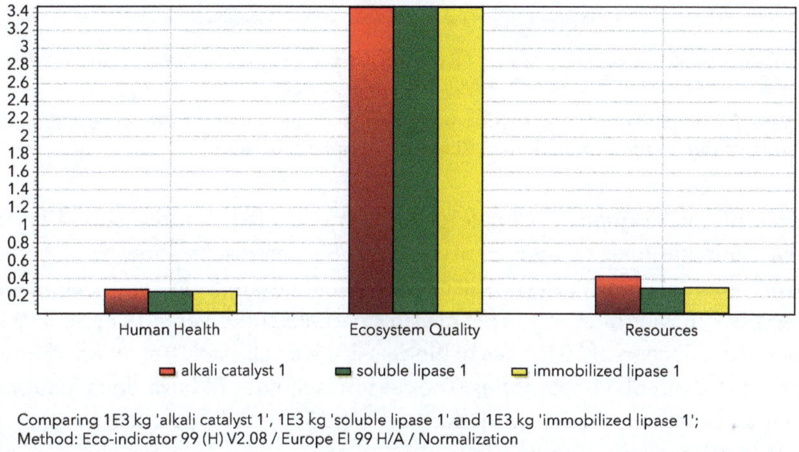

Figure 5 Comparison of the environmental impacts on human health, ecosystem, and resources due to the production of 1 tonne Jatropha biodiesel.

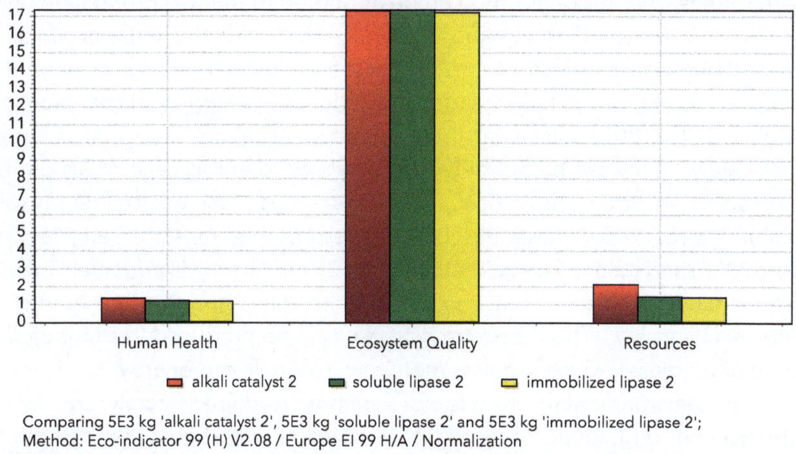

Figure 6 Comparison of the environmental impacts on human health, ecosystem, and resources due to the production of 5 tonne Jatropha biodiesel.

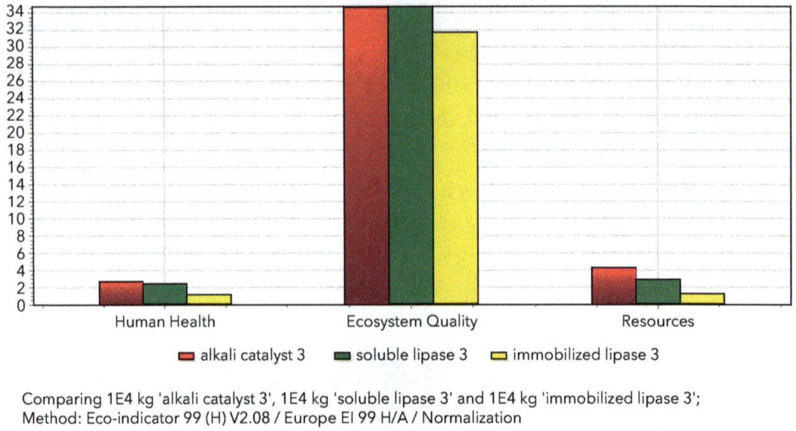

Figure 7 Comparison of the environmental impacts on human health, ecosystem, and resources due to the production of 10 tonne Jatropha biodiesel.

terms of carcinogenicity, climate change impacts (e.g. on the ozone layer), environmental toxicity, and fossil fuel use. Similar results showing a higher environmental impact using alkali catalysts were also reported with palm oil as feedstock (Ravindra et al., 2012; Kiwjaroun et al., 2009; Varanda et al., 2011). For the production of 10 tonnes of biodiesel, the alkali catalyst showed a higher impact in 8 out of 11 categories. The reason is the use of hazardous materials such as potassium hydroxide (Leung et al., 2010) and hydrochloric acids in alkali-catalyzed transesterification.

In the case of biodiesel production with soluble lipase, the impact level was almost equal to that of the alkali catalyst as production increased from 1 tonne to 10 tonnes. The reason for this can be attributed to the two production units for soluble lipase-catalyzed biodiesel production, namely the lipase production unit and biodiesel production unit. Since it is a soluble enzyme, lipase has to be produced for every batch that leads to more input of energy and raw materials (Robinson et al., 2001). However, the environmental impact is less than using an alkali catalyst. This can be attributed to the absence of toxic chemicals during the production process (Bajaj et al., 2010). The results obtained in this study with soluble lipase are comparable to LCA studies done on rapeseed oil (Harding et al., 2008) and palm oil (Jegannathan et al., 2011). The environmental impacts also increased with an increase in production from 1 tonne to 10 tonnes of biodiesel (Figs. 5–7). This is because lipase has to be produced fresh for each 1 tonne of biodiesel, which requires more raw materials and energy.

The environmental impact of immobilized-lipase-catalyzed trans-esterification for 1 tonne of biodiesel production was almost comparable to that of soluble lipase. However, an increase in production resulted in a lower environmental impact. There was a lower impact in 10 categories out of 11 for the production of 10 tonnes of biodiesel as compared to the alkali and soluble lipase. This is due to the fact that the immobilized lipase can be reused several times, thus reducing the raw material and energy input in production (Ye et al., 2006; Yaakob et al., 2013). As Fig. 8 shows, the relative environmental impact is lower in the case of immobilized lipase, showing its potential as an ecofriendly method of biodiesel production. The results are consistent with LCA studies done on rapeseed oil by Harding et al. (2008) and on palm oil by Jegannathan et al. (2011).

These results mean that biodiesel production from non-edible oils such as *J. curcas* oil is gaining momentum (Atabani et al., 2013). An LCA study conducted on Jatropha in West Africa by Ndong et al. (2009) showed that Jatropha biodiesel achieved a 72% saving in greenhouse gases as compared with conventional diesel fuel. Another study by Achten et al. (2010) found a 55% reduction in global warming potential compared to diesel fuel. It was also shown that plantation of Jatropha on wasteland improved its impact in terms of land use. Almeida et al. (2011) reported that Jatropha biodiesel production consumes eight times less non-renewable energy than conventional diesel and reduces greenhouse gas emissions by 51%.

On the other hand, Lam et al. (2009) concluded that the land area required for Jatropha to produce one tonne biodiesel is 118% higher than that required for palm. Moreover, the CO_2 sequestration of palm biodiesel is 20 times higher than that of Jatropha biodiesel. However, the energy required for palm oil extraction was 256% higher than that required for Jatropha oil extraction. If Jatropha can be grown on waste lands, it can represent a cheaper, more sustainable, non-edible biodiesel feedstock.

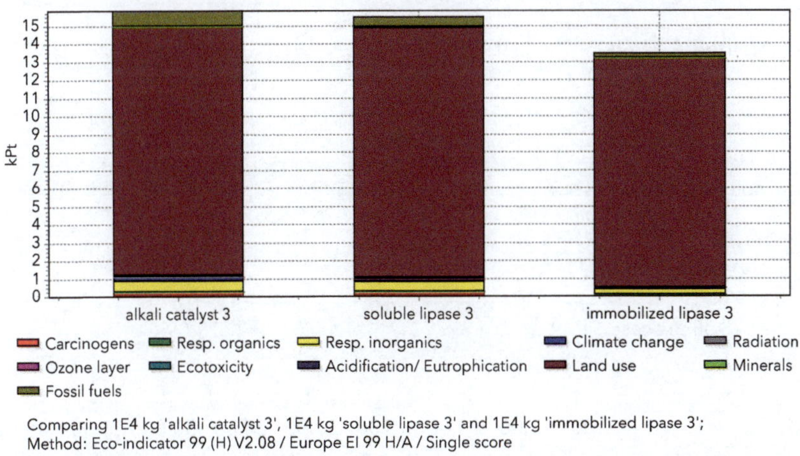

Figure 8 Comparison of the environmental impacts on each environmental category based on a single cumulative score due to the production of 10 tonne Jatropha biodiesel.

6 Economic analysis of biodiesel production

The total plant investment cost for biodiesel production from CJO (Crude Jatropha Oil) for a 1000-tonne capacity plant is shown in Fig. 9. It shows that there was an increase of 40% plant investment cost for immobilized lipase as compared to the alkali method of biodiesel production. The main factor in such an increase in plant investment is because of the installation requirements for using enzymatic transesterification. On the other hand, a marginal increase of 0.2% of plant cost for immobilized lipase as compared to the soluble lipase is due to the investment in an entrapment unit for production of immobilized lipase. Almost comparable results were also reported in studies of plant investment costs for biodiesel production from palm oil (Jegannathan et al., 2011).

The manufacturing costs of 1 tonne, 10 tonnes, 100 tonnes, and 1000 tonnes of biodiesel from CJO using three routes different techniques (namely alkali, soluble lipase, and immobilized lipase) are shown in Table 3. A comparison between the three methods of production for 1 tonne biodiesel is shown in Fig. 10. As it is clear from the figure, the most economical method of biodiesel production is by using alkali catalyst since this technique is much cheaper (Zhang et al., 2010). The cost of 1 tonne of biodiesel production using soluble lipase was found to be higher ($8032) than the alkali method ($2143). This is due to the high cost of lipase (Bajaj et al., 2010; Yücel, 2011), and for soluble-lipase-catalyzed transesterification, a fresh batch of lipase has to be produced for each step, which requires more input of energy and raw materials (Robinson et al., 2001). On the other hand, there was a drastic increase in

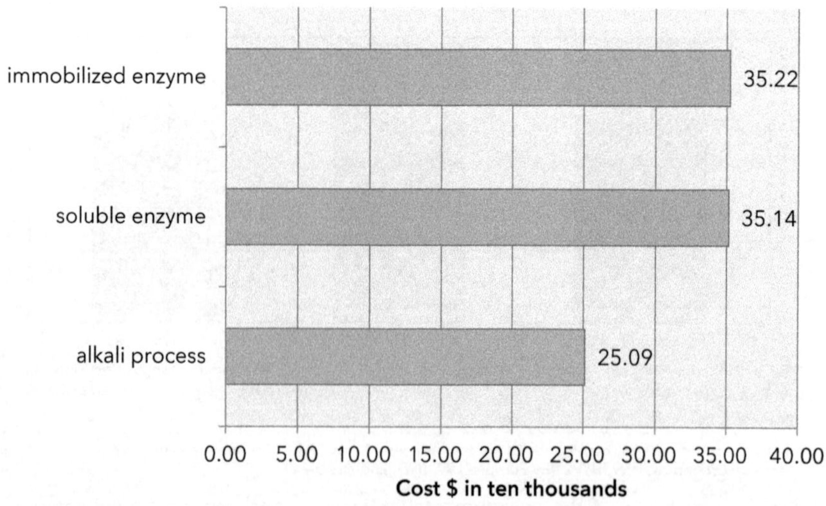

Figure 9 Plant investment costs for 1000 tonne capacity biodiesel production from CJO.

Table 3 Cost of production of biodiesel for varying capacities based on alkali, soluble lipase, and immobilized lipase catalyst

Type of catalyst	1 tonne ($)	10 tonne ($)	100 tonne ($)	1000 tonne ($)
Alkali	2143	21 430	21 43 000	21 43 000
Soluble lipase	8032	80 320	8 03 200	80 32 000
Immobilized lipase	2116 + 15 542[a] = 17 658	21 160	2 11 600 + (100*15 542) = 17 65 800	21 16 000

[a] Cost of immobilized lipase that can provide ten times reuse = $15 542.

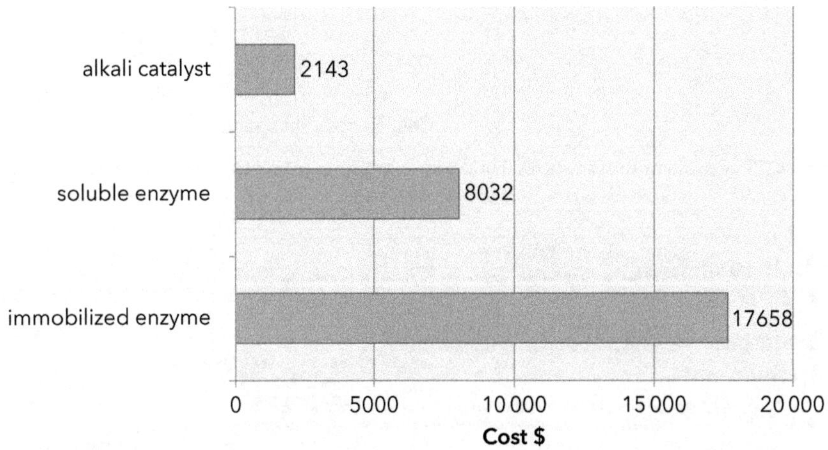

Figure 10 Manufacturing costs for 1 tonne capacity biodiesel production from CJO.

cost of production of 1 tonne of biodiesel by immobilized lipase ($17 658) as compared to alkali catalyst ($2349). This is due to the need for immobilization materials and techniques for biodiesel production. However, when immobilized lipase can be reused at least ten times, the cost of biodiesel production using immobilized lipase becomes comparable to the use of an alkali catalyst. This is clearly depicted in Fig. 11, which shows the manufacturing costs for 1000 tonnes of biodiesel production from CJO by using alkali catalyst, soluble lipase, and immobilized lipase. It can be seen that, on a large-scale, biodiesel production with soluble lipase is not at all feasible. Comparable results were obtained by Jegannathan et al. (2011) using immobilized lipase, which could be reused for five times in transesterification of palm oil for production of 1000 tonnes.

Commercially, biodiesel production using alkali catalysts is currently the most economical method of production. Biodiesel production using soluble lipase is costly because the lipase has to be manufactured for each batch of reaction and cannot be recovered. The high cost of lipase can be overcome

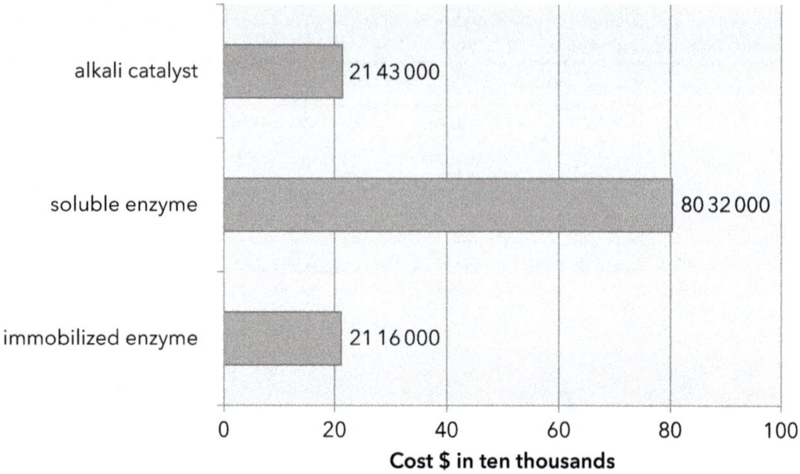

Figure 11 Manufacturing costs for 1000 tonne capacity biodiesel production from CJO.

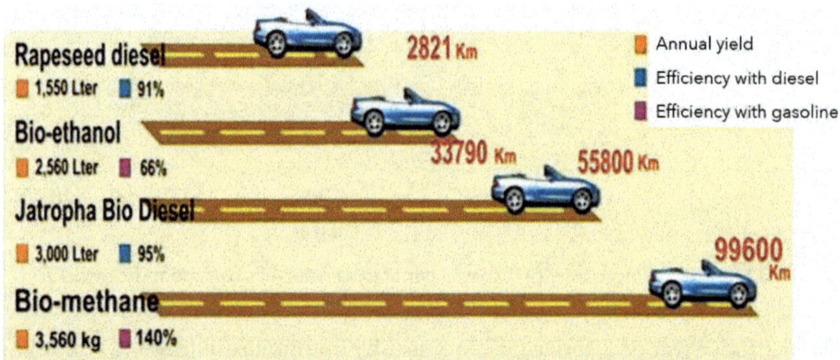

Figure 12 Economics of Jatropha biodiesel. Source: http://www.jatrophabiodiesel.org/economics.php.

by using lipase in immobilized forms. The cost of biodiesel produced by immobilized lipase can be made equivalent to the use of an alkali catalyst if the immobilized lipase can be reused at least ten times without loss of activity. The future of biodiesel production using immobilized lipase lies in the efficient design of immobilized lipase with the maximum number of reuses without loss of enzymatic activity.

The economic feasibility of CJO as biodiesel feedstock can be clearly seen from Fig. 12. The oil extracted from one hectare of *J. curcas* crops could provide biodiesel that can cover a distance of 55 800 km. Moreover, Jatropha biodiesel has 95% efficiency as compared to fossil diesel fuel. It is clear that Jatropha oil has great potential to become a cheaper and sustainable biodiesel feedstock in near future.

7 References

Abou Kheira, A. A. and Atta, N. M. M. 2009. Response of *Jatropha curcas* L. to water deficit: yield, water use efficiency and oilseed characteristics. *Biomass and Bioenergy* 33(10), 1343–50. doi:10.1016/j.biombioe.2008.05.015.

Achten, W. M. J., Mathijs, E., Verchot, L., Singh, V. P., Aerts, R. and Muys, B. 2007. *Jatropha* biodiesel fueling sustainability? *Biofuels, Bioproducts and Biorefining* 1(4), 283–91. doi:10.1002/bbb.39.

Achten, W. M. J., Almeida, J., Fobelets, V., Bolle, E., Mathijs, E., Singh, V. P., Tewari, D. N., Verchot, L. V. and Muys, B. 2010. Life cycle assessment of Jatropha biodiesel as transportation fuel in rural India. *Applied Energy* 87(12), 3652–60. doi:10.1016/j.apenergy.2010.07.003.

Agarwal, A. K. 2007. Biofuels (alcohols and biodiesel) applications as fuels for internal combustion engines. *Progress in Energy and Combustion Science* 33(3), 233–71. doi:10.1016/j.pecs.2006.08.003.

Ahmad, A. L., Yasin, N. H. M., Derek, C. J. C. and Lim, J. K. 2011. Microalgae as a sustainable energy source for biodiesel production: a review. *Renewable and Sustainable Energy Reviews* 15(1), 584–93. doi:10.1016/j.rser.2010.09.018.

Akintayo, E. T. 2004. Characteristics and composition of *Parkia biglobbossa* and *Jatropha curcas* oils and cakes. *Bioresource Technology* 92(3), 307–10. doi:10.1016/S0960-8524(03)00197-4.

Akoh, C. C., Chang, S. W., Lee, G. C. and Shaw, J. F. 2007. Enzymatic approach to biodiesel production. *Journal of Agricultural and Food Chemistry* 55(22), 8995–9005. doi:10.1021/jf071724y.

Almeida, J., Achten, W. M. J., Duarte, M. P., Mendes, B. and Muys, B. 2011. Benchmarking the environmental performance of the Jatropha biodiesel system through a generic life cycle assessment. *Environmental Science and Technology* 45(12), 5447–53. doi:10.1021/es200257m.

Al-Zuhair, S., Jayaraman, K. V., Krishnan, S. and Chan, W. H. 2006. The effect of fatty acid concentration and water content on the production of biodiesel by lipase. *Biochemical Engineering Journal* 30(2), 212–7. doi:10.1016/j.bej.2006.04.007.

Antczak, S. M., Kubiak, A., Antczak, T. and Bielecki, S. 2009. Enzymatic biodiesel synthesis - key factors affecting efficiency of the process. *Renewable Energy* 34(5), 1185–94. doi:10.1016/j.renene.2008.11.013.

Atabani, A. E., Silitonga, A. S., Ong, H. C., Mahlia, T. M. I., Masjuki, H. H., Badruddin, I. A. and Fayaz, H. 2013. Non-edible vegetable oils: a critical evaluation of oil extraction, fatty acid compositions, biodiesel production, characteristics, engine performance and emissions production. *Renewable and Sustainable Energy Reviews* 18, 211–45. doi:10.1016/j.rser.2012.10.013.

Augustus, G. D. P. S., Jayabalan, M. and Seiler, G. J. 2002. Evaluation and bioinduction of energy components of *Jatropha curcas*. *Biomass and Bioenergy* 23(3), 161–4. doi:10.1016/S0961-9534(02)00044-2.

Bajaj, A., Lohan, P., Jha, P. N. and Mehrotra, R. 2010. Biodiesel production through lipase catalyzed transesterification: an overview. *Journal of Molecular Catalysis B: Enzymatic* 62(1), 9–14. doi:10.1016/j.molcatb.2009.09.018.

Barnwal, B. K. and Sharma, M. P. 2005. Prospects of biodiesel production from vegetable oils in India. *Renewable and Sustainable Energy Reviews* 9(4), 363–78. doi:10.1016/j.rser.2004.05.007.

Becker, K. and Makkar, H. P. S. 2008. *Jatropha curcas*: a potential source for tomorrow's oil and biodiesel. *Lipid Technology* 20(5), 104–7. doi:10.1002/lite.200800023.

Berchmans, H. J. and Hirata, S. 2008. Biodiesel production from crude *Jatropha curcas* seed oil with a high content of free fatty acids. *Bioresource Technology* 99(6), 1716–21. doi:10.1016/j.biortech.2007.03.051.

Bernardes, O. L., Bevilaqua, J. V., Leal, M. C., Freire, D. M. G. and Langone, M. A. P. 2007. Biodiesel fuel production by the transesterification reaction of soybean oil using immobilized lipase. *Applied Biochemistry and Biotechnology* 137–140(1–12), 105–14. doi:10.1007/s12010-007-9043-5.

Bommarius, A. S. and Riebel, B. R. 2005. *Biocatalysis, Fundamentals and Applications*. Wiley Publishers.

Burkill, I. H. 1966. *A Dictionary of the Economic Products of the Malay Peninsula* (2nd edn.). Ministry of Agriculture and Co-operatives, Kuala Lumpur, Malaysia.

Chen, J. W. and Wu, W. T. 2003. Regeneration of immobilized *Candida antarctica* lipase for transesterification. *Journal of Bioscience and Bioengineering* 95(5), 466–9. doi:10.1016/s1389-1723(03)80046-4.

Chen, G., Ying, M. and Li, W. 2006. Enzymatic conversion of waste-cooking oils into alternative fuel biodiesel. *Applied Biochemistry and Biotechnology* 129–132, 911–21. doi:10.1385/abab:132:1:911.

Consoli, F., Allen, D., Boustead, I., Fava, J., Franklin, W. and Jensen, A. A. 1993. A code of practice. *Guidelines for Life-Cycle Assessment: A Code of Practice*. Society of Environmental Toxicology and Chemistry, SETAC Publisher.

Deng, L., Xu, X. B., Haraldsson, G. G., Tan, T. W. and Wang, F. 2005. Enzymatic production of alkyl esters through alcoholysis: a critical evaluation of lipases and alcohols. *Journal of the American Oil Chemists' Society* 82(5), 341–7. doi:10.1007/s11746-005-1076-3.

Devanesan, M. G., Viruthagiri,T. and Sugumar, N. 2007. Transesterification of Jatropha oil using immobilized *Pseudomonas fluorescens*. *African Journal of Biotechnology* 6(21), 2497–501. doi:10.5897/AJB2007.000-2396.

Du, W., Xu, Y. Y., Liu, D. H. and Zeng, J. 2004. Comparative study on lipase-catalyzed transformation of soybean oil for biodiesel production with different acyl acceptors. *Journal of Molecular Catalysis B: Enzymatic* 30(3–4), 125–9. doi:10.1016/j.molcatb.2004.04.004.

Fjerbaek, L., Christensen, K. V. and Norddahl, B. 2009. A review of the current state of biodiesel production using enzymatic transesterification. *Biotechnology and Bioengineering* 102(5), 1298–315. doi:10.1002/bit.22256.

Foidl, N., Foidl, G., Sanchez, M., Mittelbach, M. and Hackel, S. 1996. *Jatropha curcas* L. as a source for the production of biofuel in Nicaragua. *Bioresource Technology* 58(1), 77–82. doi:10.1016/S0960-8524(96)00111-3.

Francis, G., Edinger, R. and Becker, K. 2005. A concept for simultaneous wasteland reclamation, fuel production, and socio economic development in degraded areas in India: need, potential and perspectives of *Jatropha* plantations. *Natural Resources Forum* 29(1), 12–24. doi:10.1111/j.1477-8947.2005.00109.x.

Freedman, B., Pryde, E. H. and Mounts, T. L. 1984. Variables affecting the yields of fatty esters from transesterified vegetable oils. *Journal of the American Oil Chemists Society* 61(10), 1638–43. doi:10.1007/BF02541649.

Fukuda, H., Kondo, A. and Noda, H. 2001. Biodiesel fuel production by transesterification of oils. *Journal of Bioscience and Bioengineering* 92(5), 405–16. doi:10.1263/jbb.92.405.

Fukuda, H., Hama, S., Tamalampudi, S. and Noda, H. 2008. Whole-cell biocatalysts for biodiesel fuel production. *Trends in Biotechnology* 26(12), 668–73. doi:10.1016/j.tibtech.2008.08.001.

Ghaly, A. E., Dave, D., Brooks, M. S. and Budge, S. 2010. Production of biodiesel by enzymatic transesterification: review. *American Journal of Biochemistry and Biotechnology* 6(2), 54–76. doi:10.3844/ajbbsp.2010.54.76.

Gog, A., Roman, M., Tosa, M., Paizs, C. and Irimie, F. D. 2012. Biodiesel production using enzymatic transesterification- current state and perspectives. *Renewable Energy* 39(1), 10–6. doi:10.1016/j.renene.2011.08.007.

Gui, M. M., Lee, K. T. and Bhatia, S. 2008. Feasibility of edible oil vs. non-edible oil vs. waste edible oil as biodiesel feedstock. *Energy* 33(11), 1646–53. doi:10.1016/j.energy.2008.06.002.

Guinée, J. B. 2002. *Handbook on Life Cycle Assessment: Operational Guide to the ISO Standards*. Springer.

Haas, M. J., Piazza, G. J. and Foglia, T. A. 2003. Enzymatic approaches to the production of biodiesel fuels. In: Kuo, T. M. and Gardner, H. W. (Eds), *Lipid Biotechnology*. Marcel Dekker Inc., New York, NY, pp. 587–98.

Harding, K. G., Dennis, J. S., Blottnitz, H. and Harrison, S. T. L. 2008. A life cycle comparison between inorganic and biological catalysis for the production of biodiesel. *Journal of Cleaner Production* 16(13), 1368–78. doi:10.1016/j.jclepro.2007.07.003.

Heller, J. 1996. Promoting the conservation and use of underutilized and neglected crops. 1. Physic nut, *Jatropha curcas* L. International Plant Genetic Resources Institute, Rome, pp. 13–5.

Ibeto, C. N., Ofoefule, A. U. and Ezugwu, H. C. 2011. Analytical methods for quality assessment of biodiesel from animal and vegetable oils. *Trends in Applied Sciences Research* 6(6), 537–53. doi:10.3923/tasr.2011.537.553.

Jaeger, K. E. and Eggert, T. 2002. Lipases for biotechnology. *Current Opinion in Biotechnology* 13(4), 390–7. doi:10.1016/s0958-1669(02)00341-5.

Jegannathan, K. R., Abang, S., Poncelet, D., Chan, E. S. and Ravindra, P. 2008. Production of biodiesel using immobilized lipase–A critical review. *Critical Reviews in Biotechnology* 28(4), 253–64. doi:10.1080/07388550802428392.

Jegannathan, K. R., Ting, V. F. W. and Ravindra, P. 2011. Life cycle assessment of biodiesel production using alkali, soluble and immobilized enzyme catalyst processes. *Biomass and Bioenergy* 35(10), 4221–9. doi:10.1016/j.biombioe.2011.07.010.

Jeong, G. T. and Park, D. H. 2008. Lipase-catalyzed transesterification of rapeseed oil for biodiesel production with tert-butanol. *Applied Biochemistry and Biotechnology* 148, 131–9.

Jones, N. and Miller, J. H. 1992. *Jatropha curcas*: a multipurpose species for problematic sites. *Land and Resources* 1, 1–12; Washington, DC, World Bank.

Kaieda, M., Samukawa, T., Kondo, A. and Fukuda, H. 2001. Effect of methanol and water contents on production of biodiesel fuel from plant oil catalyzed by various lipases in a solvent-free system. *Journal of Bioscience and Bioengineering* 91(1), 12–5. doi:10.1263/jbb.91.12.

Kiwjaroun, C., Tubtimdee, C. and Piumsomboon, P. 2009. LCA studies comparing biodiesel synthesized by conventional and supercritical methanol methods. *Journal of Cleaner Production* 17(2), 143–53. doi:10.1016/j.jclepro.2008.03.011.

Kumar, A. and Sharma, S. 2008. An evaluation of multipurpose oil seed crop for industrial uses (*Jatropha curcas* L.): a review. *Industrial Crops and Products* 28(1), 1–10. doi:10.1016/j.indcrop.2008.01.001.

Kumar, A. and Sharma, S. 2011. Potential non-edible oil resources as biodiesel feedstock: an Indian perspective. *Renewable and Sustainable Energy Reviews* 15(4), 1791–800. doi:10.1016/j.rser.2010.11.020.

Lam, M. K., Lee, K. T. and Mohamed, A. R. 2009. Life cycle assessment for the production of biodiesel: a case study in Malaysia for palm oil versus Jatropha oil. *Biofuels, Bioproducts and Biorefining* 3(6), 601–12. doi:10.1002/bbb.182.

Leung, D. Y. C., Wu, X. and Leung, M. K. H. 2010. A review on biodiesel production using catalyzed transesterification. *Applied Energy* 87(4), 1083–95. doi:10.1016/j.apenergy.2009.10.006.

Lin, L., Cunshan, Z., Vittayapadung, S., Xiangqian, S. and Mingdong, D. 2011. Opportunities and challenges for biodiesel fuel. *Applied Energy* 88(4), 1020–31, ISSN 0306-2619, doi:10.1016/j.apenergy.2010.09.029.

Linnaeus, C. 1753. Species plantarum. In: Jatropha (Ed). *Laurentii Salvii*, pp. 1006–7. Impensis, Stockholm.

Lu, J., Nie, K., Xie, F., Wang, F. and Tan, T. 2007. Enzymatic synthesis of fatty acid methyl esters from lard with immobilized Candida sp. 99-125. *Process Biochemistry* 42(9), 1367–70. doi:10.1016/j.procbio.2007.06.004.

Lu, J., Nie, K., Wang, F. and Tan, T. 2008. Immobilized lipase *Candida* sp. 99-125 catalyzed methanolysis of glycerol trioleate: solvent effect. *Bioresource Technology* 99(14), 6070–4. doi:10.1016/j.biortech.2007.12.045.

Ma, F. and Hanna, M. A. 1999. Biodiesel production: a review. *Bioresource Technology* 70(1), 1–15. doi:10.1016/S0960-8524(99)00025-5.

Makkar, H. P. S., Becker, K., Sporen, F. and Wink, M. 1997. Studies on nutritive potential and toxic constituents of different provenances of *Jatropha curcas* L. *Journal of Agricultural and Food Chemistry* 45(8), 3152–7. doi:10.1021/jf970036j.

Marchetti, J. M., Miguel, V. U. and Errazu, A. F. 2007. Possible methods for biodiesel production. *Renewable and Sustainable Energy Reviews* 11(6), 1300–11. doi:10.1016/j.rser.2005.08.006.

Meher, L. C., Sagar, V. D. and Naik, S. N. 2006. Technical aspects of biodiesel production by transesterification – a review. *Renewable and Sustainable Energy Reviews* 10, 248–68.

Mittelbach, M. and Tratnigg, B. 1990. Kinetics of alkaline catalyzed methanolysis of sunflower oil. *Fett Wissenschaft Technologie/Fat Science Technology* 92(4), 145–8. doi:10.1002/lipi.19900920405.

Modi, M. K., Reddy, J. R. C., Rao, B. V. S. K. and Prasad, R. B. N. 2007. Lipase-mediated conversion of vegetable oils into biodiesel using ethyl acetate as acyl acceptor. *Bioresource Technology* 98(6), 1260–4. doi:10.1016/j.biortech.2006.05.006.

Mofijur, M., Masjuki, H. H., Kalam, M. A., Hazrat, M. A., Liaquat, A. M., Shahabuddin, M. and Varman, M. 2012. Prospects of biodiesel from Jatropha in Malaysia. *Renewable and Sustainable Energy Reviews* 16(7), 5007–20. doi:10.1016/j.rser.2012.05.010.

Morin, P., Hamad, B., Sapaly, G., Rocha, M. G., Oliveira, P. G., Gonzalez, W. A., Andrade, S. E. and Essayem, N. 2007. Transesterification of rapeseed oil with ethanol: I. Catalysis with homogeneous Keggin heteropolyacids. *Applied Catalysis. Part A, General* 33, 69–76.

Murugesan, A., Umarani, C., Chinnusamy, T. R., Krishnan, M., Subramanian, R. and Neduzchezhain, N. 2009. Production and analysis of bio-diesel from non-edible oils–a review. *Renewable and Sustainable Energy Reviews* 13(4), 825–34. doi:10.1016/j.rser.2008.02.003.

Ndong, R., Montrejaud-Vignoles, M., Saint Girons, O., Gabrielle, B., Pirot, R., Domergue, M. and Sablayrolles, C. 2009. Life cycle assessment of biofuels from *Jatropha curcas* in West Africa: a field study. *GCB Bioenergy* 1(3), 197–210. doi:10.1111/j.1757-1707.2009.01014.x.

Nelson, L. A., Foglia, T. A. and Marmer, W. N. 1996. Lipase-catalyzed production of biodiesel. *Journal of the American Oil Chemists' Society* 73(9), 1191–95. doi:10.1007/BF02523383.

Noureddini, H. and Zhu, D. 1997. Kinetics of transesterification of soybean oil. *Journal of the American Oil Chemists' Society* 74(11), 1457–63. doi:10.1007/s11746-997-0254-2.

Noureddini, H., Gao, X. and Philkana, R. S. 2005. Immobilized *Pseudomonas cepacia* lipase for biodiesel fuel production from soybean oil. *Bioresource Technology* 96(7), 769–77. doi:10.1016/j.biortech.2004.05.029.

Orçaire, O., Buisson, P. and Pierre, A. C. 2006. Application of silica aerogel encapsulated lipases in the synthesis of biodiesel by transesterification reactions. *Journal of Molecular Catalysis B: Enzymatic* 42(3–4), 106–13. doi:10.1016/j.molcatb.2006.08.002.

Otero, C., Berrendero, M. A., Cardenas, F., Alvarez, E. and Elson, S. W. 2005. General characterization of non-commercial microbial lipases in hydrolytic and synthetic reactions. *Applied Biochemistry and Biotechnology* 120(3), 209–24. doi:10.1385/abab:120:3:209.

Pandey, V. C., Singh, K., Singh, J. S., Kumar, A., Singh, B. and Singh, R. P. 2012. *Jatropha curcas*: a potential biofuel plant for sustainable environmental development. *Renewable and Sustainable Energy Reviews* 16(5), 2870–83. doi:10.1016/j.rser.2012.02.004.

Pehnt, M. 2006. Dynamic life cycle assessment (LCA) of renewable energy technologies. *Renewable Energy* 31(1), 55–71. doi:10.1016/j.renene.2005.03.002.

Pennington, D. W., Potting, J., Finnveden, G., Lindeijer, E., Jolliet, O., Rydberg, T. and Rebitzer, G. 2004. Life cycle assessment–part 2: current impact assessment practice. *Environment International* 30(5), 721–39. doi:10.1016/j.envint.2003.12.009.

Quirrin, M., Gartner, S. O., Pehnt, M. and Reinhardt, G. 2006. *CO2 Mitigation through Biofuels in the Transport Sector*. Technical Report. IFEU, Heidelberg.

Rahmath, A., Chan, E. S. and Ravindra, P. 2011. Biodiesel production from *Jatropha curcas*-a critical review. *Critical Reviews in Biotechnology* 31(1), 53–64. doi:10.3109/07388551.2010.487185.

Ranganathan, S. V., Narasimhan, S. L. and Muthukumar, K. 2008. An overview of enzymatic production of biodiesel. *Bioresource Technology* 99(10), 3975–81. doi:10.1016/j.biortech.2007.04.060.

Ravindra, P., Saralan, S. and Rahmath, A. 2012. LCA studies for alkaline and enzyme catalyzed biodiesel production from palm oil. *Advances in Biological Chemistry* 2, 341–52.

Rebitzer, G., Ekvall, T., Frischknecht, R., Hunkeler, D., Norris, G., Rydberg, T., Schmidt, W. P., Suh, S., Weidema, B. P. and Pennington, D. W. 2004. Life cycle assessment part 1: framework, goal and scope definition, inventory analysis, and applications. *Environment International* 30(5), 701–20. doi:10.1016/j.envint.2003.11.005.

Robinson, T., Singh, D. and Nigam, P. 2001. Solid-state fermentation: a promising microbial technology for secondary metabolite production. *Applied Microbiology and Biotechnology* 55(3), 284–9. doi:10.1007/s002530000565.

Robles, A. M., González-Moreno, P. A., Esteban-Cerdán, L. and Molina-Grima, E. 2009. Biocatalysis: towards ever greener biodiesel production. *Biotechnology Advances* 27(4), 398–408. doi:10.1016/j.biotechadv.2008.10.008.

Romijn, H. A. 2011. Land clearing and greenhouse gas emissions from Jatropha biofuels on African Miombo Woodlands. *Energy Policy* 39(10), 5751–62. doi:10.1016/j.enpol.2010.07.041.

Rowe, D. G. 2011. Beyond food versus fuel. *Nature* 474, 6–8.

Samukawa, T., Kaieda, M., Matsumoto, T., Ban, K., Kondo, A., Shimada, Y., Noda, H. and Fukuda, H. 2000. Pretreatment of immobilized *Candida antarctica* lipase for biodiesel fuel production from plant oil. *Journal of Bioscience and Bioengineering* 90(2), 180–3. doi:10.1016/S1389-1723(00)80107-3.

Shah, S. and Gupta, M. N. 2007. Lipase catalyzed preparation of biodiesel from Jatropha oil in a solvent free system. *Process Biochemistry* 42(3), 409–14. doi:10.1016/j.procbio.2006.09.024.

Shah, S., Sharma, S. and Gupta, M. N. 2004. Biodiesel preparation by lipase catalyzed transesterification of Jatropha oil. *Energy and Fuels* 18(1), 154–9. doi:10.1021/ef030075z.

Shimada, Y., Watanabe, Y., Sugihara, A. and Tominaga, Y. 2002. Enzymatic alcoholysis for biodiesel fuel production and application of the reaction oil processing. *Journal of Molecular Catalysis B: Enzymatic* 17(3–5), 133–42. doi:10.1016/S1381-1177(02)00020-6.

Silitonga, A. S., Atabani, A. E., Mahlia, T. M. I., Masjuki, H. H., Badruddin, I. A. and Mekhilef, S. 2011. A review on prospect of *Jatropha curcas* for biodiesel in Indonesia. *Renewable and Sustainable Energy Reviews* 15(8), 3733–56. doi:10.1016/j.rser.2011.07.011.

Soumanou, M. M. and Bornscheuer, U. T. 2003. Improvement in lipase-catalyzed synthesis of fatty acid methyl esters from sunflower oil. *Enzyme and Microbial Technology* 33(1), 97–103. doi:10.1016/S0141-0229(03)00090-5.

Srinivasan, S. 2009. The food v. fuel debate: a nuanced view of incentive structures. *Renewable Energy* 34(4), 950–4. doi:10.1016/j.renene.2008.08.015.

Thurmond, W. 2008. Biodiesel 2020, a global market survey. *Feedstock trends and forecasts. Multi-Client Study* (2nd edn).

Varanda, M. G., Pinto, G. and Martins, F. 2011. Life cycle analysis of biodiesel production. *Fuel Processing Technology* 92(5), 1087–94. doi:10.1016/j.fuproc.2011.01.003.

Xu, Y. Y., Du, W., Liu, D. H. and Zheng, J. 2003. A novel enzymatic route for biodiesel production from renewable oils in a solvent-free medium. *Biotechnology Letters* 25(15), 1239–41. doi:10.1023/A:1025065209983.

Yaakob, Z., Mohammad, M., Alherbawi, M., Alam, Z. and Sopian, K. 2013. Overview of the production of biodiesel from Waste cooking oil. *Renewable and Sustainable Energy Reviews* 18, 184–93. doi:10.1016/j.rser.2012.10.016.

Ye, P., Xu, Z. K., Wu, J., Innocent, C. and Seta, P. 2006. Nanofibrous poly(acrylonitrile-co-maleic acid) membranes functionalized with gelatin and chitosan for lipase immobilization. *Biomaterials* 27(22), 4169–76. doi:10.1016/j.biomaterials.2006.03.027.

Yücel, Y. 2011. Biodiesel production from pomace oil by using lipase immobilized onto olive pomace. *Bioresource Technology* 102(4), 3977–80. doi:10.1016/j.biortech.2010.12.001.

Zhang, J., Chen, S., Yang, R. and Yan, Y. 2010. Biodiesel production from vegetable oil using heterogenous acid and alkali catalyst. *Fuel* 89(10), 2939–44. doi:10.1016/j.fuel.2010.05.009.

Chapter 6

Production of biodiesel from oilseeds: canola/rapeseed

B. Brian He and Dev Shrestha, University of Idaho, USA

1 Introduction

Biodiesel is produced from chemicals called triglycerides which are the dominant components of vegetable oils and animal fats. Since biodiesel has been commercialized and widely adopted by the public, the demand for a sufficient and consistent supply of feedstock has become the decisive factor for a sustainable biodiesel industry. Vegetable oil or animal fat, either virgin and/or used, can be a biodiesel feedstock. Currently, rapeseed oil is the dominant feedstock for biodiesel production in Europe, while soybean oil is the major feedstock in the US. Canola oil is also extensively used for biodiesel production in the Pacific Northwest (PNW) of the US and Canada.

Rapeseed, also known as oilseed rape, is a member of the mustard (*Brassica*) family. Cultivars of the *Brassica* family adapt well to a wide variety of climate and soil conditions and are widely planted around the globe. The mustard family includes the oilseed crops mustard (*B. juncea*), oilseed rape (*B. napus*) and turnip rape (*B. rapa*). *B. napus* is the predominant species grown in Asia, Australia, Europe and North America, while *B. juncea* is the predominant species grown in the Indian subcontinent. The mustard family also includes vegetable crops such as cabbage and cauliflower (*B. oleracea*) and turnip (*B. rapa*). Historically, rapeseed varieties contained medium to high levels of erucic acid (30–50%) in the oil and high levels of glucosinolates (80–150 mmol/g) in the meal. Genetic modification of rapeseed has produced several new types of rapeseed including canola (or 'double zero' or 00 rapeseed), high oleic canola (or high oleic 00 rapeseed) and high erucic acid, low glucosinolate rapeseed

http://dx.doi.org/10.19103/AS.2019.0027.13

(HEAR). Varieties of canola are the Canadian modified cultivars of rapeseed in which the contents of erucic acid and glucosinolates are very low (Canola Council of Canada, 2018). Cultivars of rapeseed in Europe have also been modified to contain very low contents of erucic acid and glucosinolates and are equivalent to Canadian canola types but have traditionally been called 'doube zero' (00) rapeseed instead of canola. Canola and 00 rapeseed varieties grown in North America and Europe produce high-quality edible oil and meal. HEAR cultivars grown in North America and Europe produce a unique oil, which is used for numerous industrial purposes, and a low glucosinolate meal. Rapeseed is still being grown in Asia and the Indian subcontinent; however, the transition to canola (00 rapeseed) is in process (Hayward, 2011; Kumar et al., 2015).

For clarity and simplicity, 'canola' will be used to describe all low erucic acid, low glucosinolate rapeseed varieties while 'rapeseed' will be used to describe both traditional rapeseed (with medium to high erucic acid content and high glucosinolate content) and HEAR for the remainder of this chapter.

Canola is the major oilseed crop in Canada, with a total production of more than 21 million metric tons in 2017, a 70% increase over that of 2007. Among the total production, approximately 43–44% were crushed for oil in Canada (Statistics Canada, 2018a). According to Statistics Canada, the yield of canola has been increasingly improving through the years. In the past 5 years, the yield of canola was in the range of 2000–2400 kg/ha (1800–2100 lb/acre) (Table 1). Based on the average oil content of 42–44%, the canola oil yield is in the range of 1000–1100 L/ha (90–120 gal/acre), indicating that canola is a crop of high oil yield and a great choice for biodiesel production. In the past few years, canola oil has remained the major feedstock for biodiesel production in Canada, at an annual production of 300 million L (approx. 80 million gal).

Table 1 Canola, canola oil and biodiesel in Canada

	Calendar year				
	2013	2014	2015	2016	2017
Total production (million metric tons)[a]	18.5	16.4	18.4	19.6	21.4
Oil produced (million metric tons)[a]	2.83	3.12	3.36	3.86	4.05
Seed yield (kg/ha)[a]	2300	2000	2200	2400	2300
Oil content (%)[b]	43.5	43.6	43.5	43.6	44.1
Oil yield (L/ha)[c]	1088	948	1040	1137	1103
Biodiesel production (million L)[d]	307	269	286	267	307

[a] Data from Statistics Canada (2018a,b).
[b] Calculated from data of total seed and oil productions above.
[c] Estimated by corresponding oil contents.
[d] Raw data from USDA FAS (2018a).

The European Union (EU) is the world's largest producer of canola. According to the USDA Foreign Agricultural Services (USDA FAS), the annual production of canola in the 28 countries of the EU (EU-28) was at the level of 20–24 million metric tons (44–53 billion pounds) for the past 5 years (USDA FAS, 2018b,c). It is noticed that the yield of canola in the EU-28 is much higher than that in Canada or the US.

Canola oil was widely used for biodiesel production in the EU-28, approximately 60% of total canola oil production (Table 2). Presently, canola oil is beginning to be used in a notable quantity for renewable diesel production (USDA FAS, 2018e).

Canola production in the US is mainly in the northern states of Minnesota, Montana and North Dakota, as well as the Pacific Northwest (PNW) region of Idaho, Oregon and Washington. North Dakota produced more than 80% of the canola in the US in the past 5 years (Table 3). Although not all canola oil produced domestically is used for biodiesel production, because of the seed/oil imports from Canada, the total quantity of canola oil for biodiesel production in the US is approximately equal to that of US canola oil production. As the total biodiesel production in the US is in the range of 1300–1600 million gallons/year, canola oil contributes approximately 10% of the total feedstock (Table 3).

Just like other types of vegetable oils, canola oil is mainly composed of triglycerides (also called triacylglycerides), which are the esters of glycerol and different fatty acids. Virgin canola and rapeseed oil normally contain very low levels of free fatty acids and other heterogeneous chemicals, making them high-quality feedstocks for biodiesel production. Because of the low level of free fatty acids, soap formation is typically not a concern in converting canola/rapeseed oil to biodiesel.

Table 2 Canola and canola oil in EU-28

	Calendar year				
	2013	2014	2015	2016	2017
Total seed production (million metric tons)[a]	21.0	24.1	22.0	20.4	21.0
Oil produced (million metric tons)[a]	10.00	10.20	10.16	9.88	9.88
Oil content (%)[b]	41.8	41.6	41.8	41.7	41.7
Seed yield (kg/ha)[a]	3126	3577	3379	3119	3189
Oil yield (L/ha)[c]	1428	1626	1544	1421	1453
Oil used for biodiesel (million metric tons)[d]	5.77	6.17	5.97	5.97	6.15[e]
Total EU biodiesel production (billion L)[d]	10.4	11.2	11.5	11.6	11.98[e]

[a] Data from USDA FAS (2018b).
[b] Calculated from data of total seed and oil productions above.
[c] Estimated using oil density of 0.915 kg/L.
[d] Raw data from USDA FAS (2018d).
[e] Raw data from USDA FAS (2018e). Data includes renewable diesel.

Table 3 Canola and canola oil in the US

	Calendar year				
	2013	2014	2015	2016	2017
Total US seed production (million pounds)[a]	2210	2512	2875	3086	3119
Total ND seed production (million pounds)[a]	1.665	2.142	2.492	2.668	2.542
Seed yield (US average) (lb/acre)[a]	1748	1614	1680	1824	1588
Total US oil production (million pounds)[b]	884	1005	1150	1234	1248
Oil yield (US average) (gal/acre)[b]	89	82	85	92	80
Oil used for biodiesel (million pounds)[c]	646	1046	1057	1130	1452
Oil used for biodiesel (million gal)[d]	85	137	139	148	190
Total US. biodiesel production (million gal)[c]	1359	1271	1268	1569	1596

[a] Raw data from USDA NASS (2018a,b).
[b] Estimated by assuming that oil is extracted at 40%.
[c] Data from US EIA (2018a,b,c).
[d] Estimated using oil density of 7.63 lb/gal.

Vegetable oils are characterized by the long chains of fatty acids that are present in the triglycerides. The nomenclature for identifying fatty acids is commonly depicted as $C_{X:Y}$ where X is the number of carbon atoms and Y is the number of double bonds in the fatty acid chain. The fatty acid profiles of canola and rapeseed oils are shown in Table 4. Traditional rapeseed is not planted these days in Europe and North America but still planted in other parts of the world, such as in some Asian countries. In those regions these cultivars are often referred to as oriental rapeseed, due to their wide adaptability to climate conditions and soil types, resistibility to weeds and pests and high yield. The oil from oriental rapeseed can be also used to produce biodiesel.

The fatty acid profiles of canola and rapeseed oils are generally consistent in the ranges as shown in Table 4; however, they vary depending on the

Table 4 Fatty acid compositions of canola and rapeseed oils

	Fatty acid profiles[a,b]						
Oils	$C_{16:0}$	$C_{18:0}$	$C_{18:1}$	$C_{18:2}$	$C_{18:3}$	$C_{20:0}$	$C_{22:1}$
Canola	1-3	2-3	50-60	15-25	8-12		
Canola, high oleic	1-3	2-3	60-80	15-30	10-15		
Rapeseed, HEAR[c]	1-3	0-5	10-15	12-15	8-12	7-10	50-60
Oriental rapeseed	1-3	5-10	10-20	12-15	8-12	7-10	35-45
Soybean oil[d]	6-10	2-5	20-30	50-60	5-11		
Soybean, high oleic	2-3	2-3	80-85	3-4	3-5		

[a] Composed from various sources.
[b] The rest are other fatty acids in small quantities, making up to 100%.
[c] High erucic acid low glucosinolate.
[d] Soybean oil is listed for comparison.

conditions of weather, time, place and so on. Even for the very same cultivar planted in the same place, the fatty acid profile changes depending on the annual rainfall, use of irrigation and time of harvest. For example, the Canola Council of Canada (2018) gives an example of a fatty acid profile of a high oleic canola variety, which contains 3% palmitic acids (C16:0), 57% oleic acid (C18:1), 26% linoleic acid (C18:2) and 11% linolenic acid (C18:3).

Life cycle analysis (LCA) is a tool in evaluating the performance of a biofuel relative to its petroleum fuel counterpart, regarding reductions in greenhouse gases (GHG) emissions and energy balance of a biofuel production system. According to the Energy Independence Act (EISA) of 2007, a biofuel qualifies as an 'advanced biofuel' if it reduces GHG emissions by at least 50% compared to baseline petroleum fuel. It allows the producer to receive higher monetary incentives through tax credits or through generation of Renewable Identification Numbers (RINs) as per the Renewable Fuel Standard (RFS2) programme. Biodiesel RINs guarantee the market for a certain volume of fuel set by the EPA each year. In 2018, the mandated biomass-based diesel volume was 2.1 billion gallons. In addition to the RINs, between 2005 and 2016, the US Congress allowed biodiesel blenders to claim a credit of $1/gallon against their US federal tax liability. The biodiesel tax credit has not been reinstated for 2017 and after.

2 Current technologies and constraints of biodiesel production

Biodiesel is made from renewable plant and animal sources such as soybean oil, animal fats and waste vegetable oils. Biodiesel burns in diesel engines with the same efficiency as petroleum-based diesel fuel. Its high flash point makes it safer to use and store. Biodiesel generally has a higher cetane number than petroleum diesel fuel, in the range of 45–55 for soybean biodiesel and 49–62 for canola/rapeseed biodiesel (Mittelbach and Remschmidt, 2005). Environmentally, biodiesel is non-toxic and biodegradable. It contains very low levels of sulphur and nitrogen, and tailpipe emissions, particularly smoke, are much lower than petroleum diesel. It can be blended with petroleum diesel in any ratio and can utilize existing distribution infrastructure, all with no engine modifications. Biodiesel possesses excellent lubricity. Even low levels of biodiesel, for example 2%, with petroleum diesel fuel, will improve the lubricity of the blend to a satisfactory level and thus extend the engine life (Van Gerpen et al., 2006).

However, biodiesel has some limitations too. Biodiesel contains less energy than petroleum diesel, approximately 8% less per unit volume or 12% less per unit mass. Biodiesel generally has higher cloud and pour points, which makes it less favourable to use in low temperature environments. Biodiesel has good biodegradability, but it is also less stable chemically, thermally and oxidatively than petroleum diesel (Jain and Sharma, 2011). Biodiesel is a stronger solvent than petroleum diesel. It may not be compatible with some fuel lines and/or gasket materials in engines and could cause degradation and other compatibility problems. Biodiesel may also be incompatible with some metals and cause corrosion problems (Diaz-Ballote et al., 2009; Hu et al., 2012; Singh et al., 2012). Another drawback of biodiesel is possibly elevated nitrogen oxide emissions, a topic which is controversial.

2.1 Chemistry of biodiesel production

Biodiesel is defined as a fuel comprised of monoalkyl esters of long chain fatty acids derived from vegetable oils or animal fats, designated B100 (ASTM, 2015). The monoalkyl esters are the products of an oil or fat (chemically composed of triglycerides) reacted with an alcohol, typically methanol. The by-product of this reaction is glycerol. When biodiesel is blended with fossil diesel, the notation B*xx* is used, where *xx* stands for the percentage of biodiesel blended by volume. For example, a blend of 20% biodiesel with 80% fossil diesel is termed B20.

The chemical process to produce monoalkyl esters from oil or fat with an alcohol is known as transesterification. Methanol is the preferred alcohol for making biodiesel because it is inexpensive and more reactive compared to other alcohols such as ethanol. For transesterification to occur in a reasonable time, a catalyst is added to the oil/fat and methanol mixture. A generic form of transesterification is shown in Fig. 1.

It is shown in Fig. 1 that stoichiometrically, for every mole of triglyceride, three moles of methanol are needed, and three moles of methyl esters are produced. Additionally, a mole of by-product glycerol is formed.

Transesterification is a reversible reaction. The reaction reaches a dynamic equilibrium up to a given condition. In other words, all chemicals coexist in the system including the to-be-converted triglycerides. Since the goal is to convert

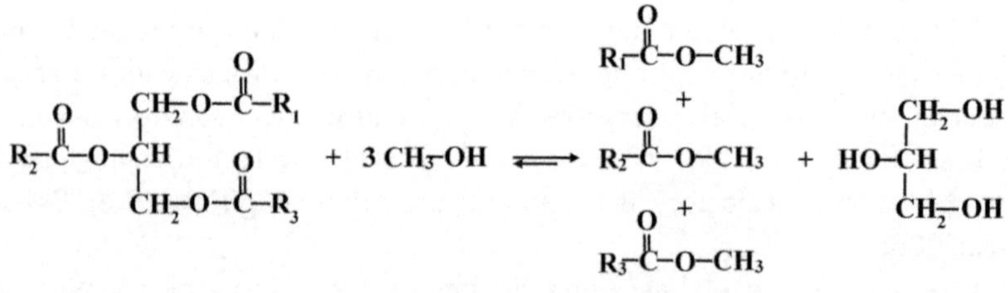

(a) triglyceride (b) methanol (c) fatty acid methyl esters (d) glycerol

Figure 1 Transesterification of oil/fat with methanol.

all triglycerides to biodiesel, the equilibrium must be shifted towards the right side of the chemical reaction.

To promote the equilibrium proceeding in the direction of the product side, one commonly used technique is to increase the molar ratio of methanol application more than its stoichiometric requirement. As a result, the equilibrium of the transesterification reaction shifts to the right and the conversion of triglycerides to esters reaches near completion. In base-catalysed transesterifications of low free fatty acid feedstock such as virgin canola or rapeseed oil, using an excess quantity of methanol of six moles for every mole of triglycerides is a common industrial practice.

Looking closely at the transesterification of Fig. 1, the chemical reaction proceeds step-wise, namely, the triglycerides are transesterified into di-glycerides first and then di-glycerides to mono-glycerides in series, as illustrated in Fig. 2. Besides triglycerides, mono- and di-glycerides can also exist as intermediates. The goal in biodiesel production is to obtain as pure a product (methyl esters) as possible. If the intermediates, either mono- or di-glycerides, are present in the final product in excess, biodiesel does not meet the ASTM 6751 standard unless further refining measures are taken.

Other possible contaminants in biodiesel include the by-product glycerol, and residual methanol and catalyst. Like the di- and mono-glycerides, these contaminants need to be removed to meet the specifications in the ASTM 6751 standard.

Methanol is a polar chemical and has limited solubility in vegetable oil and/or animal fats. Before the start of transesterification, two immiscible phases exist: the oil/fat-dominant phase and methanol-dominant phase. To allow transesterification, the oil/fat must mix with methanol at a molecular level. To

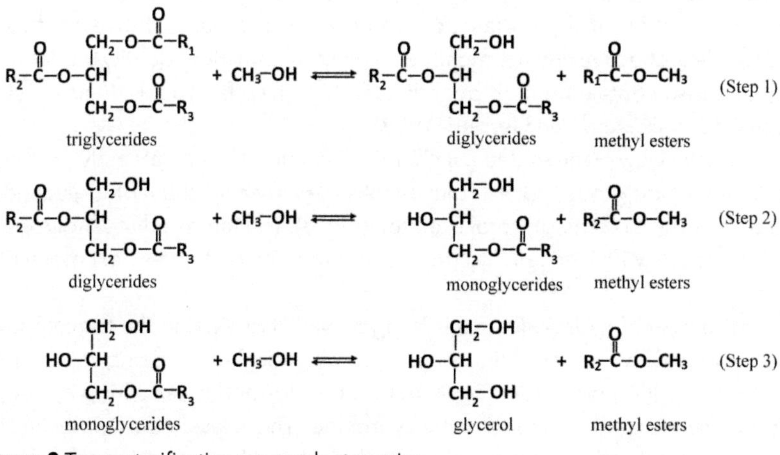

Figure 2 Transesterification proceeds step-wise.

achieve this, adequate mixing mechanically or physically by using a solvent is commonly used, although other advanced methods such as ultra-sonification-assisted mixing can be used (He and Van Gerpen, 2012). As the reaction proceeds, more and more methyl esters form, which serve as a solvent to make the oil/fat and methanol miscible in one phase, thus making the chemical reaction progress faster.

Towards the end of transesterification, the oil/fat is almost completely converted to methyl esters, a group of less polar chemicals. Meanwhile, the by-product glycerol (a polar chemical) accumulates, forming two new immiscible layers if adequate time is provided. The lower layer is mainly the heavier glycerol with unreacted methanol, most of the catalyst and other impurities. The upper layer is mainly the methyl esters, with small quantities of methanol, glycerol, catalysts and residual mono-and/or di-glycerides.

The formation of these two immiscible layers helps the methyl esters separate from the by-product layer, although separations and/or purifications are still needed to remove the residual impurities in the layer of methyl esters. As the result of excess application, methanol presence in biodiesel will negatively affect the flash point of finished biodiesel. The catalyst used for transesterification is usually the hydroxides of alkali metals of sodium (NaOH) or potassium (KOH). Residual sodium or potassium in biodiesel will cause excessive wear on engines. Residual mono- and/or di-glycerides in biodiesel will adversely affect the combustion process leading to significant carbon deposits on injectors.

Application of a catalyst is a very effective way of promoting transesterification without a long reaction time. In transesterifying high-quality feedstock, such as canola and/or rapeseed oil, the most commonly used catalyst is a strong base, either a hydroxide (-OH) or a methoxide ($-OCH_3$). A strong acid, such as sulphuric acid (H_2SO_4), can also be used in catalysing transesterification of the oil/fat. Compared to strong bases, however, strong acids are less effective and are mostly used in converting low-quality feedstocks, such as those containing high quantities of high free fatty acids, in a two-step process (Canakci and Van Gerpen, 1999).

From the view of the molecular chemical reaction, the actual catalytic moiety is the methoxide group, $-OCH_3$, that attacks the fatty acid chain of triglycerides to transesterify it off the glycerol skeleton (Fig. 3). The intermediate diglyceride salt then reacts with methanol to recover the methoxide for the next molecular reaction.

In the case of hydroxide as the catalyst, NaOH or KOH in the form of solid flakes reacts with methanol first to form methoxide before being mixed with oil/fat (Fig. 4). It is noted from Fig. 4 that water is formed in preparing a catalyst solution by reacting methanol with hydroxide. The presence of water in the system adversely affects the production of biodiesel because of potential side

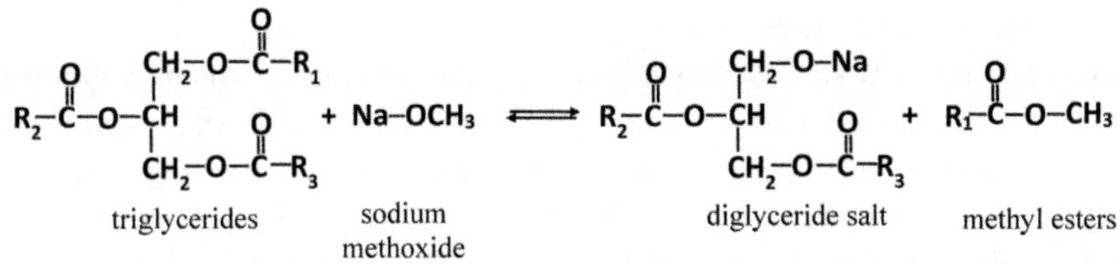

Figure 3 Methoxide moiety reacts with triglycerides to form methyl esters.

$$\underset{\text{sodium hydroxide}}{\text{Na–OH}} + \underset{\text{methanol}}{\text{HO–CH}_3} \rightleftharpoons \underset{\text{sodium methoxide}}{\text{Na–OCH}_3} + \underset{\text{water}}{\text{H}_2\text{O}}$$

Figure 4 Methoxide formation from hydroxide and methanol.

reactions of hydrolysis of methyl esters into free fatty acids, not only lowering the yield of biodiesel but also reducing its quality. Therefore, methoxide is the preferred form of catalyst.

Methoxides, either sodium methoxide ($NaOCH_3$) or potassium methoxide ($KOCH_3$), are more effective catalysts than hydroxides and are thus preferred by larger producers in continuous-flow production processes. Methoxide solutions are commercially available, typically in 25% concentrations in methanol. They are specially prepared and contain no significant amount of water. However, they are more expensive than hydroxides and require more careful handling and storage.

One drawback of using strong bases as the catalyst in transesterifying oil/fat is the side reaction of soap formation between free fatty acids contained in the oil/fat and the base catalyst. If a feedstock contains a higher level of free fatty acids, the base catalysts, either hydroxide or methoxides, can quickly react with them before effectively catalysing the transesterification. The process of soap formation is chemically termed saponification. Just like an inorganic acid reacting with a base, the organic acids (or fatty acids) react with hydroxides/methoxides to form an organic salt of fatty acid, which is commonly called soap (Fig. 5). The adverse consequence of severe soap formation is the emulsification of the reacting system, in which the biodiesel will not be able to separate from the by-product glycerol and methanol.

The rule of thumb is if an oil contains less than 3% of free fatty acids on mass basis, one would not expect serious soap formation by applying strong

$$\underset{\text{fatty acid}}{\text{R-COOH}} + \underset{\text{hydroxide}}{\text{K-OH}} \rightleftharpoons \underset{\text{soap}}{\text{R-COOK}} + \underset{\text{water}}{\text{H}_2\text{O}}$$

$$\underset{\text{fatty acid}}{\text{R-COOH}} + \underset{\text{methoxide}}{\text{K-OCH}_3} \rightleftharpoons \underset{\text{soap}}{\text{R-COOK}} + \underset{\text{methanol}}{\text{HOCH}_3}$$

Figure 5 Soap formation from free fatty acids reacting with hydroxides/methoxides.

base catalysts, either hydroxides or methoxides, in transesterification. If an oil contains more than 5% free fatty acids, serious soap formation typically occurs, which may lead to an emulsion of the transesterification system. In the latter case, a two-step process is typically applied, that is a strong acid-catalysed first step to esterify the free fatty acids followed by a base-catalysed transesterification (Canakci and Van Gerpen, 1999). Fortunately, virgin canola/rapeseed oil contains low levels of free fatty acids (Table 5); hence, soap formation is not a concern in its transesterification to biodiesel.

The common catalyst application rate is 1% of hydroxide (either NaOH or KOH) or 0.25% of methoxide (either $NaOCH_3$ or $KOCH_3$) of the oil on a mass basis. If the oil contains a small quantity of free fatty acid, the actual catalyst used needs to be adjusted to compensate the hydroxide or methoxide consumption due to soap formation. The general formula for the total catalyst application rate (in percentage of total oil on mass basis) compensating for the free fatty acids (FFA) is as follows:

$$\text{Catalyst}(\%\text{wt}) = 1\% + \%\text{FFA} \times \frac{MW_{\text{hydroxide}}}{285}$$

for hydroxide catalyst, and

$$\text{Catalyst}(\%\text{wt}) = 0.25\% + \%\text{FFA} \times \frac{MW_{\text{methoxide}}}{285}$$

Table 5 Properties of canola and rapeseed oils

	Canola[a]	Rapeseed[b]
Physical properties		
Relative density (kg/L)	0.914–0.917 at 20°C	0.912 at 38°C
Refractive index (nD 40°C)	1.465–1.467	-
Kinematic viscosity (mm²/s)	78.2 at 20°C	37.3 at 38°C
Flash point, open cup (°C)	275-290	258
Specific heat (kJ/kg at 20°C)	1.910–1.916	-
Thermal conductivity (W/mK)	0.179–0.188	-
Minor chemicals		
Phospholipids, crude oil (%)	Up to 2.5	Up to 3.5
Free fatty acids (%)	0.4-1.2	0.5–1.8
Unsaponifiables (%)	0.5–1.2	0.5–1.2
Tocopherols (ppm)	700–1200	700–1000
Chlorophylls (ppm)	5–35	5–35
Sulphur (ppm)	3–15	5–25

[a] Przybylski (2018).
[b] Demirbas (2008).

for methoxide catalyst, in which %FFA is the percentage of free fatty acid in the oil, $MW_{hydroxide}$ is the molecular weight of hydroxide (MW_{NaOH} = 40.0 kg/kmol and MW_{KOH} = 56.1 kg/kmol), $MW_{methoxide}$ is the molecular weight of methoxide ($MW_{NaOCH_3} = 54.0$ kg/kmol and $MW_{KOCH_3} = 70.1$ kg/kmol) and 285 is the pseudo-molecular weight of canola/rapeseed oil, in kg/kmol.

The use of a catalyst for the transesterification of canola/rapeseed oils helps the reaction proceed faster. However, the operating temperature helps the reaction the most. Generally, transesterification of oil takes about 8 h to reach a satisfactory conversion at room temperature. The time can be cut by half or more if the operating temperature is raised to 40°C (104°F). The time can be further reduced to 1–2 h at 60°C (140°F) for a batch operation of transesterification to reach completion. Methanol has a boiling temperature of 64.5°C (148°F), which limits the temperature that can be used for transesterification at atmospheric pressure due to the concern of severe vaporization of methanol from the reacting mixture. Therefore, the reaction temperature for transesterification of canola/rapeseed oil is 55–60°C (130–140°F).

2.2 Processes for biodiesel production

The simplest method for producing alcohol esters is to use a batch, stirred tank reactor. Agitation is necessary to ensure adequate mixing and contact between canola/rapeseed oil and methanol, which are immiscible at the beginning of the reaction. Once the reaction occurs, methyl esters (or biodiesel) are produced which serve as a solvent to bring canola/rapeseed oil and methanol into a miscible mixture, thus further promoting the reaction.

To improve the transesterification effectiveness, a two-step operation can be used: the by-product glycerol is removed once the reaction reaches a certain point, for example 90%, and the reaction proceeds with an addition of more methanol/catalyst. The advantage of this two-step operation is the reduced operating time and a higher purity product; however, the drawback is the added step of operation, which involves more intervention and thus more operating costs.

As illustrated in Fig. 6, the catalyst is first mixed with methanol to prepare a catalyst-methanol solution. If the catalyst is a hydroxide, the hydroxide flakes need to dissolve completely in methanol before being transferred to the reaction vessel.

Canola/rapeseed oil is then charged to the reaction vessel and heated to the desired operating temperature. Then, the catalyst solution is introduced to the lower layer and mixed with heated canola/rapeseed oil. The mixture is agitated adequately to provide sufficient mixing for transesterification to occur.

Upon the completion of transesterification, the reaction mixture will either remain in the reactor to separate the crude glycerol from the lighter biodiesel

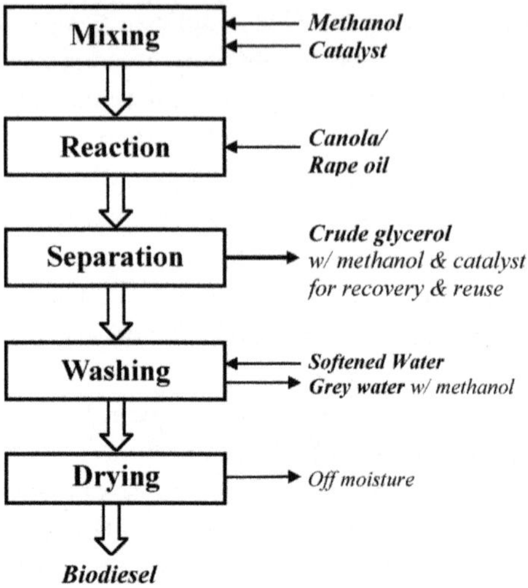

Figure 6 Process diagram of a batch transesterification system.

layer and then be discharged from the bottom, or the mixture is pumped to a centrifuge to separate the crude glycerol mechanically. In a two-step process, agitation is stopped after a certain reaction period to allow an initial separation of methyl esters and remaining oil from the heavier crude glycerol, which is discharged before additional methanol/catalyst is added and the reaction proceeds to completion.

The crude glycerol separated from the reaction mixture contains the unreacted methanol which is then recovered by a rectification/purification step before being reused back to the system. The crude glycerol is either sold directly to market (in the cases of small production facilities) or refined locally to high purity for various other applications.

After separation from the glycerol layer, the crude biodiesel layer needs further purification to remove residual methanol and catalyst, typically by softened warm (50–60°C) water washing. This polishing step can also be achieved by using a 'dry wash', that is a porous absorbent used to physically and/or chemically adsorb the impurities from the biodiesel. The advantages of dry wash include a great reduction of water usage (thus less wastewater to be treated) and a simplified operation. However, absorbents are expensive; thus, the overall cost may be higher than that of water washing.

After adequate washing, the impurities in biodiesel will have been reduced to levels meeting the specifications of the biodiesel standard ASTM D6751 in North America or EN14214 in Europe. The residual moisture in biodiesel can be high due to its hygroscopic nature (He et al., 2007) and must be reduced to 0.05% vol or lower. An effective way of reducing moisture from biodiesel is by flash vaporization under vacuum. A less effective but practical way is to simply heat the biodiesel up to an elevated temperature, for example 110°C, to drive the moisture off.

If the transesterification is incomplete for some reason, residual quantities of intermediate chemicals of di- and/or mono-glycerides may be present in the washed biodiesel, throwing it out of specification. In this scenario, one additional step, not shown in Fig. 6, is needed to remove these high-boiling-temperature chemicals, such as further reaction as described previously or by using absorbents such as an ion exchange resin.

Biodiesel production can also be achieved via a continuous-flow mode operation, to increase throughput and ensure more consistent product quality. Among the choices, the continuously stirred tank reactors (CSTRs) are the most commonly used (Fig. 7).

CSTRs vary in volume and residence time to allow for a greater extent of reaction so that the transesterification is completed before reaction mixture is discharged. CSTR can be used as a single reactor or two to three in series. When CSTRs are used in series, decanting vessels are typically utilized between the reactors to separate crude glycerol by gravity or centrifugation. Adequate mixing in CSTR is an essential element in the design, and adequate reaction time must be allowed to ensure that the reaction is complete and the composition throughout the reaction volume is consistent.

In addition to CSTR, plug-flow reactors (PFRs) are also used in continuous-flow biodiesel production. PFR is tubular. The reaction mixture moves through the reactor in which its residence time is determined by the reactor volume and the linear flow rate. Mixing is provided in the axial and radial directions by specific mechanisms such as in-line static mixers.

2.3 Processing of off-grade canola oil for biodiesel

Some canola/rapeseed oil of non-food grades may contain some level of FFA in it. In most cases, the level of FFA in virgin canola/rapeseed, even non-food

(a) (b)

Figure 7 Pilot-scale biodiesel production facility at the Biodiesel Research Laboratory, University of Idaho. (a) modular continuous-flow reactor system. (b) batch reactor system.

grades, is low, typically less than 3% wt, so that no special processing measures are needed. What one needs to do is to add extra catalyst, either hydroxides or methoxides, to compensate for the consumption due to soap formation between FFA and the catalyst.

However, if the oil is high in FFA, for example 5% wt or higher, the soap formation is severe enough to cause the reaction mixture to emulsify and lead to incapable of separating biodiesel from the by-product glycerol. Therefore, a different processing approach and/or pretreatment are needed, such as the acid- and then base-catalysed two-step process (Canakci and Van Gerpen, 2001; Van Gerpen, 2004) or supercritical processing (Bi et al., 2015; Kusdiana and Saka, 2004).

2.4 Fuel properties of biodiesel from canola and rapeseed oil

Biodiesel made from canola/rapeseed oils are reported to have better cold flow properties compared to other feedstocks. Early research conducted at the University of Idaho is summarized below to discuss the exemplary properties of biodiesel made from canola and rapeseed oils (Peterson et al., 1997).

To compare fuel characteristics of biodiesel made from a variety of feedstocks, a systematic set of performance tests were conducted. Results show that the specific gravity of canola/rapeseed biodiesel is higher than that of fossil diesel but not significantly different from the biodiesel made from canola/rapeseed and soybean; however, the viscosities of canola and rapeseed biodiesel are higher than that of soybean biodiesel (Table 6).

The pour points for biodiesel fuels were higher by 5–15°C depending on the types of feedstock, with the pour point of rapeseed biodiesel (–15°C) the closest to that of fossil diesel (–20°C). As expected, the cloud point of canola/rapeseed biodiesel is much higher than that of fossil diesel – a major drawback for it as a cold weather fuel as compared to fossil diesel. However, canola/rapeseed biodiesel has a higher flash point, making it a much safer fuel for handling and storage. Meanwhile, the cetane numbers of canola and rapeseed biodiesel (57.9 and 61.8, respectively) are superior to that of fossil diesel, attributable to the favourable combustion efficiency of biodiesel.

Other fuel properties are similar among the biodiesels made from canola, rapeseed and soybean oil. It is believed that the differences, however, are caused by the dominant fatty acid components in the vegetable oils, such as oleic acid (18:1) in canola oil (65%) and erucic acid (22:1) in rapeseed oil (approx. 30–50%), as compared to linoleic acid (18:3) in soybean oil. Therefore, the fuel properties of biodiesel are determined by the fatty acid profiles in the vegetable oils. The fatty acid profiles of oilseeds can be altered and optimized through breeding and/or genetic modification for fuel production. However, the altered profiles of canola, rapeseed or

Table 6 Comparison of canola and rapeseed methyl esters[a,b]

	Biodiesel			
	Canola	Rapeseed	Soybean	No. 2 diesel
Fuel-specific properties				
Specific gravity	0.881	0.880	0.886	0.8595
Viscosity, cSt @ 40°C	4.475	5.65	3.89	2.98
Cloud point (°C)	1	0	3	-12
Pour point (°C)	-9	-15	-3	-20
Flash point (°C)	163	79	188	87
Cetane number	57.9	61.8	54.8	46
Boiling point (°C)	334	347	339	186
Water and sediment (% vol.)	<0.005	<0.005	<0.005	<0.005
Carbon residual (% mass)	0.065	0.08	0.068	0.16
Ash (% mass)	0.002	0.002	0	0.002
Sulphur (% wt)	0.009	0.012	0.012	0.036
Heat of combustion, gross (MJ/kg)	39.9	40.54	39.77	45.42
Karl Fischer water (ppm)	1087	757	1062	38
Acid value	0.104	0.128	0.111	0.002
Iodine number	102.8	91.9	103.6	9
Fatty acid composition (%)				
Palmitic (16:0)	4.0	2.7	9.9	-
Stearic (18:0)	2.4	0.9	3.8	-
Oleic (18:1)	*65*	12.6	19.1	-
Linoleic (18:3)	17.3	12.1	*55.6*	-
Linolenic (18:3)	7.8	8.0	10.2	-
Eicosenoic (20:1)	1.3	7.4	0.2	-
Behenic (22:0)	0.4	0.7	0.3	-
Erucic (22:1)	0.1	*49.8*	0.0	-

[a] Peterson et al. (1997).
[b] Properties of soybean biodiesel are listed for comparison purpose.

soybean may not necessarily be tied to the properties of fuel made from them (Table 4).

3 Life cycle assessment of canola/rapeseed biodiesel

Canola is one of the main vegetable oil crops grown all over the world. In the US, canola production is concentrated in the Northern Plains where a drier, shorter growing season makes corn and soybean production less attractive. Most of the US production of canola is concentrated in the state of North

Dakota. Canola is also grown in the PNW as a rotational crop to wheat. A typical 3-year rotation of non-irrigated dryland farming of the Palouse region is winter wheat followed by a spring grain and then a legume or another alternative such as canola or mustard. The US consumes more canola than it produces. In year 2017–18, US produced 3.1 billion lb. of canola and imported 1.3 billion lb., mostly from Canada. Canola oil is used primarily for human consumption; however, non-food-grade canola oil has a potential market for biodiesel production.

As stated in EISA of 2007, for a biofuel to be categorized as an advanced biofuel, the fuel use must reduce the GHG emissions by at least 50% compared to the petroleum fuel it replaces. A detailed LCA was conducted by several researchers and this chapter summarizes the finding of the studies. The direct or attributional LCA covers the following stages of fuel production: (1) agricultural production, (2) crop transportation from farm to crushing plant, (3) biodiesel production from the oil and (4) biodiesel transportation and distribution. In addition to attributional LCA, EISA mandates including any significant amount of GHG emissions from indirect land use change. The indirect land use change occurs as an unintended consequence of using canola oil to produce biofuel such as the expansion of cropland to fulfil the canola oil demand in other countries.

EPA added canola and rapeseed to the list of approved biodiesel feedstock in 2010 (EPA, 2010b). The analysis used the Forestry and Agricultural Sector Optimization Model (FASOM) developed by Texas A&M University and others and the Food and Agricultural Policy and Research Institute international models as maintained by the Center for Agricultural and Rural Development (FAPRI-CARD) at Iowa State University. The midpoint of the range of LCA results indicated that the use of canola biodiesel resulted in a 50% reduction in GHG emissions compared to the diesel fuel baseline. The 95% confidence interval around that midpoint resulted in the range of a 20% reduction to a 75% reduction compared to the diesel fuel 2005 baseline.

A more detailed analysis of the Canola LCA was conducted as a joint research between University of Idaho and Washington State University for the PNW between 2012 and 2014. The study analysed scenarios for likely biofuels pathways specific to the inland northwest dryland crop production region within EPA Region 10, with locally produced canola as a feedstock for biodiesel production. Specifically, the study assessed the life cycle impacts - including crop production scenarios, crop yields, impacts on displaced crop production, carbon sequestration, nitrous oxide emissions and water use - of canola crops used to produce biodiesel for light duty and heavy-duty vehicles.

The study region is the dryland production region of the PNW, including Idaho, Oregon and Washington, and the assessment was conducted considering regional agroecological zones (AEZs), as shown in Fig. 8. Three

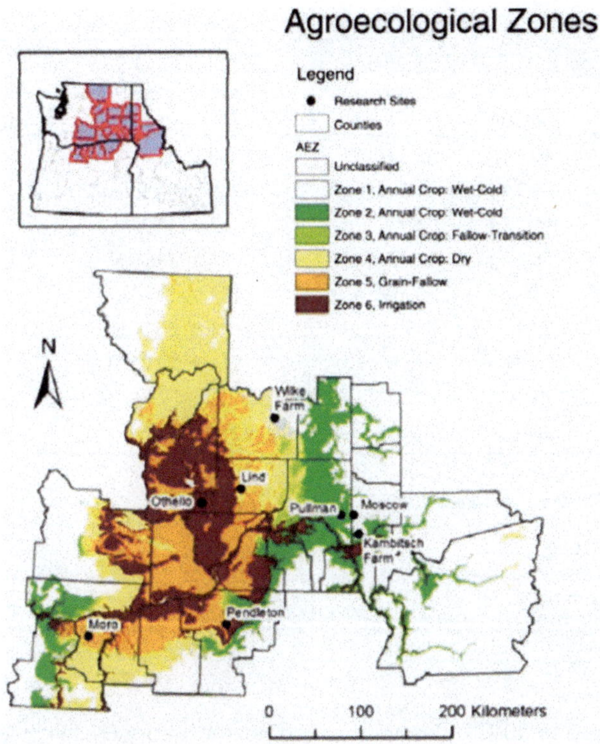

Figure 8 Agroecological zones of the inland Pacific Northwest.

AEZs were included in this assessment, following the definitions in Fig. 8: low rainfall dryland AEZ (zone 5), intermediate rainfall dryland AEZ (zone 3) and high rainfall dryland AEZ (zone 2). These three AEZs represent the most common dryland production systems in the inland Pacific Northwest (IPNW).

A computer simulation approach was used in this study, based on the cropping systems model CropSyst (Stockle et al., 2003). This is a multi-year, multi-crop model developed as an analytical tool to study the effect of climate, soils and management on cropping systems productivity, water use and hydrology, nutrient cycling and fate and the environment. A summary of energy and GHG equivalent in canola production is shown in Table 7.

Energy input for crushing and biodiesel production was measured experimentally at the University of Idaho Biodiesel Lab. A screw-press-type oil expeller (S-52 model, Hander Oil Machinery, Osaka, Japan) run by a 2.95 kW (4 hp) three-phase induction motor was used for crushing. The press had a three-stage 1 kW silicone blanket heated seed auger with an insulated jacket. The press was capable of crushing up to 50 kg/h (110 lb/h). The meal after crushing was augured to a bin using a fractional horse power motor. The seed

Table 7 Total energy input and GHG emission in spring canola production [a]

Item	Amount[i]/ha	Life cycle energy equivalent	Life cycle energy MJ/ha	GHG equivalent (gCO2e)	Total (gCO2e/ha)
Diesel	40.9 L	42.5 MJ/L[b]	1738.3	3231/L[c]	132147.9
Nitrogen	95.3 kg	51.5 MJ/kg[d]	4908.0	3592/kg[e]	342317.6
Phosphorus	11.2 kg	9.2 MJ/kg[d]	103.0	1197/kg[e]	13406.4
Potassium	16.8 kg	1.5 MJ/kg[f]	25.2	798/kg[f]	13406.4
Seeds	5.0 kg	30.4 MJ/kg[g]	152.0	189/kg[h]	945.0
Herbicide	2.2 kg	319 MJ/kg[d]	701.8	25745/kg[e]	56639.0
Insecticide	1.79 kg	325 MJ/kg[d]	581.8	29937/kg[e]	53587.2
Soil N_2O emission					330150.0[i]
	Sum		8210.1		942599.5

[a] Life cycle energy equivalents were obtained from literature.
[b] Huo et al. (2008) and Shapouri et al. (2002).
[c] DOE (2008).
[d] Pradhan et al. (2011).
[e] EPA (2010a).
[f] [S&T]^2 Consultants (2013).
[g] Shonnard et al. (2010).
[h] Sheehan et al. (1998).
[i] From the Crop Model.

was preheated to 40°C by using the seed heater before crushing. The energy used for the motor, heater and augur were recorded separately using a three-phase power data logger (Extech model 382090, Nashua, NH).

The oil content was estimated following the procedure outlined by Hammond (1991) using a Newport MKIIIA nuclear magnetic resonance (NMR) analyser (Oxford Instruments Inc, Concord, MA). The NMR analyser was calibrated with a single reference sample of known oil content, and the sample analysis was carried out as described by Hammond (1991). The procedure is based on ISO 5511-1984. The energy input in transportation and distribution of feedstock and biodiesel were obtained from the Greenhouse Gases, Regulated Emissions and Energy Use in Transportation (GREET) model (ANL, 2014)

Biodiesel was produced in a six-gallon-capacity stainless steel drum reactor (Electro-Flex Heat Inc., Bloomfield, CT) using 100% excess methanol to complete the reaction and using sodium methoxide as the catalyst. The reactor had a 250 W (1/3 hp) stirrer and a 1000 W silicone belt heater. The excess methanol from glycerol after separation was recovered using a Rotavapor (Buchi, R-114, Switzerland). A summary of energy and GHG emissions are given in Table 8.

Tables 7 and 8 contain the necessary information to conduct an LCA of canola biodiesel. A summary of the calculation is shown in Table 9.

The calculation using the conservative assumption of spring canola for the feedstock shows a 66% reduction in GHG emissions using canola oil biodiesel

Table 8 Energy input and GHG emissions from canola crushing and transesterification

Inputs	Amount per L of biodiesel[a]	Life cycle energy	Life cycle energy equivalent MJ/L of biodiesel	gCO_2e factor	Total gCO_2e/L of biodiesel
Crushing					
Electricity	0.11 kWh	10.92 MJ/kWh[b]	1.2	145.7/MJ[c]	179.3
Transesterification					
Electricity	0.212 kWh	10.92 MJ/kWh[b]	2.3	145.7/MJ[c]	233.5
Methanol	0.100 kg	33.5 MJ/kg[d]	3.4	67.7/MJ[e]	226.8
Sodium methoxide	0.020 kg	31.7 MJ/kg[f]	0.6	7.9/g[f]	13.6
		Sub total	6.3		573.9

[a] Measured values.
[b] Direct unit conversion.
[c] EPA (2016).
[d] Methanol Institute (2017).
[e] EPA (2010a).
[f] Sheehan et al. (1998).

compared to petroleum diesel. PNW spring canola-based biodiesel has a positive 3.4 net energy ratio – that is for each unit of energy spent in biodiesel production, 3.4 units of energy are available in the form of biodiesel. Similar calculations for winter canola, which has higher average productivity, indicated a net energy ratio of 3.5 and a 67% GHG reduction from use of canola biodiesel.

4 Future trends and conclusion

Current production of both biodiesel and renewable diesel are driven by the requirements of the Renewable Fuel Standard (RFS) which is mandated by the Energy Independence and Security Act of 2007. These fuels are now designated as advanced biofuels, with mandated levels that will ramp up to even higher levels in future years. Although alternative diesel requirements have been met every year, the level of lignocellulosic ethanol required by the RFS has not met expected levels and this may require the adjustment of the target production levels. The uncertainty regarding the government's continuing commitment to the RFS is one of the greatest challenges in financing alternative fuel products in the US.

The highlights of the canola biodiesel study were that the simulated crop rotations with canola were observed to have a small, generally positive impact on wheat yields. Canola production would displace some acreage of the dominant cereal grains produced in the region, but the ultimate displacement effect on a mass food value basis resulted in losses of only 10–31% depending

Table 9 Total energy input and GHG emissions from biodiesel production using spring canola

Stage	Life cycle energy MJ/L of biodiesel	GHG emissions gCO_2e /L of biodiesel	Allocation %	Allocated life cycle energy MJ/L of biodiesel	Percentage	Allocated gCO_2e /L of biodiesel	Percentage
Agriculture	13.2[a]	1519.3[a]	27.5%	3.6	38	417.8	42
Crushing	1.2[b]	179.3[b]	25.5%	0.3	4	59.3	5
Transesterification	6.3[b]	573.9[b]	89%	5.6	58	510.8	51
Biodiesel Combustion		21.7[c]	100%			21.7	2
Total (MJ/L of biodiesel)				9.5		999.6	
Biodiesel total energy output (MJ/L of biodiesel)				32.7			
Net energy ratio				3.4			
GHG emission from biodiesel production and use (gCO2e/GJ)				30 569.5			
GHG emission from 2005 baseline diesel production and use (gCO2e/GJ)[d]				90 047.4			
Net GHG reduction from biodiesel (%)				66.0%			

[a] From Table 7.
[b] From Table 8.
[c] EPA (2010a).
[d] DOE (2008).

on location and rotation, much lower than the assumed 1:1 displacement on an acreage basis. Accounting for the observed 'rotational effect' of disease and weed suppression not captured in simulations may push this trade-off closer to a net-zero displacement effect. Therefore, land use displacement or the food-for-fuel concerns should not raise significant concerns for PNW canola production.

Overall nitrous oxide emissions were slightly lower for the alternative canola rotation than the conventional wheat rotation, but the difference is too small to be significant. Soil carbon sequestration of the alternative rotation ranged from 862 to –996 kg CO_2/ha annually and is also not significantly different from the conventional wheat rotation. As seen in conventional rotations from earlier studies (Stöckle et al., 2012), the generation of nitrous oxide generally outweighs the potential benefit of increased soil carbon sequestration. The net change in total production-related GHG emissions of the alternative canola rotation over the conventional rotation ranges from 50 to 76 kg CO_2/ha annually but again is not significantly different.

From a crop production standpoint, the carbon footprint implications of shifting to alternative rotations that include canola relative to a conventional wheat rotation, while different depending on location and system, are small in comparison to the impacts of reducing tillage, as indicated by earlier studies evaluating the carbon impact of shifting tillage methods in these production systems (Stöckle et al., 2012). The potential agronomic and environmental benefits created by adding canola (or other oilseeds), especially when that addition facilitates adoption of no-till or reduced tillage, are likely far greater in significance than the carbon footprint implications of the alternative crop rotations.

Given the results of cropping system evaluation, most of the life cycle benefit of canola-based biodiesel production will be found in the processing of the seed into fuel and co-products. On average, the calculations indicated that IPNW canola-based biodiesel represents a 66% and 67% reduction in greenhouse gas emissions for spring canola and winter canola, respectively, relative to petroleum diesel and a positive net energy ratio of 3.4:1 for spring canola and 3.5:1 for winter canola.

5 Where to look for further information

5.1 Books and proceedings

- Mittelbach, M. and Remschmidt, C. 2005. *Biodiesel, the Comprehensive Handbook*. Boersedruck Ges.m.b.H., Vienna.
- Van Gerpen, J., Shanks, B., Pruszko, R. Clements, D. and Knothe, G. 2006. *Building a Successful Biodiesel Business* (2nd edn.). Biodiesel Basics, Dubuque, IA.

- Price, D. 2017. *Biodiesel Production: Processes and Technologies*. Larsen and Keller Education. ISBN-10: 1635490413; ISBN-13: 978-1635490411.
- Marcel, K. 2015. *Biodiesel: Advanced Approaches and Applications*. Clanrye International. ISBN-10: 1632400804; ISBN-13: 978-1632400802.
- Knothe, G., Krabl, J. and Van Gerpen, J. 2015. *The Biodiesel Handbook* (2nd edn.). AOCS Publishing. ISBN-10: 1893997626; ISBN-13: 978-1893997622.
- Marchetti, J. M. and Fang, Z. 2011. *Biodiesel: Blends, Properties and Applications (Energy Science, Engineering and Technology)*. Nova Science Pub Inc. ISBN-10: 1613246609; ISBN-13: 978-1613246603.
- Bart, J., Cavallaro, S. and Palmeri, N. 2010. *Biodiesel Science and Technology: From Soil to Oil*. Woodhead Publishing. ISBN-10: 1845695917; ISBN-13: 978-1845695910.
- Luque, R. and Melero, J. A. 2012. *Advances in Biodiesel Production: Processes and Technologies*. Woodhead Publishing. ISBN-10: 0857091174; ISBN-13: 978-0857091178.

5.2 Online documents

- Biodiesel Education Program, University of Idaho. Literature on biodiesel from canola, rapeseed or mustard oil. Available at: BiodieselEducation.org/Production/Canola.html
- Canola Council of Canada. 2018. Canola Encyclopedia. Available at: https://www.canolacouncil.org/canola-encyclopedia/
- US Department of Energy. *Biodiesel Handling and Use Guide* (5th edn.). 2016. DOE/GO-102016-4875. Available at: https://www.nrel.gov/docs/fy17osti/66521.pdf
- National Biodiesel Board. Biodiesel Bulletin. Available at: http://nbb.org/news/biodiesel-bulletin.
- European Biodiesel Board. Studies and Reports. Available at: http://www.ebb-eu.org/studies.php.

6 References

ANL. 2014. The greenhouse gases, regulated emissions, and energy use in transportation (GREET) model. US DOE, Argonne National Laboratory, Argonne, IL.

ASTM. 2015. D6751-15ce1 standard specification for biodiesel fuel blend stock (B100) for middle distillate fuels. ASTM International, West Conshohocken, PA.

Bi, Z., He, B. and McDonald, A. 2015. Biodiesel production from green microalgae *Schizochytrium limacinum* via *in situ* transesterification. *Energy and Fuels* 29(8), 5018–27. doi:10.1021/acs.energyfuels.5b00559.

Canakci, M. and Van Gerpen, J. 1999. Biodiesel production via acid catalysis. *Transactions of the ASAE* 42(5), 1203–10.

Canakci, M. and Van Gerpen, J. 2001. Biodiesel production from oils and fats with high free fatty acids. *Transactions of the ASAE* 44(6), 1429–36.

Canola Council of Canada. 2018. History of varietal development. Available at: https://www.canolacouncil.org/canola-encyclopedia/crop-development/history-of-varietal-development/ (accessed on 04 October 2018).

Demirbas, A. 2008. Relationships derived from physical properties of vegetable oil and biodiesel fuels. *Fuel* 87(8–9), 1743–8. doi:10.1016/j.fuel.2007.08.007.

Diaz-Ballote, L., Lopez-Sansores, J. F., Maldonado-Lopez, L. and Garfias-Mesias, L. F. 2009. Corrosion behavior of aluminum exposed to a biodiesel. *Electrochemistry Communications* 11(1), 41–4, doi:10.1016/j.elecom.2008.10.027.

DOE. 2008. Development of baseline data and analysis of life cycle greenhouse gas emission of petroleum-based fuels. DOE National Energy Technology Laboratory.

EPA. 2010a. Lifecycle Greenhouse Gas (GHG) impacts calculations. United States Environmental Protection Agency, Washington DC.

EPA. 2010b. Notice of supplemental determination for renewable fuels produced under the final RFS2 program from canola oil. Environmental Protection Agency, Washington DC.

EPA. 2016. Emissions & Generation Resource Integrated Database (eGRID). United States Environmental Protection Agency.

Hammond, E. G. 1991. Organization of rapid analysis of lipids in many individual plants. In: Linskens, H. F. and Jackson, J. F. (Eds), *Essential Oils and Waxes*. Springer, Berlin, Heidelberg, pp. 321–30.

Hayward, A. 2011. Introduction – oilseed *Brassica*. In: Edwards, D., Batley, J., Parkin, I. and Kole, C. (Eds), *Genetics Genomics and Breeding of Oilseed Brassicas*. Science Publishers/CRC Press, New York, NY, pp. 1–13. Chapter 1.

He, B. B. and Van Gerpen, J. H. 2012. Application of ultrasonication in transesterification processes for biodiesel production. *Biofuels* 3(4), 479–88. doi:10.4155/bfs.12.35.

He, B. B., Thompson, J. C., Routt, D. W. and Van Gerpen, J. H. 2007. Moisture absorption in biodiesel and its petro-diesel blends. *Applied Engineering in Agriculture* 23(1), 71–6, doi:10.13031/2013.22320.

Hu, E., Xu, Y., Hu, X., Pan, L. and Jiang, S. 2012. Corrosion behaviors of metals in biodiesel from rapeseed oil and methanol. *Renewable Energy* 37(1), 371–8, doi:10.1016/j.renene.2011.07.010.

Huo, H., Wang, M., Bloyd, C. and Putsche, V. 2008. Life-cycle assessment of energy and greenhouse gas effects of soybean-derived biodiesel and renewable fuels. *Environmental Science and Technology* 43, 750–6.

Jain, S. and Sharma, M. P. 2011. Thermal stability of biodiesel and its blends: a review. *Renewable and Sustainable Energy Reviews* 15(1), 438–48, doi:10.1016/j.rser.2010.08.022.

Kumar, A., Salisbury, P. A., Gurung, A. M. and Barbetti, M. J. 2015. Importance and origin. In: Kumar, A., Banga, S. S., Prabhu, P. D. and Priya, P. (Eds), *Brassica Oilseeds Breeding and Management*. CABI, UK, pp. 1–10. Chapter 1.

Kusdiana, D. and Saka, S. 2004. Effects of water on biodiesel fuel production by supercritical methanol treatment. *Bioresource Technology* 91(3), 289–95, doi:10.1016/S0960-8524(03)00201-3.

Methanol Institute. 2017. Physical properties of pure methanol. Methanol Institute, Alexandria, VA.

Mittelbach, M. and Remschmidt, C. 2005. *Biodiesel, the Comprehensive Handbook*. Boersedruck Ges.m.b.H., Vienna.

Peterson, C. L., Reece, D. L., Hammond, B. L., Thompson, J. C. and Beck, S. M. 1997. Processing, characterization and performance of eight fuels from lipids. *Applied Engineering in Agriculture* 13(1), 71–9, doi:10.13031/2013.21578.

Pradhan, A., Shrestha, D., McAloon, A., Yee, W., Haas, M. and Duffield, J. 2011. Energy life-cycle assessment of soybean biodiesel revisited. *Transactions of the ASABE* 54(3), 1031–9, doi:10.13031/2013.37088.

Przybylski, R. 2018. Canola oil: physical and chemical properties. Available at: https://fr.canolacouncil.org/media/515239/canola_oil_physical_chemical_properties_1.pdf (accessed on 05 October 2018).

Shapouri, H., Duffield, J. A. and Wang, M. 2002. The energy balance of corn ethanol: an update. U.S. Department of Agriculture, Washington DC.

Sheehan, J., Camobreco, V., Duffield, J., Graboski, M. and Shapouri, H. 1998. Life cycle inventory of biodiesel and petroleum diesel for use in an urban bus: a joint study. U.S. Department of Agriculture and U.S. Department of Energy, Golden, CO.

Shonnard, D. R., Williams, L. and Kalnes, T. N. 2010. Camelina-derived jet fuel and diesel: sustainable advanced biofuels. *Environmental Progress and Sustainable Energy* 29(3), 382–92, doi:10.1002/ep.10461.

Singh, B., Korstad, J. and Sharma, Y. 2012. A critical review on corrosion of compression ignition (CI) engine parts by biodiesel and biodiesel blends and its inhibition. *Renewable and Sustainable Energy Reviews* 16(5), 3401–8.

Statistics Canada. 2018a. Table 32-10-0359-01 Estimated areas, yield, production, average farm price and total farm value of principal field crops. Available at: https://www150.statcan.gc.ca/t1/tbl1/en/tv.action?pid=3210035901 (accessed on 04 October 2018).

Statistics Canada. 2018b. Table 32-10-0352-01 Crushing statistics of major oilseeds. Available at: https://www150.statcan.gc.ca/t1/tbl1/en/tv.action?pid=3210035201 (accessed on 04 October 2018).

$(S\&T)^2$ Consultants, Inc. 2013. GHGenius model 4.01 volume 2: data and data sources. $(S\&T)^2$ Consultants, Delta, BC, Canada.

Stockle, C. O., Donatelli, M. and Nelson, R. 2003. CropSyst, a cropping systems simulation model. *European Journal of Agronomy* 18(3–4), 289–307, doi:10.1016/S1161-0301(02)00109-0.

Stöckle, C., Higgins, S., Kemanian, A., Nelson, R., Huggins, D., Marcos, J. and Collins, H. 2012. Carbon storage and nitrous oxide emissions of cropping systems in eastern Washington: a simulation study. *Journal of Soil and Water Conservation* 67(5), 365–77.

US EIA. 2018a. Monthly biodiesel production report July 2018. Table 3. U.S. inputs to biodiesel production. Available at: https://www.eia.gov/biofuels/biodiesel/production/archive/2018/2018_07/biodiesel.pdf (accessed on 05 October 2018).

US EIA. 2018b. Monthly biodiesel production report December 2016. Table 3. U.S. inputs to biodiesel production. Available at: https://www.eia.gov/biofuels/biodiesel/production/archive/2016/2016_12/biodiesel.pdf (accessed on 05 October 2018).

US EIA. 2018c. Monthly biodiesel production report December 2014. Table 3. U.S. inputs to biodiesel production. Available at: https://www.eia.gov/biofuels/biodiesel/production/archive/2014/2014_12/biodiesel.pdf (accessed on 05 October 2018).

USDA FAS. 2018a. Canada biofuels annual 2017. Available at: https://gain.fas.usda.gov/Recent%20GAIN%20Publications/Biofuels%20Annual_Ottawa_Canada_4-6-2018.pdf (accessed on 05 October 2018).

USDA FAS. 2018b. Oilseeds and products annual 2017. Available at: https://gain.fas.usda.gov/Recent%20GAIN%20Publications/Oilseeds%20and%20Products%20Annual_Vienna_EU-28_3-31-2017.pdf (accessed on 05 October 2018).

USDA FAS. 2018c. Oilseeds and products annual 2015. Available at: https://gain.fas.usda.gov/Recent%20GAIN%20Publications/Oilseeds%20and%20Products%20Annual_Vienna_EU-28_3-31-2015.pdf (accessed on 05 October 2018).

USDA FAS. 2018d. EU biofuels annual 2015. Available at: https://gain.fas.usda.gov/Recent%20GAIN%20Publications/Biofuels%20Annual_The%20Hague_EU-28_7-15-2015.pdf (accessed on 09 October 2018).

USDA FAS. 2018e. EU biofuels annual 2018. Available at: https://gain.fas.usda.gov/Recent%20GAIN%20Publications/Biofuels%20Annual_The%20Hague_EU-28_7-3-2018.pdf (accessed on 09 October 2018).

USDA NASS. 2018a. Crop production 2017 summary. Available at: http://usda.mannlib.cornell.edu/usda/current/CropProdSu/CropProdSu-01-12-2018.pdf (accessed on 05 October 2018).

USDA NASS. 2018b. Crop production 2015 summary. Available at: http://usda.mannlib.cornell.edu/usda/nass/CropProdSu//2010s/2016/CropProdSu-01-12-2016.pdf (accessed on 05 October 2018).

Van Gerpen, J. 2004. Biodiesel processing and production. *Fuel Processing Technology* 86(10), 1097–107. doi:10.1016/j.fuproc.2004.11.005.

Van Gerpen, J., Shanks, B., Pruszko, R., Clements, D. and Knothe, G. 2006. *Building a Successful Biodiesel Business* (2nd edn.). Biodiesel Basics, Dubuque, IA.

Chapter 7

Sustainable use of *Miscanthus* for biofuel

Paul Robson, University of Aberystwyth, UK; Astley Hastings, University of Aberdeen, UK; John Clifton-Brown, University of Aberystwyth, UK; and Jon McCalmont, University of Exeter, UK

1 Introduction

Highly productive terrestrial plants have the potential to deliver significant amounts of biomass for thermal or chemical conversion to displace fossil fuels and deliver sustainable energy for the future. Of the available candidates for use as bioenergy crops, C4 species have some of the highest potential and recorded productivities of all terrestrial plants (Morison et al., 2000); however, the majority of C4 species are of tropical or subtropical origin and are poorly 'adapted' to growing in cool and temperate climates where C3 species tend to dominate. An exception to this is *Miscanthus*, a perennial rhizomatous grass genus that appears to be more suited to growing under cooler conditions (Jones, 2011). *Miscanthus* spp. have a wide native range covering most of SE Asia from the Kharbarosk Krai in NE Asian Russia to the tropical Philippines, and from the Himalaya to Taiwan; they have adapted to a variety of different climates and, therefore, contain a considerable diverse genetic potential to help create hybrids with improved combinations of traits. Despite this diversity, *Miscanthus* is largely undomesticated and the main commercial type grown is

http://dx.doi.org/10.19103/AS.2019.0027.15

a sterile triploid *M.* × *giganteus* (Greef and Deuter, 1993) thought to arise as a natural hybrid between a diploid *M. sinensis* and a tetraploid *M. sacchariflorus* (Linde-Laursen, 1993). The cultivation of *M.* × *giganteus* has demonstrated good potential as a sustainable biomass crop because it combines a number of useful attributes, some of which will be explored in more detail later, including high dry matter yields, perennial growth, efficient use of nitrogen and water, and good disease resistance. This combination of characteristics has meant that *Miscanthus* is of particular interest for growth in the temperate regions of Asia, North America and Europe. A large amount of published research has focussed on crop trials within these regions, which will, therefore, be the focus of this review. However, interest in the crop is increasing in other regions of the world such as Africa and many of the considerations around sustainability discussed here will likely be transferrable.

As a perennial crop, *Miscanthus* may be grown continuously from a single cultivation for many years and stands have been harvested for 20 years or more on a yearly cycle. Full harvestable economic yield is usually achieved after 2–3 years. By this time, the crop has established sufficient rhizome to produce a dense canopy to maximise seasonal light interception, and consequently yield, with a typical planting density of around 2 plants m^{-2} (Atkinson, 2009). The earliest replicated plot yield trials growing *M.* × *giganteus* in Europe were planted in 1983 in Hornum, Denmark (Jørgensen, 1995) and produced regular autumn yields of 10 to 25 tonnes of dry matter per hectare (Mg DM ha^{-1}) (Lewandowski et al., 2000). A spring harvest allowed improved biomass quality as a solid fuel, particularly lower moisture contents, but at the expense of recoverable biomass with yields up to 30% less than autumn yields being reported in European trials (Lewandowski et al., 2000). Commercial interest in *Miscanthus* developed in the USA around 2000 and potential spring harvest yields of 30 t DM ha^{-1} y^{-1} demonstrated the huge potential to produce lignocellulosic biomass from *Miscanthus* in mid-west USA (Heaton et al., 2008). However, yields of *M.* × *giganteus* showed high inter-annual and inter-site variability, ranging between sites from around 10 to 35 Mg DM ha^{-1} and were inversely correlated with latitude (Lesur et al., 2013). Yield of *Miscanthus* was the most influential factor in its potential for economic return and soil water availability the most crucial input in determining yield (Wang et al., 2012).

In its native Asia, *Miscanthus* has a number of uses: the young tips are eaten by humans and the straw is used for thatching, building or animal bedding. *Miscanthus* was used for the conversion of biomass to liquids using gasification and the Fischer-Tropsch process and a pilot plant was operational in Germany from 1998 to 2004 producing biodiesel (Kittelman, 2014). Other uses of *Miscanthus* explored at this time included direct thermal conversion by combustion and use as a raw material for the paper and pulp industries. The majority of *Miscanthus* currently grown in Europe is used for thermal conversion, that is burnt to generate power or combined heat and power (CHP) (Baxter et al., 2014), but projects are developing its use across a range of different end uses (Ogunsona et al., 2017; Wagner et al., 2017).

The potential for biomass crop cultivation to compete unfavourably with food production has been discussed widely (Valentine et al., 2012), but one potential benefit of perennial *Miscanthus* is that it may be grown on marginal, poorer quality agricultural land that would otherwise not be economic or suitable for cultivation of food crops. But in considering the potential impact of *Miscanthus* cultivation, it is important to know what area may be practically available for its production. Such considerations may be particularly important in smaller countries where it will be vital to know if there is sufficient land available to generate a sufficient quantity of biomass crop for sustainable commercial usage. As an example, in the UK, Lovett et al. (2014) used a 'constraint mapping' approach to calculate the potential land available for biomass crops. This GIS mapping approach identified and excluded areas where biomass crops could or should not be grown such as roads, rivers, slopes >15%, areas of cultural heritage and national parks, woodlands, peat soils and natural habitats; these were combined to produce 'prohibition' areas at a spatial resolution of 1 ha. Also excluded was land graded 1 and 2, the highest quality land, which was regarded as essential for the production of food crops. The less productive land grades 3, 4 and 5 make up the vast majority of the UK agricultural area. What remained was an 8.5 Mha area in the UK that could be considered suitable for deployment of these modern energy crops. It is important to consider these significant crop changes carefully as conversion from different land management practices to *Miscanthus* plantation may have either a positive or a negative impact on greenhouse gas (GHG) emissions, physical impacts on radiated energy from albedo change, and further consequences for biodiversity and soil hydrology (Jørgensen et al., 2014). A sensible approach would likely exclude areas of permanent pasture or established woodland (Pogson et al., 2016; Richards et al., 2017).

The 2007 Biomass Strategy (DEFRA, 2007) set a target of 0.35 Mha, that is less than 5% of the potentially available land identified in Lovett et al. (2014). Modelling studies (Lovett et al., 2009; Hastings et al., 2014) show that the mean yield of *Miscanthus* on grade 3b, 4 and 5 land outside the excluded areas would be around 10 Mg DM ha^{-1} yr^{-1}, and 0.35 Mha could produce up to 70 PJ energy, equivalent to 1.67 Mt of oil or 1.17% of the total UK energy. In comparison, across the larger areas available in the US with *Miscanthus* yielding up to 30 Mg ha^{-1}, 12 million ha or 9.3% of the current US cropland would be sufficient to provide the equivalent of around one-fifth of the current US gasoline use (Heaton et al., 2008). The available areas vary by country; for example, grassland in Ireland accounts for 60% of the agricultural land. This is currently used largely for livestock production, but it is possible that future shifts towards diets with less meat and milk or sustainable intensification of extensively managed grasslands could release significant pasture land for energy crops.

As a source of energy for biofuels, biomass should of course contain more potential energy than is required for growth, harvesting and processing of the crop (i.e. provide an energy output/input ratio significantly greater than 1 (Felten et al., 2013)). Improved sustainability is derived from a number of factors including energy balance, carbon mitigation and environmental impacts. The growth of any crop and perturbation of any living system by removal of biomass will have an effect on factors that impact sustainability. In a study of the cradle-to-farm gate GHG budget of *Miscanthus* plantations in the UK, Robertson et al. (2017) showed that the main variable was CO_2 rather than other significant GHGs such as methane (CH_4) and nitrous oxide (N_2O). Despite reporting relatively low yields at the study site, these results showed that *Miscanthus* can make a considerable contribution to GHG mitigation in energy production from steam turbines compared with coal and natural gas. However, use in a CHP facility would improve the efficiency of fuel chemical energy conversion to useful energy from 35% to nearly 80%. The importance of specific components of a sustainability assessment is likely to vary between sites because of complex soil/plant/atmosphere interactions and crop management and transport requirements that impact positive and negative effectors of GHG balance. For example, higher local air temperature may achieve a combination of positive and negative impacts. Positive impacts may occur from higher yield and, therefore, greater above-ground fixed carbon or a drier crop and, therefore, an increased lower heating value. Negative impacts may include higher soil respiration due to increased soil temperatures and subsequent loss of fixed carbon. The complexity of such interactions necessitates the use of systems models that integrate spatially explicit climate and soil data together with plant growth and biomass utilisation models to make sustainability predictions at the macro scale. When applied across large areas of land such as Europe, systems models reveal that the sustainability of *Miscanthus* cultivation can be expected to vary dramatically, driven by a complex pattern of changes in yield, soil organic content and carbon intensity from *Miscanthus* cultivation (Fig. 1). Such complexities make definitive conclusions difficult, but there are a number of characteristics that can be used to improve the sustainability of *Miscanthus* some of which are discussed later, starting with the energy balance.

2 The energy balance

The balance of energy output to energy input is clearly an important consideration for biomass crops that are used for bioenergy. *Miscanthus* is a low-input, fast-growing perennial energy grass and as such is an attractive

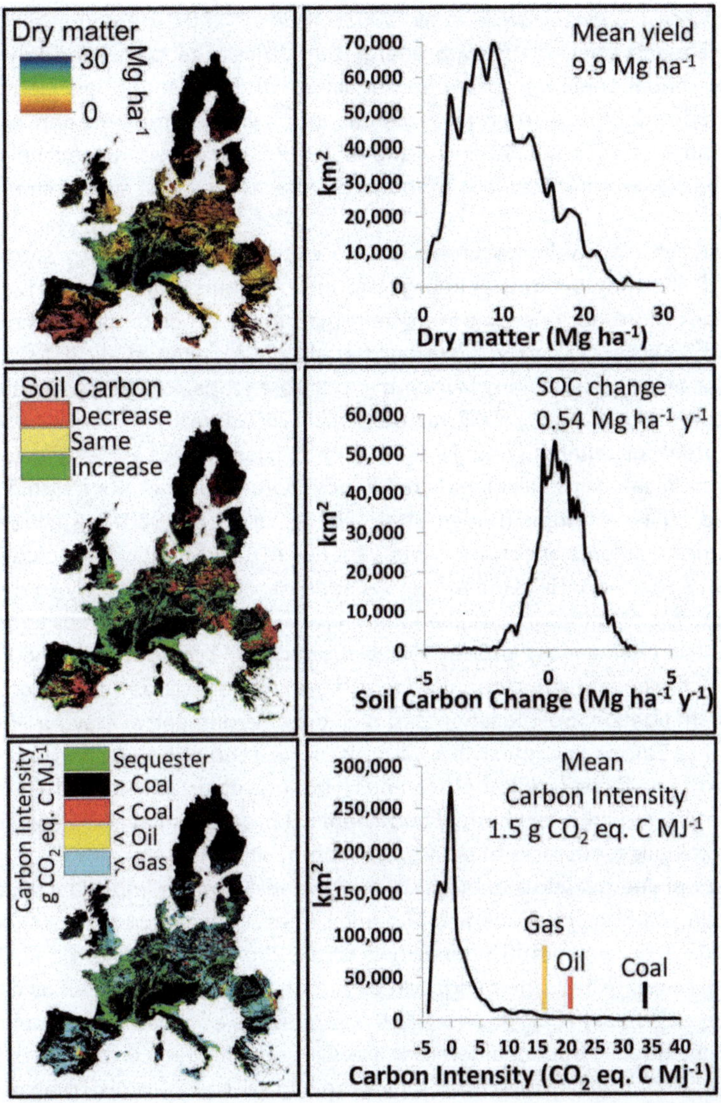

Figure 1 Output from the MiscanFor Model (Hastings et al., 2009) using the Harmonized World Soil Data (IIASA) and Climate Research Unit 4.2 climate data of the world as input for period 2000–10. Top panel: map of mean harvest yield for the period 2000–10 in Mg ha^{-1} $year^{-1}$ with a histogram of yields for EU27 + Switzerland. Scale red = 0 to green = 30. Black area is non-cropland and high organic soils. Middle panel: Map of Soil Organic Carbon change of *Miscanthus* planted on arable land, green = sequestration, yellow = no change, red = loss, Histogram shows SOC change in EU27 in Mg ha^{-1} $year^{-1}$. Black area is non-cropland and high organic soils. Bottom Panel: Map of carbon intensity of *Miscanthus* fuel for combustion in furnace, grown on arable land; scale green = sequester C, turquoise = less than gas (16), yellow = less than oil (22), red = less than coal (33), and black area is non-cropland and high organic soils and more than coal (33); histogram shows carbon intensity of crop grown in EU27 on arable land in g CO_2-eq. C MJ^{-1}.

biomass crop (Lewandowski et al., 2003b; Harvey, 2007; Heaton et al., 2008, 2010; Zhuang et al., 2013) with energy output/input ratios around ten times that of annual energy crops (Felten et al., 2013). The energy balances for oil seed rape (OSR), maize and *Miscanthus* crops were compared and output/input ratios of 4.7 ± 0.2, 5.5 ± 0.2 and 47.3 ± 2.2 calculated, respectively, with only the low-input *Miscanthus* found to be effectively a CO_2 sink (Felten et al., 2013).

Land is a valuable resource and land cultivated for bioenergy should be utilised efficiently. In terms of energy-production intensity, *Miscanthus* biomass produces more net energy per hectare than many other bioenergy crops at around 200–250 GJ ha^{-1} yr^{-1} (Hastings et al., 2012; Felten et al., 2013). These values compare particularly favourably to arable crops, for example maize for biogas (98 GJ ha^{-1} yr^{-1}), OSR for biodiesel (25 GJ ha^{-1} yr^{-1}) and wheat and sugar beet for ethanol (7 to 15 GJ ha^{-1} yr^{-1}) (Hastings et al., 2012). Energy-production intensity calculated for woody perennials can vary significantly between different areas (Bauen et al., 2010). Tallis et al. (2013) showed that in the right circumstances, even old varieties of short-rotation coppice (SRC) willow can exceed 150 GJ ha^{-1} yr^{-1}. A sensible approach is likely to include the strategic planting of combinations of crops across different areas to deliver efficient overall energy production, but adequate consideration should be given to the energy inputs required to achieve different yields and the resulting carbon mitigation. Soil, location and climatic environment will favour specific crops; in a European context, this will typically be a choice between woody SRC/short rotation forest (SRF) or *Miscanthus* (Hastings et al., 2014). In addition, the equipment required for thermal and chemical processes have to be available to use specific feedstocks of a particular composition.

A high energy ratio and biomass energy density will be important factors in mitigating carbon emissions, and *Miscanthus* has significant potential to reduce fossil fuel CO_2 emission (Clifton-Brown et al., 2004, 2007; Hillier et al., 2009; Hastings et al., 2009). The mitigation of carbon is usually expressed as carbon dioxide equivalent (CO_2-eq.) to allow comparisons across different fuel types and to include the impact of fertiliser use. Carbon mitigation is the sum of fossil fuel carbon emissions displaced minus the carbon cost of growing an energy crop and producing a useable fuel plus any soil carbon sequestration or loss. Depending on soil type, climate and previous land use, soil carbon changes can either add or detract from the overall carbon mitigation. When such factors were modelled across available land in the UK, most of the land could produce *Miscanthus* biomass with a carbon index of 1.12 g CO_2-C equivalent per MJ energy in the furnace. The carbon index value for *Miscanthus* production was substantially lower than coal (33), oil (22), liquefied natural gas (21), Russian gas (20) and North Sea gas (16) (McCalmont et al., 2017). These values were calculated for use of different feedstock for thermal conversion to electricity, but other potential uses of *Miscanthus* are considered in a later section including recent life cycle assessment (LCA) studies.

3 Nutrient-use efficiency

Nutrient-use efficiency is related to the particular use of the crop. It is important that the crop has the nutrients that are needed to grow efficiently, but for biomass crops, the harvest is of fixed carbon and, therefore, many of the nutrients utilised in a growing crop are not required (or are even undesirable) in the harvested product. The composition of the harvested crop such as the carbon content, the form of the carbon and the presence of other compounds determines the ease of conversion. The presence of potential high-value bi-products may contribute to the economic sustainability of the conversion process. Thermal conversion of biomass for heat or energy ideally utilises dry biomass with a low water content and a high carbon content. *Miscanthus* has a high C:N ratio (average of 142.6) at spring harvest (Heaton et al., 2009). After senescence, when most of the N has been repartitioned to the rhizome and stored for the next growing season and the majority of leaf biomass has fallen to the ground, the remaining long woody stems can be an ideal source of not only carbon for fuel production/burning, but also bioplastics and other products. Low N content at harvest time means reduced nitrogen oxide (NOx) emissions during burning, while the high C content implying low levels of easily metabolised sugars and proteins makes the harvested crop a fairly poor food source, which contributes to low levels of herbivory and biotic stress.

Lewandowski and Schmidt (2006) compared triticale and reed canary grass to *Miscanthus* and showed far higher N-use efficiency in *Miscanthus*. Maximum yields were observed with no fertiliser, but with existing soil N at 50 kg N ha^{-1}; higher applications of N fertilisation (above 114 kg N ha^{-1} yr^{-1}) were detrimental to crop performance, particularly where soil water was in short supply. This very low demand for added fertiliser was investigated by Christian et al. (2006), who used ^{15}N isotope-enriched nitrogen fertiliser applied at 60 kg N ha^{-1} to study uptake during the establishment phase following planting. Only around 20% of the N taken up by the developing crop had come from the fertiliser; 80% had come from mineralisation of soil organic matter of the former grassland or atmospheric deposition. There is growing evidence that high nutrient efficiency may, in part, derive from bacterial nitrogen fixation associated with *Miscanthus* (Davis et al., 2010; Dohleman et al., 2012). Nitrogenase activity has been found in both rhizomes and surrounding soil bacteria (Eckert et al., 2001; Miyamoto et al., 2004) with isotope analysis revealing high levels of biologically fixed nitrogen in *Miscanthus* biomass, particularly in the first year of establishment (Keymer and Kent, 2014). One trade-off to this low nitrogen requirement is that emissions and leaching can initially arise following planting into highly fertilised land or grassland killed in preparation for conversion (Christian and Riche, 1998; Behnke et al., 2012; Holder et al., 2018b) as *Miscanthus* is unlikely to utilise all the available nutrients in the first year.

The potential for high nitrogen-use efficiency means that the need for regular agronomic amendments that would otherwise reduce a favourable energy balance and detract from other sustainability criteria can be largely avoided. Cadoux et al. (2012) reviewed nutrient offtake in mature *Miscanthus* harvests in 27 studies over 10 countries and found a median content of 4.9 g N $(kg\ DM)^{-1}$ when harvested in the early spring. Given a typical UK offtake of 10 to 15 Mg DM $ha^{-1}\ yr^{-1}$, the annual export of organic nitrogen from a site in harvest material would range between 49 and 73.5 kg N ha^{-1}. Accounting for an atmospheric N deposition rate of 35–50 kg N $ha^{-1}\ yr^{-1}$, Goulding et al. (1998) suggests that *Miscanthus* is unlikely to benefit greatly from inputs of N unless it was being established in very low fertility soils. For example, an optimum application of 100 kg N ha^{-1} was seen to give significant yield benefits on a low fertility sandy loam soil in S. England (Shield et al., 2014). Lewandowski et al. (2000) reviewed 19 *Miscanthus* field trials across Europe and reported that there was little response to N fertiliser after the second or third year, though there was some suggestion that early rhizome development may benefit from a low level of application where soils may be low in available N to begin with. Christian et al. (2008) followed a *Miscanthus* crop for 14 years treated with 0, 60 and 120 kg N $ha^{-1}\ yr^{-1}$ and concluded that there was no yield response to the application of N fertiliser though monitoring of soil fertility and offtake did suggest, in these soils at least, a benefit from additions of phosphate (7 kg P $ha^{-1}\ yr^{-1}$) and potassium (100 kg K $ha^{-1}\ yr^{-1}$). A large number of studies have been aggregated into a database to demonstrate yield and sustainability across different crops and ecosystems (LeBauer et al., 2018). To avoid negative impacts on sustainability, any increase in yield from fertiliser application would need to be sufficient to offset any associated increased GHG emissions. An LCA of a mature *Miscanthus* trial showed that yield improvements and the associated increased displacement of fossil fuels associated with different fertiliser application rates were unlikely to be sufficient to offset resulting increased soil N_2O emissions (Roth et al., 2015).

The entire above-ground biomass of *Miscanthus* is harvested, and therefore, the cropped biomass reflects the perennial growth cycle of the plant. This cycle begins with the remobilisation of nutrients from below-ground rhizome in late spring, which are used for early growth, the crop then develops a mature canopy and the rhizome is replenished through summer before finally the canopy senesces, typically followed by leaf loss throughout the winter. The latter two processes contribute to the recycling of nutrients so that they are available for the next growth cycle, but also negatively impact harvested biomass yield, which may decline by up to 30% (Lewandowski and

Heinz, 2003). This creates a strong correlation towards the end of the growth year between yield and crop quality, with yield declining as some aspects of quality such as moisture and C:N ratio improve. The interaction between yield and quality is of varying importance depending on how the harvested biomass is utilised. A spring harvest of *Miscanthus* greatly improves the suitability of biomass for combustion because the material is usually fully senesced, reducing concentrations of moisture, ash and alkali metals at the expense of dry matter yield (Lewandowski et al., 2003a). An ideal biomass composition for thermal conversion comprises low moisture content, strongly bonded complex molecules such as lignified cell walls and low amounts of N to limit NOx emissions. In addition, the presence of compounds that may contribute to the increased ash slagging in boilers, such as silica and potassium, should be low; such compounds lower the slagging temperature and hence reduce thermal efficiency. Heating values of *Miscanthus* biomass have been found to vary due to senescence and/or some other process occurring during the winter period, which were not identified, but might include nutrient leaching and leaf loss (Lewandowski and Kicherer, 1997). Mos et al. (2013) concluded that the change in heating value of *Miscanthus* is likely due to variation in the lignin content of crops harvested at different times of the year. The harvest time and associated variation in senescence also impacted the quality and stability of bio-oil made by fast pyrolysis processing of *Miscanthus* biomass, with summer harvests producing lower yields of bio-oil. Other harvest dates in September and February produced similar quality bio-oil despite the crop being less senesced in September than February (Mos et al., 2013). Crop yield declined from September to February suggesting an earlier harvest may be preferable, but harvesting *Miscanthus* early may impact on the recycling of nutrients to the rhizome and soil during senescence and leaf-drop that occurs over winter. The resulting depletion of nutrients may, in turn, affect long-term yields unless these nutrients were replaced, for example, using post-processing residue; however, such additional applications would inevitably increase the overall carbon footprint of the crop.

Miscanthus is also used to produce liquid fuels, such as ethanol, which have a different ideal biomass composition requirement. In a study looking at the environmental sustainability of ethanol production from a range of second-generation feedstocks, Falano et al. (2014) concluded that the global warming potential of ethanol from *Miscanthus* biomass is more than 80% lower than petrol/gasoline from fossil sources. A modelling study on biofuel production demonstrated that cellulosic ethanol from *Miscanthus* will have a considerably lower environmental impact than current first-generation biofuels (Gabrielle et al., 2014). Conversion of biomass to liquid fuels ideally utilises biomass rich in simple sugars that are easily accessed and energetically favourable for enzymatic or catalytic conversion. Le Ngoc Huyen et al. (2010) showed that harvesting *Miscanthus* at an earlier stage improved saccharification efficiency with alkali pre-treated biomass. Harvesting before full senescence results in more nutrients being harvested, which are therefore likely to be above the levels that are replaced by natural annual cycling; however, in a similar approach to that discussed in the preceding paragraph, the digestate can be used as a soil amendment and fertiliser to create a circular nutrient cycle though any resulting impacts on soil trace gas production, such as N_2O, would need to be considered.

Genotypic variation in senescence and, therefore, composition such as moisture content has been identified across a broad genotypic panel of *Miscanthus* (Robson et al., 2012). Cell wall composition varied between *Miscanthus* growing at different sites and genotypic variation in cell wall composition has been linked to yields of pyrolysis products (Hodgson et al., 2010, 2011). Such studies suggest that there is potential to breed for improvements in composition targeted towards conversion efficiency of *Miscanthus* for each end use, but such genotypic improvements may have different impacts across different climatic regions or require specific agronomies.

4 Water use/water-use efficiency

Water is a limiting factor for yield in *Miscanthus,* as in all vegetation; one major factor that affects the sustainability of bioenergy crops is the use of fresh water for growth. Plants in general use roughly one litre of water to produce between 2 and 6 g of biomass; therefore, the water use attributed to plant growth for bioenergy may seem very large. However, any land used to grow bioenergy feedstocks would usually support other plant growth, such as pasture, forest, scrub-land or other crops, if it was not used for bioenergy. So the real question is the difference between the water consumption of the bioenergy feedstock ecosystem and the one it replaces. Evapotranspiration and water-use efficiency (WUE) and the amount of biomass harvested per unit of water consumed varied significantly depending on crop and ecosystem (Rockstrom et al., 1999) and therefore the extent of the impact and issues of sustainability will be location specific. As such, it is more realistic to focus on the relative performance within a particular environment and how best to increase WUE to obtain 'more crop per drop' within sustainable limits of a particular location.

Plant growth is governed by the rate of photosynthesis, a process that consumes a small proportion of the total amount of water that is transpired and produces carbohydrates as one of the products. The efficiency with which water is utilised varies among the three photosynthesis processes used by plants, of which two, known as C3 and C4, are the most common in higher plants. In general, C3 plants, which encompass many grasses, flowering plants and woody trees, use more water per unit biomass than C4 plants. In suitable climates, C4 species should exhibit higher efficiencies of radiation, nutrient and water use than C3 species, and hence attain higher productivity. However, C4 photosynthesis is more typical of tropical and subtropical species. In cooler temperate climates, most C4 species, for example maize (*Zea mays)*, fail to achieve high productivity due to high thermal requirements for growth and impaired photosynthesis at low temperatures (Miedema et al., 1987; Sage et al., 2010). *Miscanthus* is one of a seemingly small number of C4 species that appear to be well adapted to temperate environments. In a comparison with another temperate C4 grass (*Spartina cynosuroides), Miscanthus* achieved higher WUE, up to 9.1 g kg^{-1}, and retained the theoretical high WUE of C4 species, when grown under temperate conditions (Beale et al., 1999).

Despite highly efficient WUE, producing more biomass may have a significant impact on soil hydrology cycles. *Miscanthus* used more water across the season (954 mm) than comparable plots of maize (611 mm) and switchgrass (*Panicum virgatum*) (764 mm) at the same location. This resulted from increased latent heat flux (λET) transferring water to the atmosphere and a longer growing season (Hickman et al., 2010), though *Miscanthus* produced more biomass per unit of water used. WUE was very similar between *Miscanthus* at ca. 19 kg ha^{-1} mm^{-1} and maize at ca. 18.6 kg ha^{-1} mm^{-1}, and WUE in switchgrass was about half that of maize and *Miscanthus* (Hickman et al., 2010), illustrating the considerable variation that exists even among C4 grass species. Changes in water use may impact ecological functions of land areas and if production of utilisable biomass is an aim, then options for high and low biomass crops to limit impacts in hydrologically sensitive areas may be required. This may include a long-season, slow-growing *Miscanthus* that combines high WUE and provides soil cover to reduce evaporative loss and competition from hydrologically inefficient C3 species. It should be noted that high water use may be a benefit in some regions prone to flooding. A long-season canopy may have additional benefits related to excess rainfall. A recent study measured evapotranspiration by eddy covariance in a *Miscanthus* plantation and found it was higher than expected during winter months. The canopy interception of water by *Miscanthus* was similar to that of the mixed deciduous forest in the latter months of the year, suggesting that the potential for flooding may be reduced by lessening soil water recharge by rainfall (Holder et al., 2018a).

Whatever be the relative merits of high and low biomass crops, it is likely that a requirement will be for the biomass to be produced with high WUE. Malinowska et al. (2017) examined WUE across a broad genotypic range of *Miscanthus* growing under different water treatments; they identified that bigger and faster growing plants tended to have lower WUE, suggesting the general trend is in opposition to the more desirable interaction. However, against this trend, there were individuals that produced high biomass under control treatments and high WUE under the different water treatments, although the absolute values achieved would need to be confirmed by larger scale field trials. The same study also identified the climatic regions associated with high WUE in *Miscanthus* suggesting where favourable phenotypes might be found and further accessions could be collected with the unusual, but highly desirable link between high biomass production and high WUE (Malinowska et al., 2017).

In addition to the water used by the plant to grow, as in any energy system, the water used in the conversion of the feedstock into a suitable fuel and then into energy must also be considered. If coal is considered in comparison to biomass fuel, for example, the coal must be mined and washed before use. This uses water and also produces a lot of acidic waste water containing iron and other heavy metals, which requires processing before disposal or it will damage water courses and aquifers. Similarly, oil and gas production also results in large quantities of associated subsurface brines that require disposal. Unconventional oil and gas production requires the use of hydrologic fracking, which not only consumes a large quantity of fresh water, but also produces saline and chemical-laden brine in the flow-back process, which needs careful processing and disposal (Hastings, 2018). Detailed comparisons of water use via different conversion routes have not been produced for *Miscanthus*, but such comparisons are likely to be similar among different feedstock. Water efficiency in the use of the fuel also needs consideration as many thermal power stations do not recover the 54–65% of heat that is not converted into electricity, but lose it to the atmosphere, mostly by evaporative cooling in cooling towers. To improve efficiency further, any thermal production of electricity should also provide the waste heat for either space heating or industrial processes through a process known as CHP.

5 Carbon flux

Land disturbance resulting from cultivation of any crop will lead to impacts on the main drivers of soil GHG emissions, that is soil oxygen status, microbial diversity, carbon-to-nitrogen input ratios and hydrological status. Changes of crop type will further perturb the system by changing nutrient availability and mineralisation rates. All these impacts will have a direct bearing on GHG emissions, particularly CO_2 and CH_4 from the decomposition of organic matter and N_2O from nitrification/denitrification processes in the soil. For perennial energy crops primarily aimed at mitigating climate change, such as *Miscanthus*, quantifying these impacts is essential in determining if the crop is a source or sink of GHG, and if it provides useful gains over fossil fuel use. There have been many studies looking at soil carbon fluxes during transitions from conventional crops to *Miscanthus,* many taking advantage of the fact that, being a C4 species as opposed to the C3 crops that are being replaced, soil carbon derived from

Miscanthus will carry a distinctive isotopic signal (e.g. Clifton-Brown et al., 2007). Among others, Zimmermann et al. (2012) found that in the short term, land disturbance resulted in relatively rapid losses of existing C3 soil carbon to the atmosphere. However, the large turnover of root and leaf biomass in *Miscanthus* lead to carbon sequestration rates of 0.6 Mg ha^{-1} yr^{-1} (following arable crops) and 0.9 Mg ha^{-1} yr^{-1} (following grassland); thus, the lost soil carbon was replaced within 4–5 years. Early work by Hansen et al. (2004) found that after a typical *Miscanthus* crop lifetime of around 16 years, 31% of the carbon in the upper soil layers was derived from the *Miscanthus* inputs. This evidence for rapid turnover and replacement of C3-derived soil carbon with C4 has been reported in several other studies; for examples, see (Clifton-Brown et al., 2007; Dondini et al., 2009; Zimmermann et al., 2013; Zatta et al., 2014). Previous land use and the soil carbon levels associated with it appear to be key determinants of the levels of carbon sequestration that might be expected following cultivation of *Miscanthus*. Changes from annual crop species to perennial *Miscanthus* production are likely to see increases in soil carbon due the removal of annual soil disturbance and the build-up of large inputs through overwinter leaf-drop of around 30% of aboveground biomass and large inputs from roots (Lewandowski et al., 2000; Zhu et al., 2018). Changes from perennial grassland systems are not so clear cut, with studies more likely to show maintenance of soil carbon stocks at similar levels to starting status (Anderson-Teixeira et al., 2009; Zatta et al., 2014). Recent literature syntheses (Qin et al., 2016; Zang et al., 2018) have confirmed that these overall trends with changes to perennial *Miscanthus* from annual cropping show net sequestration, while cultivations into previous grasslands showed no significant change.

Less well studied or reported are the other important greenhouse gases in agriculture, CH_4 and N_2O. While CH_4 is not considered to be an important GHG in temperate cropland, as opposed to during animal production (Snyder et al., 2009), N_2O is of a far more significant concern, with agriculture being the largest global contributor to atmospheric concentrations. A recent review paper on the sustainability of energy crop production (Whitaker et al., 2018) called for more work on soil N_2O emissions under energy crops and literature is now beginning to develop, particularly around the critical impact periods during the cropping cycle: cultivations at the beginning and end of crop life and during fertilisation events. Recent work has reported on the soil N_2O flux at the beginning of *Miscanthus* establishment into grassland (Holder et al., 2018b) and at the end of the cropping cycle and reversion back to grassland (McCalmont et al., 2018). Both studies revealed significant, short-term spikes in soil N_2O emissions driven by soil disturbance associated with crop cultivations; they showed that integrations of these spikes into annual sum estimates added significantly to overall LCA of the GHG costs of crop production. For the land-use conversion period, Holder et al. (2018b) suggested that directly attributable

N_2O costs would add 4.13 Mg ha^{-1} CO_2-eq to a previously calculated LCA carbon cost of the biomass production (Hastings et al., 2017) of 9.49 Mg h^{-1} CO_2-eq (calculated over ten years' production), an increase of 44% over the original estimate which had not considered N_2O emission in its calculation. Similarly, McCalmont et al. (2018) found that the reversion N_2O costs added around 50% to the overall global warming potential cost of biomass energy production. However, while these figures are significant and need to be considered in GHG accounting for these systems, it should be noted that total CO_2-eq. costs associated with biomass crop production still remain far lower than the equivalent energy produced through more conventional fossil fuels such as coal. McCalmont et al. (2018) calculated that even including the N_2O cost of reverting back to grassland at the end of crop life, the CO_2 eq. cost of energy produced through *Miscanthus* would be 18 times lower than the equivalent produced through coal. Of course, one key question regarding sustainability of energy crop production and the impact of land-use change (LUC) remains, and it is the perennial problem of attribution of costs and benefits in LCA studies. If the GHG cost of establishing an energy crop is attributed to the biomass produced, then should the GHG cost of its reversion to a more conventional crop also be attributed, or would this be better apportioned to the following crop?

In a modelling exercise for the UK, the relative emissions of a suite of bioenergy crops potentially grown in the UK were analysed for their impact on soil GHG emissions, considering three initial LUC scenarios: from forestry, grassland and arable. These crops included wheat, sugar beet, OSR, *Miscanthus*, SRC willow and poplar and SRF for many species. This analysis showed that where *Miscanthus* yields were the highest, it produced the least GHG emissions. For all conversions from arable emissions were negative, grassland conversions slightly positive and replacing forestry with all crops resulted in significant GHG emissions (Pogson et al., 2016; Richards et al., 2017).

6 Life cycle assessment (LCA) for different end uses

Miscanthus can be used in many pathways for the production of energy. It can be used as a direct fuel for thermal conversion or as a feedstock for conversion to liquid fuels or biogas. *Miscanthus* can also be used as a reinforcing fibre for plastic- or cement-based components for either structural engineering use or insulation. The latter structural uses retain the carbon for a longer-term storage/utilisation, than when biomass is used as a fuel, and therefore have great potential for reducing atmospheric GHG levels. Inherent in energy use is the emission of carbon, with the possible exception of Bioenergy with Carbon Capture and Storage (BECCS) discussed later, and therefore careful consideration must be given to the balance of carbon and environmental impact in comparison to other energy sources across the cycle of use. Such considerations invariably will also include a particular context of economic and/or political requirements. Here, we will only consider energy production, although structural and insulation materials can be used to improve energy efficiency. As the two main policy drivers for using bioenergy are to reduce GHG emissions and enhance energy security, each production pathway should be compared using metrics that quantify the potential to achieve these policy objectives and provide a comparison to other energy systems, be they fossil based or intermittent renewables such as solar, wind, tidal and wave generation.

LCA is commonly used to evaluate the environmental impact of an energy production pathway. This approach defines the boundaries of a process and calculates all of the inputs and outputs to the system, including energy use and waste products, and estimates their impact or cost to the environment. LCAs for various uses of *Miscanthus* have been calculated, from being a biomass fuel to a feedstock for producing biogas, biodiesel, methanol and ethanol, to assess if it does indeed reduce emissions compared to alternative fuels within these specific utilisation pathways (Styles and Jones, 2007; Hastings et al., 2012). Of the variables that have been incorporated into an LCA model, the establishment rate of the *Miscanthus* stand and canopy longevity are crucial inputs. The length of time over which energy and equivalent carbon costs accrued at planting are amortised has a significant impact on the overall carbon (and economic) balance. Many models assume a 15-year cycle with 3 years to achieve the maximum yield, but *Miscanthus* plantations may be productive for much longer and establish faster and, thereby, further improve carbon and energy returns.

An early LCA examined the impact of replacing 30% of peat and 10% coal for electricity production in Ireland by co-firing with *Miscanthus*. The reduction in CO_2 emitted was estimated at 1.9 Mt CO_2 eq. yr^{-1} and represented 2.8% of Ireland's 2004 GHG emissions, but it was calculated to require just 1.7% of agricultural land area using a conservative estimate of yield (Styles and Jones, 2007). The reduction in CO_2 emissions from incorporating *Miscanthus* into peat or coal electricity generation was so significant that even very low-yield estimations continued to generate significant GHG savings, suggesting models have considerable flexibility in the assumptions of benefit. Felten et al. (2013) compared different biomass systems using rapeseed, maize and *Miscanthus*. Compared to equivalent fossil fuel-related energy supply, the potential reduction in CO_2-equivalents ranged between 30–76% for electrical energy from maize biomass, 29–82% for biodiesel from rapeseed, and 96–117% for *Miscanthus* chips. Interestingly, the authors concluded that, in their study, CO_2-neutrality was only reached by the *Miscanthus* cropping system and was related to an additional credit from carbon sequestration in soil during the cultivation period; thus, this cropping system acted as a CO_2-sink (Felten et al., 2013). A later paper (Robertson et al., 2017) further concluded that *Miscanthus* cropping could be carbon neutral (incorporating soil N_2O emissions as well as CO_2) even without net carbon sequestration to the soil.

The consideration of GHG emissions should not focus solely on comparisons with fossil fuel alternatives that are displaced by the use of biomass crops, but should also include consideration of the impacts of the cropping systems that are displaced. Dondini et al. (2009) compared soil carbon across a soil depth profile in an arable crop to *Miscanthus* conversion and showed that the total amount of soil organic carbon (SOC) was higher under *Miscanthus* than under the arable crop and that this difference was largely due to the input of new carbon. The *Miscanthus* system gained 25.4 Mg C ha^{-1} across the soil profile and this increase in carbon storage within the soil appeared to be largely due to the decrease in soil tillage (Dondini et al., 2009), an important advantage in utilising perennial crops. An estimate of yearly carbon mitigation by *Miscanthus* over 15 years ranged from 5.2 to 7.2 Mg C ha^{-1} yr^{-1} depending on time of harvest (Clifton-Brown et al., 2007). Zatta et al. (2014) studied a grassland-to-*Miscanthus* conversion and showed that after an initial decline in SOC due to soil disturbance resulting from cultivation at planting, SOC levels rapidly recovered due to input of new carbon from *Miscanthus*. In this instance, the input of more carbon from *Miscanthus* did not produce a significant increase in SOC and the authors suggested the additional carbon may have been utilised through increased microbial respiration, so called soil priming, thus illustrating the potential for complex interactions and the need for empirical data to test assumptions used in models.

The GHG emissions including those associated with LUC were calculated comparing bioenergy feedstock and a rotational food production system it displaced (Styles et al., 2015). Only *Miscanthus* and rotational maize offered GHG savings when indirect LUC (iLUC) impacts were considered, and the percentage of displaced production that was directly replaced had a substantial impact on the estimated global warming potential. However, the GHG benefits for rotational maize were offset by impacts on ecosystem services, and of the six bioenergy crop systems investigated, *Miscanthus* was shown to offer the greatest benefits in the provision of ecosystem services (Styles et al., 2015). Tonini et al. (2012) used sensitivity analysis to show that uncertainties around LUC could have a significant impact on LCA results. They compared four conversion pathways (anaerobic digestion, gasification, small-scale CHP, and large-scale co-firing with coal) for ryegrass (*Lolium perenne*), willow (*Salix* spp) and *Miscanthus* and found that only large-scale co-firing of *Miscanthus* and willow offered real GHG savings compared to fossil fuel alternatives. The impacts of LUC and any positive effects could be localized and consideration should be given to where production might be displaced to and the impacts of any further consequential LUCs thereby incurred. The consequences of LUC and iLUC are complex, particularly the latter, and the impacts and values assigned have been the subject of recent debate and discussion (Muñoz et al., 2014; Jepson and Caldas, 2017).

The production of biogas from biomass has received much attention recently and serves as a good example of the considerations in assessing the potential impacts and possible benefits of deploying *Miscanthus* as a source of biomass for biofuels. *Miscanthus* has been researched as a potential substrate for the production of biogas, particularly in Germany where biogas production has increased significantly in recent years (Kiesel and Lewandowski, 2017; Kiesel et al., 2017a). The cultivation of biomass to provide substrate for anaerobic digestion makes up a significant proportion of the environmental impact and GHG cost of biogas (Hijazi et al., 2016). In the mono-digestion of maize (AD with maize alone), it was the cultivation that had the largest environmental impact due to diesel fuel use in agriculture and emissions linked to fertiliser use (Lijó et al., 2014). *Miscanthus* has been researched as a more environmentally sustainable alternative to the use of maize in biogas production. LCA demonstrated biogas production using *Miscanthus*, as compared with Maize, resulted in a reduction across five impact categories such as climate change, terrestrial acidification and eutrophication of water (Kiesel et al., 2017b).

One of the benefits of dedicated perennial *Miscanthus* crops is the ability to be grown on marginal lands where conventional annual crops would not provide an economically useful return. Marginal lands are usually defined as such due to limitations in productivity, which may occur for a number of reasons; these can include abiotic stresses such as low water availability or salinity. *Miscanthus* grown on marginal land was tested for biogas production and showed a cost advantage over similar conversions to biogas using maize (Wagner et al., 2019). Overall yields of *Miscanthus* were critical in the competitiveness of gas production per hectare. Potential reductions in yield of growing Miscanthus on marginal land may be offset by the fact that otherwise unutilised uneconomical land is being brought back in to production. Enhancing *Miscanthus*' capacity to tolerate marginal soil conditions through breeding or imaginative valorisation of other potential benefits such as remediating denuded or contaminated soils and incorporating this into long-term rotations with food crops may improve economic viability. Cultivation on overworked and nutrient-denuded soils may even increase overall productivity of associated food crops as soils improved by using *Miscanthus* as a long-term break crop are brought back into conventional food production.

7 Traits and/or agronomy for improved sustainability

Yield is an important trait for improvement in *Miscanthus,* and biomass crops in general; provided increased yield is not achieved through energy-intensive means, higher yield improves the energy balance and the economics of the crop. Yield is limited by abiotic stress and water availability in particular. Areas of high solar radiation needed to fuel high yields are often associated with low precipitation. Growing *Miscanthus* on marginal land subject to individual or combinations of stress that limit conventional agriculture, as discussed earlier, generates a highly desirable mix of sustainability criteria, reduces competition with food crops and possibly regenerates land. Therefore, a key requirement for future development of sustainable *Miscanthus* through breeding is to conserve and enhance this tolerance to abiotic stress and increase yields accordingly. Efforts are being made to understand and reduce the impact of abiotic stress on *Miscanthus* including the screening of germplasm under single or combinations of different stresses (Ezaki et al., 2008; Jones et al., 2015; Kalinina et al., 2017; Malinowska et al., 2017; Stavridou et al., 2017; van der Weijde et al., 2017; Fonteyne et al., 2018).

Miscanthus is a versatile biomass feedstock and is utilised in a number of ways across many regions of the world. As discussed above, GHG savings need to be calculated against existing technologies and alternative utilisations, and the greatest savings and optimum conversion pathways will vary from region to region. This potential complexity was illustrated by a study across five European countries. At all five sites, the highest energy savings were achieved by CHP generation via combustion (Meyer et al., 2017). The GHG savings were more complex: the highest savings were achieved by heat and power production in Portugal (42.7 t CO_2-eq ha^{-1} yr^{-1}); however, at other European locations (Sweden, Denmark, Germany, England), bioethanol production gave the highest GHG savings. However, there was some uncertainty over the ability to generate the GHG savings attributed to the use of fermentation residues for heat in bioethanol production, which had a significant contribution to the GHG savings in the scenario discussed. The study concluded that the improved GHG savings were primarily associated with increased yield and that composition was of comparatively lower importance, but would impact other sustainability characteristics such as emissions (Meyer et al., 2017).

The economic viability, and competitiveness with less sustainable crops, of growing *Miscanthus* on marginal land for biogas-based electricity production has been shown to be limited by yield (Wagner et al., 2019), so improvements in this area will be an important focus for ongoing research. However, there is also potential for manipulating *Miscanthus* cell wall structure and composition to optimise biomass for particular end uses (Slavov et al., 2013). Lignin, in particular, is regarded as one of the main factors impeding saccharification by enzymatic hydrolysis as it prevents enzymes accessing the hemicellulose and cellulose in plant cell walls (Zeng et al., 2014). Biomass quality parameters depend on the intended technology for conversion; for example, low lignin content may improve enzymatic conversion, whereas the high energy contained within lignin bonds means that a high lignin content is favourable for thermochemical conversion (Welker et al., 2015). In addition to yield and composition improvements, process optimisation has the potential to increase sustainability and economic viability by reducing the energy required across post-harvesting treatments. The development of such improvements is at an early stage because novel *Miscanthus* cultivars and their usage are still relatively new; however, improvements are already being demonstrated. For example, in the production of biogas, there was a considerable cost of pre-treatment of biomass for anaerobic digestion, but ensiling *Miscanthus* biomass greatly reduced the need for pre-treatment (Mangold et al., 2019). A study of 50 diverse *Miscanthus* genotypes demonstrated that drought tolerance and cell wall composition were only weakly correlated, suggesting that the potential exists for both traits to be improved independently (van der Weijde et al., 2017). Some traits may be highly aliased and act in opposition but, where possible, future domestication and breeding efforts are likely to target the combined improvement of important traits such as biomass yield, cell wall composition and abiotic stress resilience.

A particular challenge is perhaps one of identifying how to domesticate *Miscanthus* quickly enough to maximise the potential global benefits across a short enough time scale to contribute significantly to the fight against climate change. Breeding programmes are well established across several biomass crops and experiments have demonstrated the potential of next-generation sequencing to generate markers and speedup breeding cycles using genome-wide association studies (Slavov et al., 2014; Davey et al., 2017) and optimised selection indices to test strategies for efficient improvement of multiple traits (Slavov et al., 2018).

8 Conclusion and future trends

Miscanthus often compares favourably in terms of GHG and energy efficiency with not just the fossil fuel alternatives, but also other potential cropping systems that may be used for biomass/bioenergy. Towards the aim of achieving carbon-negative biofuel, of particular note is the sequestration of carbon into soils associated with these more stable perennial, rhizomatous systems in many studies. The composition of *Miscanthus* biomass may be optimised for different end uses though the environmental and economic sustainability of such end uses varies across different growing regions. Broadly, it may be assumed that because dry biomass with high C:N ratios, and thus excellent conservation of nutrients within the plant/soil system, is better suited to thermal conversion, it is these routes that will achieve the highest savings in GHG (Fig. 2). This may be further enhanced by the implementation of BECCS (Kemper, 2015), a technology that has not yet been thoroughly explored in the *Miscanthus*

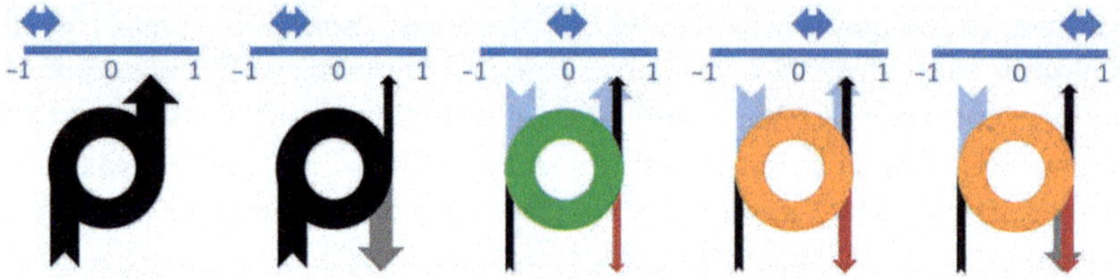

Figure 2 Carbon flow (below) and approximate greenhouse gas balance (above) achievable from different fuel use scenarios including fossil fuel, green or senesced biomass with and without carbon capture and storage (CCS). Source: adapted from Kemper (2015).

literature, but has great potential for achieving carbon-negative biofuel. Implementing BECCS from bioenergy crops at a large scale faces a number of challenges such as transport of CO_2 and the co-location of marginal lands and carbon storage basins (Turner et al., 2018). Modelling the potential for BECCS highlights the contributions of biomass residues and dedicated biomass crop growth on marginal land and improving yield. Notwithstanding these contributions, other sustainability characteristics discussed previously plus the importance of good governance and the need for effective incentives are also significant factors in modelling the potential of this new technology (Vaughan et al., 2018). The implementation of BECCS would see *Miscanthus* capture carbon from the atmosphere as part of its annual cycle of growth, a portion of that carbon would be sequestered in the soil through the perennial flux of photosynthate from the crop and, as part of the thermal conversion process, CO_2 that would be otherwise re-emitted to the atmosphere would be captured and sequestered in long-term storage under sea or ground. This GHG-removal technology is a key component in possible future emission scenario pathways to limit global warming to 1.5°C (IPCC, 2018). The technology readiness to implement BECCS is largely at the demonstration phase but the capture, transport and potential storage capacity suggest that this will be an important part of reducing atmospheric CO_2 levels in the future (Royal Society, 2018) and a key contributor to making biomass crop use even more sustainable.

The growth and harvesting of biomass is of particular significance to overall energy balance, and there are a number of crop characteristics that contribute to an improved energy ratio. Many of these are exemplified in the current commercial production of *Miscanthus*, but the tremendous geographical range over which *Miscanthus* species are present and the ability to generate wide hybrids between diverse species present a significant opportunity to improve characteristics and deliver high levels of sustainability along with improved yield and composition. There is further potential to enhance tolerance for growth under marginal conditions to reduce competition with food crops and potentially remediate soils. The unusual cold tolerance of *Miscanthus* provides an additional bonus in allowing the potential benefits of WUE associated with C4 crops to be utilised in temperate regions.

At present, the cultivation of bioenergy crops is often seen as separate to food crops in potential competition for land. Perhaps more emphasis should be given to the potential to incorporate *Miscanthus* within land-use rotations with food crops. Improvements in genetics and agronomy are attempting to produce *Miscanthus* plantations that produce economic returns in the second growth year making shorter term plantations more economically attractive and a negative GHG balance achievable in fewer perennial cycles. The perennial rhizomatous nature of *Miscanthus*, and associated reductions in soil disturbance over time, gives it the potential to improve degraded soils by increasing soil carbon, organic matter and earthworm diversity (Kahle et al., 2001; Hansen et al., 2004; Felten and Emmerling, 2011). Perhaps in the future *Miscanthus* will be incorporated into land-use rotations as a longer term, low-intensity break crop providing a range of alternative bio-products, increasing regional fuel security and socio/economic benefits while enhancing ecosystem services.

The focus of this chapter has been to highlight the research that demonstrates *Miscanthus* embodies a range of attributes that make it an ideal sustainable biomass crop for biofuels. Added to this, the diversity identified within global collections of *Miscanthus* species represents a wealth of potential for future improvements. Thus far, the impact and potential of *Miscanthus* have mostly been demonstrated using a small number of high-yielding clones and experiments with diverse genotypes. These experiments have identified improvements in yield, resilience against stress and improved yield quality. Fewer studies report the tremendous potential of hybridisations between diverse *Miscanthus* genotypes. In the space of a few years, the focus of research using *Miscanthus* has been on delivering sustainable solutions, the development of underpinning biological knowledge, agronomy, environmental modelling, LCA and on demonstrating the potential of new genotypes. The main question now is can the investment be made to deliver the potential within *Miscanthus* fast enough to produce the necessary impact.

9 Acknowledgements

The authors thank the many colleagues who contributed the information and papers upon which this chapter is based and the authors acknowledge the support of CERES Inc. and Terravesta Ltd. through the GIANTLINK project (LK0863). This chapter was funded by several projects, the UK Biotechnology and Biological Sciences Research Council (grant numbers BBS/E/W/10963A01A and BB/CSP1730/1) and the Natural Environment Research Council (grant numbers NERC - NE/M019691/10 and NERC - NE/P019951/1) and UKERC/FFR2/3.

10 Where to look for further information

There is a video, FAQs and literature available at the *Miscanthus* Breeding site http://www.miscanthusbreeding.org/.

The context and challenges of climate change and sustainable global decarbonisation including the potential contributions of carbon-negative biomass crops are discussed in many reviews including two recently published by IPCC (2018) and the Royal Society (2018). There are a number of reviews that discuss the current status of developing bioenergy crops (Clifton-Brown et al., 2019) and *Miscanthus* at commercial scale (Clifton-Brown et al., 2017) and also reviews of the relative costs and benefits of *Miscanthus* cultivation and a general consideration of land availability, land use, land conversion sustainability and competition for land with food crops (Hastings et al., 2009; Fazio and Monti, 2011; Valentine et al., 2012; Holland et al., 2015; McCalmont et al., 2017).

11 References

Anderson-Teixeira, K. J., Davis, S. C., Masters, M. D. and Delucia, E. H. 2009. Changes in soil organic carbon under biofuel crops. *GCB Bioenergy* 1(1), 75–96. doi:10.1111/j.1757-1707.2008.01001.x.

Atkinson, C. J. 2009. Establishing perennial grass energy crops in the UK: a review of current propagation options for *Miscanthus*. *Biomass and Bioenergy* 33(5), 752–9. doi:10.1016/j.biombioe.2009.01.005.

Bauen, A. W., Dunnett, A. J., Richter, G. M., Dailey, A. G., Aylott, M., Casella, E. and Taylor, G. 2010. Modelling supply and demand of bioenergy from short rotation coppice and *Miscanthus* in the UK. *Bioresource Technology* 101(21), 8132–43. doi:10.1016/j.biortech.2010.05.002.

Baxter, X. C., Darvell, L. I., Jones, J. M., Barraclough, T., Yates, N. E. and Shield, I. 2014. *Miscanthus* combustion properties and variations with *Miscanthus* agronomy. *Fuel* 117, 851–69. doi:10.1016/j.fuel.2013.09.003.

Beale, C. V., Morison, J. I. L. and Long, S. P. 1999. Water use efficiency of C_4 perennial grasses in a temperate climate. *Agricultural and Forest Meteorology* 96(1–3), 103–15. doi:10.1016/S0168-1923(99)00042-8.

Behnke, G. D., David, M. B. and Voigt, T. B. 2012. Greenhouse gas emissions, nitrate leaching, and biomass yields from production of *Miscanthus* × *giganteus* in Illinois, USA. *Bioenergy Research* 5(4), 801–13. doi:10.1007/s12155-012-9191-5.

Cadoux, S., Riche, A. B., Yates, N. E. and Machet, J. M. 2012. Nutrient requirements of *Miscanthus* × *giganteus*: conclusions from a review of published studies. *Biomass and Bioenergy* 38, 14–22. doi:10.1016/j.biombioe.2011.01.015.

Christian, D. G. and Riche, A. B. 1998. Nitrate leaching losses under *Miscanthus* grass planted on a silty clay loam soil. *Soil Use and Management* 14(3), 131–5. doi:10.1111/j.1475-2743.1998.tb00136.x.

Christian, D. G., Poulton, P. R., Riche, A. B., Yates, N. E. and Todd, A. D. 2006. The recovery over several seasons of 15N-labelled fertilizer applied to *Miscanthus × giganteus* ranging from 1 to 3 years old. *Biomass and Bioenergy* 30(2), 125–33. doi:10.1016/j.biombioe.2005.11.002.

Christian, D. G., Riche, A. B. and Yates, N. E. 2008 Growth, yield and mineral content of *Miscanthus × giganteus* grown as a biofuel for 14 successive harvests. *Industrial Crops and Products* 28(3), 320–7. doi:10.1016/j.indcrop.2008.02.009.

Clifton-Brown, J. C., Stampfl, P. F. and Jones, M. B. 2004. *Miscanthus* biomass production for energy in Europe and its potential contribution to decreasing fossil fuel carbon emissions. *Global Change Biology* 10(4), 509–18. doi:10.1111/j.1529-8817.2003.00749.x.

Clifton-Brown, J. C., Breuer, J. and Jones, M. B. 2007. Carbon mitigation by the energy crop, *Miscanthus*. *Global Change Biology* 13(11), 2296–307. doi:10.1111/j.1365-2486.2007.01438.x.

Clifton-Brown, J., Hastings, A., Mos, M., McCalmont, J. P., Ashman, C., Awty-Carroll, D., Cerazy, J., Chiang, Y., Cosentino, S., Cracroft-Eley, W., Scurlock, J., Donnison, I. S., Glover, C., Gołąb, I., Greef, J. M., Gwyn, J., Harding, G., Hayes, C., Helios, W., Hsu, T., Huang, L. S., Jeżowski, S., Kim, D., Kiesel, A., Kotecki, A., Krzyzak, J., Lewandowski, I., Lim, S. H., Liu, J., Loosely, M., Meyer, H., Murphy-Bokern, D., Nelson, W., Pogrzeba, M., Robinson, G., Robson, P., Rogers, C., Scalici, G., Schuele, H., Shafiei, R., Shevchuk, O., Schwarz, K., Squance, M., Swaller, T., Thornton, J., Truckses, T., Botnari, V., Vizir, I., Wagner, M., Warren, R., Webster, R., Yamada, T., Youell, S., Xi, Q., Zong, J. and Flavell, R. 2017. Progress in upscaling *Miscanthus* biomass production for the European bio-economy with seed-based hybrids. *GCB Bioenergy* 9(1), 6–17. doi:10.1111/gcbb.12357.

Clifton-Brown, J., Harfouche, A., Casler, M. D., Dylan Jones, H., Macalpine, W. J., Murphy-Bokern, D., Smart, L. B., Adler, A., Ashman, C., Awty-Carroll, D., Bastien, C., Bopper, S., Botnari, V., Brancourt-Hulmel, M., Chen, Z., Clark, L. V., Cosentino, S., Dalton, S., Davey, C., Dolstra, O., Donnison, I., Flavell, R., Greef, J., Hanley, S., Hastings, A., Hertzberg, M., Hsu, T. W., Huang, L. S., Iurato, A., Jensen, E., Jin, X., Jørgensen, U., Kiesel, A., Kim, D. S., Liu, J., McCalmont, J. P., McMahon, B. G., Mos, M., Robson, P., Sacks, E. J., Sandu, A., Scalici, G., Schwarz, K., Scordia, D., Shafiei, R., Shield, I., Slavov, G., Stanton, B. J., Swaminathan, K., Taylor, G., Torres, A. F., Trindade, L. M., Tschaplinski, T., Tuskan, G. A., Yamada, T., Yeon Yu, C., Zalesny, R. S., Zong, J. and Lewandowski, I. 2019. Breeding progress and preparedness for mass-scale deployment of perennial lignocellulosic biomass crops switchgrass, *Miscanthus*, willow, and poplar. *GCB Bioenergy* 11(1), 118–51. doi:10.1111/gcbb.12566.

Davey, C. L., Robson, P., Hawkins, S., Farrar, K., Clifton-Brown, J. C., Donnison, I. S. and Slavov, G. T. 2017. Genetic relationships between spring emergence, canopy phenology, and biomass yield increase the accuracy of genomic prediction in *Miscanthus*. *Journal of Experimental Botany* 68(18), 5093–102. doi:10.1093/jxb/erx339.

Davis, S. C., Parton, W. J., Dohleman, F. G., Smith, C. M., Del Grosso, S. D., Kent, A. D. and DeLucia, E. H. 2010. Comparative biogeochemical cycles of bioenergy crops reveal nitrogen-fixation and low greenhouse gas emissions in a *Miscanthus × giganteus* agro-ecosystem. *Ecosystems* 13(1), 144–56. doi:10.1007/s10021-009-9306-9.

DEFRA. 2007. Department of Food and Rural Affairs (DEFRA) UK biomass strategy. Available at: http://www.globalbioenergy.org/uploads/media/0705_Defra_-_UK_Biomass_Strategy_01.pdf (accessed on 17 October 2018).

Dohleman, F. G., Heaton, E. A., Arundale, R. A. and Long, S. P. 2012. Seasonal dynamics of above- and below-ground biomass and nitrogen partitioning in *Miscanthus × giganteus* and *Panicum virgatum* across three growing seasons. *GCB Bioenergy* 4(5), 534–44. doi:10.1111/j.1757-1707.2011.01153.x.

Dondini, M., Van Groenigen, K.-J., Del Galdo, I. and Jones, M. B. 2009. Carbon sequestration under *Miscanthus*: a study of 13C distribution in soil aggregates. *GCB Bioenergy* 1(5), 321–30. doi:10.1111/j.1757-1707.2009.01025.x.

Eckert, B., Weber, O. B., Kirchhof, G., Halbritter, A., Stoffels, M. and Hartmann, A. 2001. *Azospirillum doebereinerae* sp. nov., a nitrogen-fixing bacterium associated with the C_4-grass *Miscanthus*. *International Journal of Systematic and Evolutionary Microbiology* 51(1), 17–26. doi:10.1099/00207713-51-1-17.

Ezaki, B., Nagao, E., Yamamoto, Y., Nakashima, S. and Enomoto, T. 2008. Wild plants, *Andropogon virginicus* L. and *Miscanthus sinensis* Anders, are tolerant to multiple stresses including aluminum, heavy metals and oxidative stresses. *Plant Cell Reports* 27(5), 951–61. doi:10.1007/s00299-007-0503-8.

Falano, T., Jeswani, H. K. and Azapagic, A. 2014. Assessing the environmental sustainability of ethanol from integrated biorefineries. *Biotechnology Journal* 9(6), 753–65. doi:10.1002/biot.201300246.

Fazio, S. and Monti, A. 2011. Life cycle assessment of different bioenergy production systems including perennial and annual crops. *Biomass and Bioenergy* 35(12), 4868–78. doi:10.1016/j.biombioe.2011.10.014.

Felten, D. and Emmerling, C. 2011. Effects of bioenergy crop cultivation on earthworm communities-a comparative study of perennial (*Miscanthus*) and annual crops with consideration of graded land-use intensity. *Applied Soil Ecology* 49, 167–77. doi:10.1016/j.apsoil.2011.06.001.

Felten, D., Fröba, N., Fries, J. and Emmerling, C. 2013. Energy balances and greenhouse gas-mitigation potentials of bioenergy cropping systems (*Miscanthus*, rapeseed, and maize) based on farming conditions in Western Germany. *Renewable Energy* 55, 160–74. doi:10.1016/j.renene.2012.12.004.

Fonteyne, S., Muylle, H., Lootens, P., Kerchev, P., Van Den Ende, W., Staelens, A., Reheul, D. and Roldán-Ruiz, I. 2018. Physiological basis of chilling tolerance and early-season growth in *Miscanthus*. *Annals of Botany* 121(2), 281–95. doi:10.1093/aob/mcx159.

Gabrielle, B., Gagnaire, N., Massad, R. S., Dufossé, K. and Bessou, C. 2014. Environmental assessment of biofuel pathways in Ile de France based on ecosystem modeling. *Bioresource Technology* 152, 511–8. doi:10.1016/j.biortech.2013.10.104.

Goulding, K. W. T., Bailey, N. J., Bradbury, N. J., Hargreaves, P., Howe, M., Murphy, D. V., Poulton, P. R. and Willison, T. W. 1998. Nitrogen deposition and its contribution to nitrogen cycling and associated soil processes. *New Phytologist* 139, 49–58.

Greef, J. M. and Deuter, M. 1993. Syntaxonomy of *Miscanthus × giganteus* GREEF et DEU. *Angewandte Botanik* 67, 87–90.

Hansen, E. M., Christensen, B. T., Jensen, L. S. and Kristensen, K. 2004. Carbon sequestration in soil beneath long-term *Miscanthus* plantations as determined by ^{13}C abundance. *Biomass and Bioenergy* 26(2), 97–105. doi:10.1016/S0961-9534(03)00102-8.

Harvey, J. 2007. A versatile solution-growing *Miscanthus* for bioenergy. *Renewable Energy World* 10, 86.

Hastings, A. 2018. Bio-energies impact on natural capital and ecosystem services: a comparison of biomass and coal fuels. In: Qin, Z., Mishra, U. and Hastings, A. (Eds), *Bioenergy and Land Use Change*. The American Geophysical Union (AGU) and Wiley & Sons, Hoboken, NJ, pp. 83-97.

Hastings, A., Clifton-Brown, J., Wattenbach, M., Mitchell, C. P., Stampfl, P. and Smith, P. 2009. Future energy potential of *Miscanthus* in Europe. *GCB Bioenergy* 1(2), 180–96. doi:10.1111/j.1757-1707.2009.01012.x.

Hastings, A., Yeluripati, J., Hillier, J. and Smith, P. 2012. Biofuel crops and greenhouse gasses. In: Singh, B. P. (Ed.), *Biofuel Crop Sustainability*. John Wiley & Sons Inc., Ames, IA, pp. 383–406. Chapter 12. ISBN 978-0-470-96304-3.

Hastings, A., Tallis, M. J., Casella, E., Matthews, R. W., Henshall, P. A., Milner, S., Smith, P. and Taylor, G. 2014. The technical potential of Great Britain to produce ligno-cellulosic biomass for bioenergy in current and future climates. *GCB Bioenergy* 6(2), 108–22. doi:10.1111/gcbb.12103.

Hastings, A., Mos, M., Yesufu, J. A., McCalmont, J., Schwarz, K., Shafei, R., Ashman, C., Nunn, C., Schuele, H., Cosentino, S., Scalici, G., Scordia, D., Wagner, M. and Clifton-Brown, J. 2017. Economic and environmental assessment of seed and rhizome propagated *Miscanthus* in the UK. *Frontiers in Plant Science* 8, 1058. doi:10.3389/fpls.2017.01058.

Heaton, E. A., Dohleman, F. G. and Long, S. P. 2008. Meeting US biofuel goals with less land: the potential of *Miscanthus*. *Global Change Biology* 14(9), 2000–14. doi:10.1111/j.1365-2486.2008.01662.x.

Heaton, E. A., Dohleman, F. G. and Long, S. P. 2009. Seasonal nitrogen dynamics of *Miscanthus* × *giganteus* and *Panicum virgatum*. *GCB Bioenergy* 1(4), 297–307. doi:10.1111/j.1757-1707.2009.01022.x.

Heaton, E. A., Dohleman, F. G., Miguez, A. F., Juvik, J. A., Lozovaya, V., Widholm, J., Zabotina, O. A., McIsaac, G. F., David, M. B., Voigt, T. B., Boersma, N. N. and Long, S. P. 2010. *Miscanthus*. A promising biomass crop. *Advances in Botanical Research* 56, 76–137.

Hickman, G. C., Vanlooche, A., Dohleman, F. G. and Bernacchi, C. J. 2010. A comparison of canopy evapotranspiration for maize and two perennial grasses identified as potential bioenergy crops. *GCB Bioenergy* 2, no. doi:10.1111/j.1757-1707.2010.01050.x.

Hijazi, O., Munro, S., Zerhusen, B. and Effenberger, M. 2016. Review of life cycle assessment for biogas production in Europe. *Renewable and Sustainable Energy Reviews* 54, 1291-300. doi:10.1016/j.rser.2015.10.013.

Hillier, J., Whittaker, C., Dailey, G., Aylott, M., Casella, E., Richter, G. M., Riche, A., Murphy, R., Taylor, G. and Smith, P. 2009. Greenhouse gas emissions from four bioenergy crops in England and Wales: integrating spatial estimates of yield and soil carbon balance in life cycle analyses. *GCB Bioenergy* 1(4), 267–81. doi:10.1111/j.1757-1707.2009.01021.x.

Hodgson, E. M., Lister, S. J., Bridgwater, A. V., Clifton-Brown, J. and Donnison, I. S. 2010. Genotypic and environmentally derived variation in the cell wall composition of *Miscanthus* in relation to its use as a biomass feedstock. *Biomass and Bioenergy* 34(5), 652–60. doi:10.1016/j.biombioe.2010.01.008.

Hodgson, E. M., Nowakowski, D. J., Shield, I., Riche, A., Bridgwater, A. V., Clifton-Brown, J. C. and Donnison, I. S. 2011. Variation in *Miscanthus* chemical composition and implications for conversion by pyrolysis and thermo-chemical bio-refining for fuels and chemicals. *Bioresource Technology* 102(3), 3411-8. doi:10.1016/j.biortech.2010.10.017.

Holder, A. J., McCalmont, J. P., McNamara, N. P., Rowe, R. and Donnison, I. S. 2018a. Evapotranspiration model comparison and an estimate of field scale *Miscanthus* canopy precipitation interception. *GCB Bioenergy* 10(5), 353-66. doi:10.1111/gcbb.12503.

Holder, A. J., McCalmont, J. P., Rowe, R., McNamara, N. P., Elias, D. and Donnison, I. S. 2018b. Soil N_2O emissions with different reduced tillage methods during the establishment of *Miscanthus* in temperate grassland. *GCB Bioenergy* 11, 539-49.

Holland, R. A., Eigenbrod, F., Muggeridge, A., Brown, G., Clarke, D. and Taylor, G. 2015. A synthesis of the ecosystem services impact of second generation bioenergy crop production. *Renewable and Sustainable Energy Reviews* 46, 30-40. doi:10.1016/j.rser.2015.02.003.

IPCC. 2018. Summary for policymakers. In: Masson-Delmotte, V., Zhai, P., Pörtner, H. O., Roberts, D., Skea, J., Shukla, P. R., Pirani, A., Moufouma-Okia, W., Péan, C., Pidcock, R., Connors, S., Matthews, J. B. R., Chen, Y., Zhou, X., Gomis, M. I., Lonnoy, E., Maycock, T., Tignor, M. and Waterfield, T. (Eds), *Global Warming of 1.5°C. An IPCC Special Report on the Impacts of Global Warming of 1.5°C above Pre-Industrial Levels and Related Global Greenhouse Gas Emission Pathways, in the Context of Strengthening the Global Response to the Threat of Climate Change, Sustainable Development, and Efforts to Eradicate Poverty*. World Meteorological Organization, Geneva, Switzerland, 32pp.

Jepson, W. and Caldas, M. 2017. Changing energy systems and land-use change. *Journal of Land Use Science* 12(6), 405-6. doi:10.1080/1747423X.2017.1408889.

Jones, M. B. 2011. C_4 species as energy crops. In: *C_4 Photosynthesis and Related CO_2 Concentrating Mechanisms*. Advances in Photosynthesis and Respiration (vol. 32). Springer, Dordrecht, The Netherlands, pp. 379-97.

Jones, M. B., Finnan, J. and Hodkinson, T. R. 2015. Morphological and physiological traits for higher biomass production in perennial rhizomatous grasses grown on marginal land. *GCB Bioenergy* 7(2), 375-85. doi:10.1111/gcbb.12203.

Jørgensen, U. 1995. Low cost and safe establishment of *Miscanthus*. In: Chartier, P., Beenackers, A. A. C. M. and Grassi, G. (Eds), *Biomass for Energy, Environment, Agriculture and Industry*. Elsevier, Oxford, pp. 541-7.

Jørgensen, S. V., Cherubini, F. and Michelsen, O. 2014. Biogenic CO_2 fluxes, changes in surface albedo and biodiversity impacts from establishment of a *Miscanthus* plantation. *Journal of Environmental Management* 146, 346-54. doi:10.1016/j.jenvman.2014.06.033.

Kahle, P., Beuch, S., Boelcke, B., Leinweber, P. and Schulten, H. R. 2001. Cropping of *Miscanthus* in Central Europe: biomass production and influence on nutrients and soil organic matter. *European Journal of Agronomy* 15(3), 171-84. doi:10.1016/S1161-0301(01)00102-2.

Kalinina, O., Nunn, C., Sanderson, R., Hastings, A. F. S., van der Weijde, T., Özgüven, M., Tarakanov, I., Schüle, H., Trindade, L. M., Dolstra, O., Schwarz, K. U., Iqbal, Y., Kiesel, A., Mos, M., Lewandowski, I. and Clifton-Brown, J. C. 2017. Extending *Miscanthus* cultivation with novel germplasm at six contrasting sites. *Frontiers in Plant Science* 8, 563. doi:10.3389/fpls.2017.00563.

Kemper, J. 2015. Biomass and carbon dioxide capture and storage: a review. *International Journal of Greenhouse Gas Control* 40, 401-30. doi:10.1016/j.ijggc.2015.06.012.

Keymer, D. P. and Kent, A. D. 2014. Contribution of nitrogen fixation to first year *Miscanthus × giganteus*. *GCB Bioenergy* 6(5), 577–86. doi:10.1111/gcbb.12095.

Kiesel, A. and Lewandowski, I. 2017. *Miscanthus* as biogas substrate - cutting tolerance and potential for anaerobic digestion. *GCB Bioenergy* 9(1), 153-67. doi:10.1111/gcbb.12330.

Kiesel, A., Nunn, C., Iqbal, Y., Van der Weijde, T., Wagner, M., Özgüven, M., Tarakanov, I., Kalinina, O., Trindade, L. M., Clifton-Brown, J. and Lewandowski, I. 2017a. Site-specific management of *Miscanthus* genotypes for combustion and anaerobic digestion: a comparison of energy yields. *Frontiers in Plant Science* 8, 347. doi:10.3389/fpls.2017.00347.

Kiesel, A., Wagner, M. and Lewandowski, I. 2017b. Environmental performance of *Miscanthus*, switchgrass and maize: can C_4 perennials increase the sustainability of biogas production? *Sustainability* 9(1), 5. doi:10.3390/su9010005.

Kittelman, H. 2014. Carbo-V® biomass gasification technology. Status after application of sound engineering practices. Presented at *IAE Workshop*, Karlesruhe, 4 November 2014. Available at: www.ieatask33.org/app/webroot/files/file/2014/WS2/Kittelmann.pdf.

LeBauer, D., Kooper, R., Mulrooney, P., Rohde, S., Wand, D., Long, S. P. and Dietze, M. C. 2018. BETYdb: a yield, trait, and ecosystem service database applied to second-generation bioenergy feedstock production. *GCB Bioenergy* 10(1), 61–71. doi:10.1111/gcbb.12420.

Le Ngoc Huyen, T., Rémond, C., Dheilly, R. M. and Chabbert, B. 2010. Effect of harvesting date on the composition and saccharification of *Miscanthus × giganteus*. *Bioresource Technology* 101(21), 8224–31. doi:10.1016/j.biortech.2010.05.087.

Lesur, C., Jeuffroy, M. H., Makowski, D., Riche, A. B., Shield, I., Yates, N., Fritz, M., Formowitz, B., Grunert, M., Jorgensen, U., Laerke, P. E. and Loyce, C. 2013. Modeling long-term yield trends of *Miscanthus × giganteus* using experimental data from across Europe. *Field Crops Research* 149, 252–60. doi:10.1016/j.fcr.2013.05.004.

Lewandowski, I. and Heinz, A. 2003. Delayed harvest of *Miscanthus* – influences on biomass quantity and quality and environmental impacts of energy production. *European Journal of Agronomy* 19(1), 45–63. doi:10.1016/S1161-0301(02)00018-7.

Lewandowski, I. and Kicherer, A. 1997. Combustion quality of biomass: practical relevance and experiments to modify the biomass quality of *Miscanthus × giganteus*. *European Journal of Agronomy* 6(3–4), 163–77. doi:10.1016/S1161-0301(96)02044-8.

Lewandowski, I. and Schmidt, U. 2006. Nitrogen, energy and land use efficiencies of *Miscanthus*, reed canary grass and triticale as determined by the boundary line approach. *Agriculture, Ecosystems and Environment* 112(4), 335–46. doi:10.1016/j.agee.2005.08.003.

Lewandowski, I., Clifton-Brown, J. C., Scurlock, J. M. O. and Huisman, W. 2000. *Miscanthus*: European experience with a novel energy crop. *Biomass and Bioenergy* 19(4), 209–27. doi:10.1016/S0961-9534(00)00032-5.

Lewandowski, I., Clifton-Brown, J. C., Andersson, B., Basch, G., Christian, D. G., Jørgensen, U., Jones, M. B., Riche, A. B., Schwarz, K. U., Tayebi, K. and Teixeira, F. 2003a. Environment and harvest time affects the combustion qualities of *Miscanthus* genotypes. *Agronomy Journal* 95(5), 1274–80. doi:10.2134/agronj2003.1274.

Lewandowski, I., Scurlock, J. M. O., Lindvall, E. and Christou, M. 2003b. The development and current status of perennial rhizomatous grasses as energy crops in the US and Europe. *Biomass and Bioenergy* 25(4), 335–61. doi:10.1016/S0961-9534(03)00030-8.

Lijó, L., González-García, S., Bacenetti, J., Fiala, M., Feijoo, G. and Moreira, M. T. 2014. Assuring the sustainable production of biogas from anaerobic mono-digestion. *Journal of Cleaner Production* 72, 23–34. doi:10.1016/j.jclepro.2014.03.022.

Linde-Laursen, I. 1993. Cytogenetic analysis of *Miscanthus 'Giganteus'*, an interspecific hybrid. *Hereditas* 119, 297–300.

Lovett, A. A., Sünnenberg, G. M., Richter, G. M., Dailey, A. G., Riche, A. B. and Karp, A. 2009. Land use implications of increased biomass production identified by gis-based suitability and yield mapping for miscanthus in England. *Bioenergy Research* 2(1–2), 17–28. doi:10.1007/s12155-008-9030-x.

Lovett, A., Sünnenberg, G. and Dockerty, T. 2014. The availability of land for perennial energy crops in Great Britain. *GCB Bioenergy* 6(2), 99–107. doi:10.1111/gcbb.12147.

Malinowska, M., Donnison, I. S. and Robson, P. R. H. 2017. Phenomics analysis of drought responses in Miscanthus collected from different geographical locations. *GCB Bioenergy* 9(1), 78–91. doi:10.1111/gcbb.12350.

Mangold, A., Lewandowski, I., Möhring, J., Clifton-Brown, J., Krzyżak, J., Mos, M., Pogrzeba, M. and Kiesel, A. 2019. Harvest date and leaf:stem ratio determine methane hectare yield of miscanthus biomass. *GCB Bioenergy* (in press). doi:10.1111/gcbb.12549.

McCalmont, J. P., Hastings, A., McNamara, N. P., Richter, G. M., Robson, P., Donnison, I. S. and Clifton-Brown, J. 2017. Environmental costs and benefits of growing *Miscanthus* for bioenergy in the UK. *GCB Bioenergy* 9(3), 489–507. doi:10.1111/gcbb.12294.

McCalmont, J. P., Rowe, R., Elias, D., Whitaker, J., McNamara, N. P. and Donnison, I. S. 2018. Soil nitrous oxide flux following land-use reversion from *Miscanthus* and SRC willow to perennial ryegrass. *GCB Bioenergy* 10(12), 914–29. doi:10.1111/gcbb.12541.

Meyer, F., Wagner, M. and Lewandowski, I. 2017. Optimizing GHG emission and energy-saving performance of *Miscanthus*-based value chains. *Biomass Conversion and Biorefinery* 7(2), 139–52. doi:10.1007/s13399-016-0219-5.

Miedema, P., Post, J. and Groot, P. 1987. The effects of low temperature on seedling growth in maize genotypes. *Agricultural Research Reports Wageningen: Pudoc* 926, 124.

Miyamoto, T., Kawahara, M. and Minamisawa, K. 2004. Novel endophytic nitrogen-fixing clostridia from the grass *Miscanthus sinensis* as revealed by terminal restriction fragment length polymorphism analysis. *Applied and Environmental Microbiology* 70(11), 6580–6. doi:10.1128/AEM.70.11.6580-6586.2004.

Morison, J. I. L., Piedade, M. T. F., Müller, E., Long, S. P., Junk, W. J. and Jones, M. B. 2000. Very high productivity of the C_4 aquatic grass *Echinochloa polystachya* in the Amazon floodplain confirmed by net ecosystem CO_2 flux measurements. *Oecologia* 125(3), 400–11. doi:10.1007/s004420000464.

Mos, M., Banks, S. W., Nowakowski, D. J., Robson, P. R. H., Bridgwater, A. V. and Donnison, I. S. 2013. Impact of *Miscanthus × giganteus* senescence times on fast pyrolysis bio-oil quality. *Bioresource Technology* 129, 335–42. doi:10.1016/j.biortech.2012.11.069.

Muñoz, I., Schmidt, J. H., Brandão, M. and Weidema, B. P. 2014. Avoiding the streetlight effect: Rebuttal to 'Indirect land use change (iLUC) within life cycle assessment (LCA) - scientific robustness and consistency with international standards' by Prof. Dr. Matthias Finkbeiner, pp. 1–13.

Ogunsona, E. O., Misra, M. and Mohanty, A. K. 2017. Sustainable biocomposites from biobased polyamide 6,10 and biocarbon from pyrolyzed *Miscanthus* fibers. *Journal of Applied Polymer Science* 134(4), 44221. doi:10.1002/app.44221.

Pogson, M., Richards, M., Dondini, M., Jones, E. O., Hastings, A. and Smith, P. 2016. ELUM: a spatial modelling tool to predict soil greenhouse gas changes from land conversion to bioenergy in the UK. *Environmental Modelling and Software* 84, 458–66. doi:10.1016/j.envsoft.2016.07.011.

Qin, Z., Dunn, J. B., Kwon, H., Mueller, S. and Wander, M. M. 2016. Soil carbon sequestration and land use change associated with biofuel production: empirical evidence. *GCB Bioenergy* 8(1), 66–80. doi:10.1111/gcbb.12237.

Richards, M., Pogson, M., Dondini, M., Jones, E. O., Hastings, A., Henner, D. N., Tallis, M. J., Casella, E., Matthews, R. W., Henshall, P. A., Milner, S., Taylor, G., McNamara, N. P., Smith, J. U. and Smith, P. 2017. High-resolution spatial modelling of greenhouse gas emissions from land-use change to energy crops in the United Kingdom. *GCB Bioenergy* 9(3), 627–44. doi:10.1111/gcbb.12360.

Robertson, A. D., Whitaker, J., Morrison, R., Davies, C. A., Smith, P. and McNamara, N. P. 2017. A *Miscanthus* plantation can be carbon neutral without increasing soil carbon stocks. *GCB Bioenergy* 9(3), 645–61. doi:10.1111/gcbb.12397.

Robson, P., Mos, M., Clifton-Brown, J. and Donnison, I. 2012. Phenotypic variation in senescence in *Miscanthus*: towards optimising biomass quality and quantity. *BioEnergy Research* 5(1), 95–105. doi:10.1007/s12155-011-9118-6.

Rockstrom, J., Gordon, L., Folke, C., Kalkenmark, M. and Engwall, M. 1999. Linkages among water vapor flows, food production, and terrestrial ecosystem services. *Conservation Ecology* 3(2), 5. doi:10.5751/ES-00142-030205.

Roth, B., Finnan, J. M., Jones, M. B., Burke, J. I. and Williams, M. L. 2015. Are the benefits of yield responses to nitrogen fertilizer application in the bioenergy crop *Miscanthus × giganteus* offset by increased soil emissions of nitrous oxide? *GCB Bioenergy* 7(1), 145–52. doi:10.1111/gcbb.12125.

Royal Society. 2018. Greenhouse gas removal. Available at: https://royalsociety.org/~/media/policy/projects/greenhouse-gas-removal/royal-society-greenhouse-gas-removal-report-2018.pdf (accessed October 2018).

Sage, R. F., Kocacinar, F. and Kubien, D. S. 2010. C_4 photosynthesis and temperature. In: Raghavendra, A. and Sage, R. (Eds), *C_4 Photosynthesis and Related CO_2 Concentrating Mechanisms*. Advances in Photosynthesis and Respiration (vol. 32). Springer, Dordrecht, pp. 161–95. Chapter 10.

Shield, I. F., Barraclough, T. J. P., Riche, A. B. and Yates, N. E. 2014. The yield and quality response of the energy grass *Miscanthus × giganteus* to fertiliser applications of nitrogen, potassium and sulphur. *Biomass and Bioenergy* 68, 185–94. doi:10.1016/j.biombioe.2014.06.007.

Slavov, G. T., Allison, G. and Bosch, M. 2013. Advances in the genetic dissection of plant cell walls: tools and resources available in *Miscanthus*. *Frontiers in Plant Science* 4, 217. doi:10.3389/fpls.2013.00217.

Slavov, G. T., Nipper, R., Robson, P., Farrar, K., Allison, G. G., Bosch, M., Clifton-Brown, J. C., Donnison, I. S. and Jensen, E. 2014. Genome-wide association studies and prediction of 17 traits related to phenology, biomass and cell wall composition in the energy grass *Miscanthus sinensis*. *The New Phytologist* 201(4), 1227–39. doi:10.1111/nph.12621.

Slavov, G. T., Davey, C. L., Bosch, M., Robson, P. R. H., Donnison, I. S. and Mackay, I. J. 2018. Genomic index selection provides a pragmatic framework for setting and refining multi-objective breeding targets in *Miscanthus*. *Annals of Botany*. doi:10.1093/aob/mcy187.

Snyder, C. S., Bruulsema, T. W., Jensen, T. L. and Fixen, P. E. 2009. Review of greenhouse gas emissions from crop production systems and fertilizer management effects. *Agriculture, Ecosystems and Environment* 133(3–4), 247–66. doi:10.1016/j.agee.2009.04.021.

Stavridou, E., Hastings, A., Webster, R. J. and Robson, P. R. H. 2017. The impact of soil salinity on the yield, composition and physiology of the bioenergy grass *Miscanthus × giganteus*. *GCB Bioenergy* 9(1), 92–104. doi:10.1111/gcbb.12351.

Styles, D. and Jones, M. B. 2007. Energy crops in Ireland: quantifying the potential life-cycle greenhouse gas reductions of energy-crop electricity. *Biomass and Bioenergy* 31(11–12), 759–72. doi:10.1016/j.biombioe.2007.05.003.

Styles, D., Gibbons, J., Williams, A. P., Dauber, J., Stichnothe, H., Urban, B., Chadwick, D. R. and Jones, D. L. 2015. Consequential life cycle assessment of biogas, biofuel and biomass energy options within an arable crop rotation. *GCB Bioenergy* 7(6), 1305–20. doi:10.1111/gcbb.12246.

Tallis, M. J., Casella, E., Henshall, P. A., Aylott, M. J., Randle, T. J., Morison, J. I. L. and Taylor, G. 2013. Development and evaluation of ForestGrowth-SRC a process-based model for short rotation coppice yield and spatial supply reveals poplar uses water more efficiently than willow. *GCB Bioenergy* 5(1), 53–66. doi:10.1111/j.1757-1707.2012.01191.x.

Tonini, D., Hamelin, L., Wenzel, H. and Astrup, T. 2012. Bioenergy production from perennial energy crops: a consequential LCA of 12 bioenergy scenarios including land use changes. *Environmental Science and Technology* 46(24), 13521–30. doi:10.1021/es3024435.

Turner, P. A., Mach, K. J., Lobell, D. B., Benson, S. M., Baik, E., Sanchez, D. L. and Field, C. B. 2018. The global overlap of bioenergy and carbon sequestration potential. *Climatic Change* 148(1–2), 1–10. doi:10.1007/s10584-018-2189-z.

Valentine, J., Clifton-Brown, J., Hastings, A., Robson, P., Allison, G. and Smith, P. 2012. Food vs. fuel: the use of land for lignocellulosic 'next generation' energy crops that minimize competition with primary food production. *GCB Bioenergy* 4(1), 1–19. doi:10.1111/j.1757-1707.2011.01111.x.

van der Weijde, T., Huxley, L. M., Hawkins, S., Sembiring, E. H., Farrar, K., Dolstra, O., Visser, R. G. F. and Trindade, L. M. 2017. Impact of drought stress on growth and quality of *Miscanthus* for biofuel production. *GCB Bioenergy* 9(4), 770–82. doi:10.1111/gcbb.12382.

Vaughan, N. E., Gough, C., Mander, S., Littleton, E. W., Welfle, A., Gernaat, D. E. H. J. and Van Vuuren, D. P. 2018. Evaluating the use of biomass energy with carbon capture and storage in low emission scenarios. *Environmental Research Letters* 13(4). doi:10.1088/1748-9326/aaaa02.

Wagner, M., Kiesel, A., Hastings, A., Iqbal, Y. and Lewandowski, I. 2017. Novel miscanthus germplasm-based value chains: a life cycle assessment. *Frontiers in Plant Science* 8, 990. doi:10.3389/fpls.2017.00990.

Wagner, M., Mangold, A., Lask, J., Petig, E., Kiesel, A. and Lewandowski, I. 2019. Economic and environmental performance of miscanthus cultivated on marginal land for biogas production. *GCB Bioenergy* 11(1), 34–49. doi:10.1111/gcbb.12567.

Wang, S., Wang, S., Hastings, A., Pogson, M. and Smith, P. 2012. Economic and greenhouse gas costs of *Miscanthus* supply chains in the United Kingdom. *GCB Bioenergy* 4(3), 358–63. doi:10.1111/j.1757-1707.2011.01125.x.

Welker, C. M., Balasubramanian, V. K., Petti, C., Rai, K. M., De Bolt, S. and Mendu, V. 2015. Engineering plant biomass lignin content and composition for biofuels and bioproducts. *Energies* 8(8), 7654–76. doi:10.3390/en8087654.

Whitaker, J., Field, J. L., Bernacchi, C. J., Cerri, C. E. P., Ceulemans, R., Davies, C. A., DeLucia, E. H., Donnison, I. S., McCalmont, J. P., Paustian, K., Rowe, R. L., Smith, P., Thornley, P. and McNamara, N. P. 2018. Consensus, uncertainties and challenges for perennial bioenergy crops and land use. *GCB Bioenergy* 10(3), 150–64. doi:10.1111/gcbb.12488.

Zang, H., Blagodatskaya, E., Wen, Y., Xu, X., Dyckmans, J. and Kuzyakov, Y. 2018. Carbon sequestration and turnover in soil under the energy crop *Miscanthus*: repeated ^{13}C natural abundance approach and literature synthesis. *GCB Bioenergy* 10(4), 262-71. doi:10.1111/gcbb.12485.

Zatta, A., Clifton-Brown, J., Robson, P., Hastings, A. and Monti, A. 2014. Land use change from C_3 grassland to C_4 *Miscanthus*: effects on soil carbon content and estimated mitigation benefit after six years. *GCB Bioenergy* 6(4), 360–70. doi:10.1111/gcbb.12054.

Zeng, Y., Zhao, S., Yang, S. and Ding, S. Y. 2014. Lignin plays a negative role in the biochemical process for producing lignocellulosic biofuels. *Current Opinion in Biotechnology* 27, 38–45. doi:10.1016/j.copbio.2013.09.008.

Zhu, X., Liang, C., Masters, M. D., Kantola, I. B. and DeLucia, E. H. 2018. The impacts of four potential bioenergy crops on soil carbon dynamics as shown by biomarker analyses and DRIFT spectroscopy. *GCB Bioenergy* 10(7), 489–500. doi:10.1111/gcbb.12520.

Zhuang, Q., Qin, Z. and Chen, M. 2013. Biofuel, land and water: maize, switchgrass or *Miscanthus*? *Environmental Research Letters* 8(1). doi:10.1088/1748-9326/8/1/015020.

Zimmermann, J., Dauber, J. and Jones, M. B. 2012. Soil carbon sequestration during the establishment phase of *Miscanthus* × *giganteus*: a regional-scale study on commercial farms using ^{13}C natural abundance. *GCB Bioenergy* 4(4), 453-61. doi:10.1111/j.1757-1707.2011.01117.x.

Zimmermann, J., Dondini, M. and Jones, M. B. 2013. Assessing the impacts of the establishment of miscanthus on soil organic carbon on two contrasting land-use types in Ireland. *European Journal of Soil Science* 64, 747–56.

Chapter 8

Sustainable use of switchgrass for biofuel

John Fike, Virginia Tech, USA; Vance Owens, South Dakota State University, USA; David Parrish, Virginia Tech, USA; and Rana Genedy, Cairo, Egypt

1 Introduction

Energy cropping systems have been promoted on the basis of delivering energy independence and national security, as an engine of economic development for rural economies and as a means of addressing the challenge of climate change. Researchers who consider that bioenergy - and specifically a switchgrass-to-biofuel system is feasible - are working to answer the question, 'Where will switchgrass fit on the agricultural landscape?' (Mitchell et al., 2012) and this is often discussed in the context of greater sustainability. However, and given the current state of industry development, it may be as fitting to ask a more basic question: 'Will switchgrass *be* on the agricultural landscape?' To clarify, 'Will switchgrass be grown at any scale relevant for meeting a significant portion of US energy system needs?' Although we will not and cannot answer questions of whether dedicated energy-cropping biofuel systems will be developed on a grand scale, this chapter considers several of the issues currently affecting bioenergy system development in a switchgrass-based context; it does not address the production of other bioproducts from either switchgrass directly or from byproducts of conversion processes. Although we recognize that bioproducts will play an important role in economics of switchgrass-to-biofuel

http://dx.doi.org/10.19103/AS.2019.0027.16

systems, we nonetheless are limiting this discussion to biofuel development. Individual sections will consider issues related to adaptation and productivity, agronomics, logistics, processing and conversion, environmental sustainability and economics (including competing crops). Although some of these issues may challenge systems development, the use of switchgrass (*Panicum virgatum* L.) and other dedicated biofuels systems may present opportunities that are not possible with current monocultural cropping systems. While the jury is out, there is cause for optimism at least on some fronts.

Challenges for bioenergy development span the spectrum of environmental, technological, economic and sociopolitical domains, and most of these issues intersect and interact. Many of these challenges are beyond the scope of a chapter on sustainable use of switchgrass for biofuel. However, readers must understand that independent of agronomy and productivity, the feasibility and sustainability of switchgrass or any species used in dedicated energy systems will be intrinsically linked to this broader set of factors. As an example, Federal policy of the Energy Independence and Security Act of 2007 (Pub.L. 110–140) set in place mandates for United States biofuel production of 30 billion gallons (ethanol equivalent) by 2020, with not more than half (15 billion gallons) of this renewable fuel to be derived from conventional, 'first generation' feedstock sources such as corn (*Zea mays* L.) or other grains. However, and despite a policy framework that ostensibly supports renewables, production of cellulosic and other advanced biofuels has fallen far (almost 30%) short of government targets (USEIA, 2018). In response to this slow growth, the US EPA used its waiver authority to lower the cellulosic biofuel volume standards (USEIA, 2018), a move that many in the industry felt reduced overall demand for research to drive the growth in cellulosic biofuels. This direct evidence of a lack of industry development begs the questions, 'What are the factors surrounding the low – dare we say disappointing – level of biofuel production system development in the US? And how might they be overcome?'

2 Switchgrass and environmental sustainability metrics

Development of a perennial, grass-based bioenergy industry faces several issues, and all impact sustainability in some manner. Thus, before proceeding, we digress with a discussion of what we consider to be 'sustainable'. Metaphorically, sustainability often is described as a three-legged stool. Each 'leg' – economic viability, environmental soundness and social acceptability – is critical to the overall 'standability' and stability of the stool (i.e. the practice being considered). First-generation biofuels, based on corn or other grain production systems, have been questioned for their sustainability across all three criteria. Although they have provided opportunity for corn growers and alcohol refiners – and were expected to jumpstart the second generation biofuel industry - grain-based biofuel systems may present high (if unaccounted) social costs. Soil erosion, pesticide and nutrient losses to the environment, loss of species diversity, reductions in soil health and consequences for water quality are among the concerns (e.g. see Fike et al., 2007). Although this type of system may make economic sense at the grower level, the production may be subsidized directly or indirectly by federal policy (e.g. through the renewable fuels standard) and the many environmental costs typically are externalized. In either case, society pays for these systems and their impacts.

While there is plenty of debate about the merit of first-generation biofuels, we take the position that sustainable energy cropping with perennial species can and should positively affect a number of environmental sustainability metrics - and generate positive economic returns. In the specific case of switchgrass, with proper management, the species should positively affect the environment through improvements to key indicators such as greenhouse gas (GHG) emissions, soil organic C (SOC), water quantity and quality, and biodiversity (e.g. wildlife habitat), and we briefly consider those factors here.

2.1 Greenhouse gases

Although energy cropping systems are inherently intended to reduce GHG, concerns have been raised that biofuels production will not meet this objective (e.g. see Searchinger et al., 2008). Even in cases where a net benefit may be predicted to occur in terms of reduced fossil-based GHG, the CO_2 cost associated with converting land from one system to another may require years to overcome. Along with CO_2, N_2O and CH_4, appropriate system boundaries must be considered and adequately accounted to evaluate the GHG balance of energy cropping systems (Smith and Searchinger, 2012; Smith, 2012). Greater effort to understand greenhouse gas emissions in a life-cycle context is warranted. For example, a 5-year study on 10 farms in Nebraska, South Dakota and North Dakota, indicated average GHG emissions from switchgrass-based ethanol would be 94% lower than estimated GHG emissions from gasoline (Schmer et al., 2008). However, N_2O emissions and their contribution to the switchgrass system apparently were not evaluated.

Perennials typically have much lower direct N_2O emissions than annuals, and emissions are lower in low-fertility production systems (Oates et al., 2016). Switchgrass position on the landscape also matters. For example, in a multi-year study with switchgrass grown across landscape positions, Mbonimpa et al. (2015) found that CO_2 fluxes were higher on footslope than shoulder slope positions while nitrogen (N) rate (0, 56 and 112 kg ha^{-1}) had no impact on CO_2 flux. However, their group later reported that N_2O fluxes were often higher with the higher N rate at this same location (Lai et al., 2018b). Based on these evaluations, they recommended a N rate of no higher than 56 kg N ha^{-1} in their environment, both because there was no yield advantage above this rate and in order to reduce greenhouse gas emissions.

2.2 Soil carbon

Perennial grass crops, in general, are considered soil conserving and soil improving, and in this regard, switchgrass is something of a superstar. The species is broadly adapted, with capacity to grow on a range of sites and soils, and the plant's dense foliage and massive, fibrous root system protects against erosion, improves water infiltration and may sink carbon at depth. Several studies have demonstrated that switchgrass grown for bioenergy can increase soil C levels (Ameen et al., 2019; Frank et al., 2004; Luo et al., 2002; Ma et al., 2000), but this will depend on site conditions and management (Chatterjee et al., 2018; Das et al., 2018). Switchgrass can be used by intensive livestock operations as a means to reduce manure loads. In fact, Lee et al. (2007) found that applying manure to switchgrass increased its ability to sequester C more than switchgrass to which equal rates of N fertilizer were applied. Increases in soil C are not achieved quickly, however. Ma et al. (2000) reported that N fertilizer and harvest frequency during switchgrass establishment did not change soil organic carbon (SOC) content in the short term (over 2-3 years). Compared with fallowed soil, switchgrass sequestered more SOC after 10 years, with increases of 44.8 and 28.2% at depths between 0–15 cm and 15–30 cm (Ma et al., 2000). In contrast, a 9-year study in the Great Plains found that SOC increased under switchgrass to depths of 150 cm; in the 0–30 cm layer, SOC was approximately doubled with applied N (Follett et al., 2012). Landscape position is another factor influencing soil properties, with higher SOC contents at the footslope position than at the midslope and shoulder positions throughout the soil profile to a depth of 100 cm in mature stands (Lai et al., 2018a). Switchgrass – as an energy crop, forage or conservation planting – is likely to be a good candidate on the basis of increasing soil C, particularly where SOC levels are low.

2.3 Water quantity

With regard to water resources, we consider dryland production systems only. We choose this tack simply because irrigating to produce biomass for energy production will face strong headwinds, except perhaps in a few very limited cases. In many regions, neither surface nor groundwater supplies are sufficient to sustain irrigation for bioenergy. In many agricultural settings, groundwater removal currently exceeds recharge. This practice is unsustainable, and even if the annual deficit is very small, it is only a matter of time until the water table will be too depressed to allow pumping. That said, producing switchgrass or other water-efficient biomass crops may be the future for agricultural regions with economies currently based on irrigated annual crop production. In some cases, elaborate schemes to replace alfalfa (*Medicago sativa*) and other water-intensive crops with switchgrass have been floated in arid regions of the United States. However, the payoff for the landowner would likely not be for biomass, but rather for rights to the 'saved' water, which could then be bought by thirsty urban areas.

Studies of water-use efficiency (WUE) seem equivocal regarding whether switchgrass is (VanLoocke et al., 2012) or is not (Hamilton et al., 2015) more water-use efficient than corn, which may depend on the calculation and if grain is included. Switchgrass is less water-use efficient than *Miscanthus × giganteus* (Miscanthus; Hamilton et al., 2015; VanLoocke et al., 2012) however, which may make the issue of efficiency relative to corn a moot point, at least in regions where Miscanthus is well-adapted. Regardless of which species predominates, seasonal water supply will be a driver for bioenergy systems, as switchgrass' high summer growth rates can result in end-of-season soil water depletion (Hendrickson et al., 2013).

2.4 Water quality

Any energy cropping system, whether based on switchgrass or other species, must not do damage to the quality of the water resource. Many grasses have an inherent advantage over annual crop species in this regard as a result of their perenniality (always covering the soil) and extensive root system, two characteristics that drastically reduce soil runoff and erosion. Reduced sediment runoff generally coincides with reduced nutrient runoff and reduced eutrophication of water bodies.

Several studies have shown the value of switchgrass filter strips for removing nutrients and contaminants (Lee et al., 1998; Mersie et al., 1999; Sanderson et al., 2001). A study of filter strips indicated that switchgrass supports reductions in runoff velocity and increased retention by soils which can reduce herbicide loss to the environment (Mersie et al., 1999). Switchgrass reduces NO_3^- leaching as stands age, regardless of landscape position or applied N rate (Lai et al., 2018b), helping to reduce environmental problems associated with N fertilizer application. Greater reductions in sediment, nutrient and herbicide losses would be expected where switchgrass is planted in landscape scale. Productivity of bioenergy crop systems could be supported by the application of waste products (Liu et al., 2014, 2015; Sanderson et al., 2001). Although the impacts of long-term repeated applications would need monitoring, it is likely this could be balanced with the effects of repeated harvest and removal (Lee et al., 2007; Oliveira et al., 2017).

2.5 Biodiversity

Converting large expanses of land to monoculture plantings of any species can be disconcerting given the potential negative consequences to ecosystem diversity and wildlife habitat, especially in sensitive sites such as wetlands, and this is particularly pertinent for annual row crops. There is general recognition that perennial species often provide better habitat than annuals, particularly under appropriate management schemes (e.g. proper harvest timing, employment of mixtures, etc.). Expansion of agricultural crops in the Prairie Pothole Region, USA, a region critical for migrating waterfowl, threatens the grasslands and wetlands that provide critical habitat and other ecosystem services (Johnston, 2014). From 2006–2011, nearly 530 000 ha were converted from grass-dominated land cover to agricultural row crops (Wright and Wimberly, 2013). Conversion away from grassland poses a serious threat to waterfowl breeding generally, and in the Prairie Pothole Region specifically.

Planting native species such as switchgrass has been proposed as a method of increasing grassland bird abundance, although there is recognition that large, monotypic stands may prove less than ideal. There is some concern about the quality of upland switchgrass ecotypes as habitat for ground-nesting birds, because these types can form dense stands that may prove impenetrable to fledglings. Harvest timing may provide one way to address this or other issues associated with diverse fauna (Roth et al., 2005). In fields with lowland switchgrass ecotypes, habitat limitations for ground-nesting birds may also be addressed through planting management. Lowland switchgrass maintains more of a bunch-type structure, maintaining high productivity with wide (up to 90 cm) row planting (Foster et al., 2013; Ma et al., 2001). Thus, wide-row configurations could meet the needs of both wildlife and biomass producer communities.

Despite the apparent potential for switchgrass to provide wildlife habitat, large monotypic stands may still not deliver the diversity needed across ecosystems. On the other hand, if switchgrass use were to displace a monoculture crop, there should be no net loss of biodiversity; indeed, with good management, biodiversity would increase. The diversity of these switchgrass-based systems could be further improved by adding other grass and forb species based on site suitability (Roth et al., 2005; Tilman et al., 2006b; Zilverberg et al., 2014, 2015).

In natural stands across the landscape, switchgrass is only dominant in a narrow ecological niche of North America's tallgrass prairie. This suggests that individual species or mixtures of species may be better adapted at various landscape positions and could support both greater biomass production and species diversity than switchgrass grown in monocultures. Evidence of this response was reported from a South Dakota study in which switchgrass monocultures were less productive than mixed stands at most landscape positions (Zilverberg et al., 2015). However, even switchgrass in monoculture is likely to have greater yield and yield stability when the stand is comprised of a mixture of different cultivars (DeGraff et al., 2016).

3 Challenges in becoming a primary feedstock resource

The previous discussion provides some basis for the general consensus that biomass-to-bioenergy systems based on perennial grasses are more sustainable than first generation, grain-based bioenergy systems – at least on one leg of the sustainability stool. Annual systems also face concerns about the negative effects of extensive residue removal and concomitant reductions in soil organic C, as well as the increased risk from an over-reliance on annual monocultures (Chavas and Holt, 1990). Perennial-based energy production could diversify farm production systems. They also would have a smaller environmental footprint in terms of lower energy inputs and reduced soil erosion and they would confer ecosystem services such as filtering water, improving soil health, sequestering carbon and providing wildlife habitat. However, the absence of wide-spread, second generation biofuel systems speaks volumes about the current state of the progress in perennial, biomass-based renewable energy, despite the potential environmental benefits. While environmentally based payments might improve this situation, paying for some combination of environmental benefits may still be insufficient to compete with the economic returns from existing, annual-based cropping systems (Noe et al., 2016). This is context and model-specific, however. For example, Noe et al. (2016) used Monte-Carlo simulation modeling to compare the value of corn-soy rotations with feedstocks derived from low-input, high-diversity prairies. However, the output from constructed prairies may underrepresent the potential productivity and conservation services of well-managed switchgrass stands on marginal lands (Schmer et al., 2008). Woodbury et al. (2018), too, suggested that switchgrass could be used to replace corn in the Chesapeake Bay drainage provided a \$25 Mg^{-1} ecosystem service payment (for N loss mitigation) was added to a \$111 $Mg\text{-}ha^{-1}$ farm-gate price for biomass. Bonner et al. (2016) suggested switchgrass incorporation into unprofitable subfields would increase supplies of biomass by over a third, with environmental values estimated at \$180 ha^{-1}. Of course, these models are only as good as their assumptions about biomass production, the degree and value of nutrient loss mitigation, and biomass-to-energy conversion (vis-a-vis energy gain and fossil C displacement), among other factors. The feasibility of pushing forward bioenergy systems with ecosystems service payments is likely to remain in question until society arrives at adequate ways to account and pay for ecosystem services in a direct, market-driven manner.

3.1 Competing feedstock supplies

Despite their promise for proving environmental benefits, and despite substantial federal, state and private investment, critics contend that biomass-to-biofuel systems built around dedicated, perennial feedstocks rarely make sufficient economic sense under existing market and policy realities (Foreman, 2018). The currently existing cellulosic energy industry in the United States largely has been built on the potential to collect corn stover (the residue from grain harvest) and convert it to liquid fuel (viz. ethanol). Such annual-crop-based systems have been thought economically viable because the harvest residues, considered an abundant by-product of the grain production system, represent a low-cost, readily available resource. The inputs needed to grow the stover essentially are heavily discounted or written off as the cost of producing the grain. In contrast, the accounting for a dedicated, perennial feedstock is entirely attributed to its production, resulting in higher total costs, despite the lower input requirements associated with these production systems.

Given the available supplies of stover throughout the US Midwest, some contend that switchgrass and other perennial species would only compete with crop residues in the southern Great Plains and southeastern United States (Dumortier, 2016). More recent efforts to understand and model yield potential of a variety of feedstocks (Lee et al., 2018) resulted in the development of production maps that suggest switchgrass may not be the feedstock of choice in much of the United States. Rather, the model indicates Miscanthus is likely to be more productive than switchgrass over much of the eastern half of the continental United States (Fig. 1). Switchgrass may be better suited to the Southeast and Northern Plains east of the 100th meridian, while Miscanthus appears well suited to much of the US Corn Belt. Switchgrass production in these regions would make additional sense in that this would fit with a common land use (beef cattle production on pasture) and allow farmers to direct the resource to the most profitable option (cattle, bioenergy or mixed use) depending on the value of these commodities (Biermacher et al., 2017). While more productive in other parts of the Southeast, Miscanthus may be less appealing in that region unless and until there are other markets to compete for the material.

While the model from Lee et al. (2018) would suggest only a limited role for switchgrass production in the Great Plains, Mitchell et al. (2016) offer a different perspective. While acknowledging that stover is likely to be the region's primary feedstock, they note that production of perennial species such as switchgrass has proven profitable on marginal croplands and that this practice can enhance ecosystem services. In much of this region, corn production is made possible with center pivot irrigation systems. Fields typically are square, and water is applied over the field in a circular pattern. The unirrigated corners of such

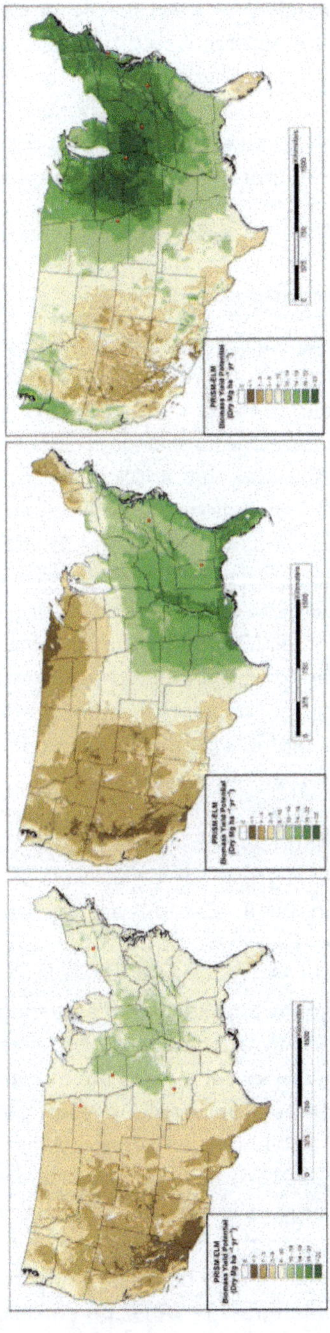

Figure 1 Yield potential maps for upland (left) and lowland (middle) switchgrass ecotypes and for Miscanthus × giganteus (right) developed from PRISM-ELM models. Source: adapted from Lee et al. (2018).

fields typically are planted with perennial, drought-tolerant crops. Uden et al. (2013) modeled the availability of biomass supply for ethanol production in an agricultural 'fuelshed', assuming corn would be produced under center pivot and switchgrass in the unirrigated field corners. The authors concluded that the combined harvest of corn stover and switchgrass (from the field corners) could provide enough biomass to support year-round production of cellulosic ethanol, increasing ethanol output while keeping marginal land under perennial cover.

3.2 Logistics and conversion issues

For any energy crop, sufficient acres and sufficient biomass must be produced within a given area to warrant the investment required to establish a proces facility. Studies from the heart of the Corn Belt note that grain production is not a net positive on all crop lands at the subfield level (Bonner et al., 2014; Brandes et al., 2016). This has fed the idea that application of precision technologies and multi-criteria decision analysis techniques could make it feasible to convert such marginal crop lands to switchgrass to improve both profits and environmental outcomes (Bonner et al., 2016; Brandes et al., 2018a,b). Using only moderate price points for switchgrass and for corn and soybeans (*Glycine max*) in rotation, Brandes et al. (2018b) suggested that more than 4% of the existing annual crop land in Iowa, USA, could be converted to switchgrass at break even or higher returns per hectare. Such data lend hope to the idea that switchgrass for biomass-to-biofuels production could be a viable, competitive land use to increase environmental sustainability.

Such assumptions must face and address the hard realities of logistics, however. The cost of hauling low-density materials increases with increasing refinery size, and this effect is greater at lower levels of land conversion (for a more thorough treatment of the interactions among feedstock yield, available acres and refinery size, see Fike et al. (2007)). If this were the only supply of biomass, such low density (4.3%) of available acres likely would be insufficient to attract a large biorefinery to a region, even assuming 100% conversion of these economically unproductive lands. In a Midwestern context, however, where millions of acres of corn stover already are available, this application of switchgrass at first look seems a potential means of providing revenue from marginal crop land while it also helps to meet ecosystem service goals (Bonner et al., 2016; Brandes et al., 2018b). In some cases, it may be more feasible to utilize a distributed bioenergy collection and pretreatment system to densify switchgrass prior to hauling to the refinery. Over seven Great Plains states, perhaps as much as 13.6 M ha of land not suited for crop production could be repurposed to growing switchgrass or other perennials, protecting soils and water under bioenergy crops (Li and Chen, 2018). However, added equipment costs, limited site accessibility, lost efficiency due to irregularity of shape and unevenness of terrain, and the need to work around these 'ecosystem islands' are all part of the constraints that will slow the adoption of such land use (Soldavini and Tyner, 2018; Fred Circle, pers. comm.). Whether this will remain so once production is based on algorithms, positioning technology and autonomous systems remains to be seen, but this is clearly an area of interest.

Perennial feedstock systems could offer advantage over crop residues in terms of harvest timing since they have the potential to be harvested twice per year in some geographic locations, but again, large acreage would need to be available for the energy system. In such a scenario, multiple harvests of switchgrass per season (while likely increasing nutrient removal and thus long-term input requirements), would increase use of hauling and storage infrastructure, and thus reduce the costs associated with their use. Dedicated biomass production likely also would be advantageous to stover in terms of having better (lower) moisture levels and a wider harvest window (Shinners et al., 2007) given the relatively narrow time available for residue collection.

Another issue, harvest method, relates to logistics and has important implications for in-field handling and preprocessing systems at the biorefinery. Detailed treatment of such issues is tangential to the scope of this chapter – but this is a relevant concern for most herbaceous biomass feedstocks. In a nutshell, researchers have explored the best way to gather and package biomass for transport, and baling systems appear to be at the forefront because unlike chopped biomass systems, baling offers discontinuity between in-field harvest and postharvest transport, and bales may be easier to manage for handling and storage.

Although crop production and logistics have important implications for the success of a biomass-to-bioenergy system, and while it is best to consider a system in its entirety, it should be informative to consider the progress of feedstock conversion itself – that is, independent of the costs and challenges of growing and moving any presumed feedstock. While continued advances have brought the turning of 'straw' into a liquid 'gold' closer to reality, many companies working to process stover to ethanol have struggled to be profitable. Moreover, most of the currently operational systems that turn crop residues into fuel are utilizing only a limited portion of the available cellulosic biomass in their conversion process. A number of technical challenges exist regarding the processing and conversion of cellulosic materials to liquid fuels. While a detailed exploration of these factors is well beyond the scope of a chapter on switchgrass sustainability, the issues clearly have consequence for the development of a functioning, profitable biomass-based bioenergy system. That said, while the next big breakthrough in conversion seems always 'just around the corner', there is a promise that these systems will be propelled forward when one gets traction.

4 Switchgrass production and agronomics in the context of sustainability

As noted before, switchgrass or any energy crop must yield sufficient tonnage over sufficient acres to make economic sense. However, energy crops should not be viewed through the lens of that single metric, yield, but rather within the context of an energy system to which it may be the central component (Fike et al., 2007). Clearly, several other potential dedicated energy crops yield more than switchgrass, but each crop faces its own 'Achilles' heel' with respect to energy system development. For example, while sterile miscanthus hybrids are more productive in parts of the United States, they have higher establishment costs given that they require vegetative propagation. This is also true for energycane (*Saccharum* spp.), which has a more limited geographic range, and for sorghum (*Sorghum bicolor*), which must be replanted annually. In the case of miscanthus, some scientists are working to overcome the planting limitations through the development of hybrid *M.* × *giganteus* seed, or even fertile *M. sinensis* with functional sterility. However, these systems still present the opportunity for the escape of individuals that would then pose the risk of broader invasion (Quinn et al., 2010). Invasiveness is a significant concern for other productive species such as giant reed (*Arundo donax*) and napier/elephant grass (*Pennisetum purpureum*). Several of these species also can be difficult to dry, which can increase challenges and costs of harvest, storage, and load/haul logistics and may constrain the process system to biochemical conversion.

Switchgrass does well with respect to many of the metrics that one might desire for herbaceous energy crops (Fike et al., 2007). The species is readily established from seed, produces high biomass yields, is adaptable with breeding, provides a number of conservation benefits, and in its native range east of the Rocky Mountains (Hitchcock, 1935), it is not generally considered an invasive threat. Given our earlier discussion of sustainability features that accrue to switchgrass, we will move on to brief considerations of switchgrass production from an agronomic perspective. Interested readers may find more detailed information on switchgrass production for biomass in Parrish and Fike (2005) and Fike et al. (2014) among other sources.

4.1 Ecotypes and cultivars

Switchgrass is generally categorized into upland or lowland 'ecotypes'. Upland types typically have shorter stature, with thicker roots, longer root internodes and rhizomes of sufficient length to support sod formation. Lowlands generally are much taller, with thicker stems. Their shorter rhizomes result in a bunch form, which as noted before, makes it better for wildlife from the perspective of habitat structure. Less distinct hybrids of these two forms can also be found

(Zhang et al., 2011). Although there is a considerable overlap in geographic range, upland switchgrass is more commonly found from midlatitude to points north, while lowland ecotypes are better adapted from midlatitudes to points south. Both types can be found within a habitat (Hultquist et al., 1997), but each has unique traits that promote fitness for survival in the different growing environments from which it would have evolved in eastern North America.

Historically, available switchgrass cultivars have largely been selections from some local population chosen on the basis of their productivity, typically for forage and conservation. More recent cultivar registrations have included descriptions of use to include suitability for bioenergy production. As noted above, the ability to provide feedstock resources suitable for more than one market or enterprise may be a strength and selling point of the species. While breeding programs are resulting in new feedstock materials coming on line, a dual purpose approach is typically considered as the end use.

4.2 Establishment

Switchgrass has in the past had a reputation both for being difficult to establish and for requiring a long (3-year) period to reach full productivity (e.g. Mitchell et al., 2012). As our understanding of the plant and its required agronomic practices has improved, its notoriety for being difficult to work with has dissipated. Stand frequencies of 40% or greater are considered adequate for bioenergy stands (Schmer et al., 2006), with 50% of full stand yield possible in the establishment year. A suite of factors for successful establishment, including site selection, seedbed preparation, planting depth, row spacing, seeding rate, seed dormancy, time of planting and seed inocula can affect stand success and are discussed in some detail elsewhere (Fike et al., 2014). In a sustainability context, the fact that switchgrass can be established from seed presents a significant advantage for the crop relative to other high-yielding but vegetatively propagated species. This allows for no-till planting and reduces establishment costs, added pluses in the consideration of energy crop sustainability (Fike et al., 2007).

4.3 Fertility

Fertility management is another important consideration for any energy cropping system. Switchgrass (and other native grasses from North America's tallgrass prairies) have capacity to be productive with limited fertility inputs. (This, coupled with suitability to marginal lands and lower economic and environmental costs of production, supported its development as a model bioenergy crop). Generally, N is the most needed - and thus most expensive - nutrient input; it also has the greatest potential for negative impacts on the environment and on life cycle outcomes (Mitchell and Schmer, 2012). The carbon balance for biomass-to-bioenergy production systems is driven by the fossil fuel use and greenhouse gas emissions associated with synthesizing N fertilizer and the potential losses to the environment as N_2O (a highly potent greenhouse gas).

In biomass systems, the general recommendation is to apply N at a replacement rate to prevent overfertilization and soil N accumulation. While this is straightforward in principle, substantial variation in response to N has been seen across research studies (Parrish and Fike, 2005). For example, in a review of switchgrass aboveground biomass responses to added N, the range of response relative to control (zero N) treatment was from 0 to 6.2 Mg ha^{-1} (Brejda, 2000). More recent field-scale research found that on some marginal sites, typically with greater soil moisture, N application could even reduce switchgrass yield (Fike et al., 2017). Such variation reflects differences in the plant's inherent demand and capacity for N recycling, as well as factors such as soil type, soil N status, precipitation, atmospheric N deposition and timing of the biomass harvest. These factors, in turn, have downstream consequences for feedstock quality and biofuel yield, further affecting the overall greenhouse gas balance for the system. Long-term, switchgrass stand health and productivity for bioenergy production will require some level of nutrient replacement (Lemus et al., 2008), although these inputs typically would not be necessary on an annual basis (Springer, 2017).

The relationship and value of N fertility to overall system sustainability in terms of greenhouse gas balance has been the subject of debate among agronomists and ecologists. Some have argued that low-input, high-diversity systems (which mimic native prairies) should be grown in place of monocultural cropping systems on marginal lands (e.g. Tilman et al., 2006a). However, the productivity of such sites may be constraining for biofuel system development, because significant economic, energetic and environmental costs would be incurred by harvesting and transporting a low-yielding, low bulk-density biomass from widely dispersed marginal sites to a biorefinery. The value of such plantings may be increased, however, if other ecosystem services (such as wildlife habitat, pollinator services, erosion reduction) are fully accounted.

Another consideration for fertility application is their effect on plants and soils. Changes in shoot:root ratios and soil C (Garten et al., 2011; Heggenstaller et al., 2009) may affect both GHG emissions and C sequestration of the system. Optimizing fertility management to provide both high aboveground biomass production and SOC accretion is an ongoing area of study, however, and likely will require more effort to deliver greater clarity given the complex interactions among numerous bio-geochemical factors that drive these responses. In studies in which exogenous N increased aboveground yields, the added CO_2 captured in biomass and the fossil carbon displaced in the fuel system had net favorable effects on greenhouse gas balance and would also be expected to reduce competition for lands that are currently used for food production (Schmer et al., 2008; Nikièma et al., 2011). However, Kibet et al. (2016) found no response of root mass or soil organic C with applied fertility, while Valdez et al. (2017) reported that C and N stocks were reduced by added N fertilization. Lower soil C stocks likely were driven by reduced levels of C partitioned to roots, changes in root composition and decomposability, and attendant effects on the soil microbial community and consequent mineralization rate.

The N needed to support switchgrass production during the growing season may be relatively high on a mass basis, but the concentration of tissue N declines into the growing season (Waramit et al., 2011). By season's end, much of the N used for aboveground biomass production is returned to the plant's roots and rhizomes (Beaty et al., 1978; Garten et al., 2011; Lemus et al., 2008), and end-of-season N concentrations typically range from 5 to 8 g kg^{-1} for plants harvested after senescence (Fike et al., 2006a,b; Guretzky et al., 2011; Madakadze et al., 1999).

Although the case can be made for adding fossil fuel-based N fertilizers, biologically derived N sources may increase the productivity of these systems. Adding legumes is one potential means of minimizing exogenous N inputs, with yields comparable to moderate levels of applied N (Ashworth et al., 2016; Warwick et al., 2016). Species compatibility is an important consideration for creating mixed stands, however, as some species combinations may reduce switchgrass production (Bow et al., 2008; Butler et al., 2013; Springer et al., 2001).

Associative N fixation may be another means to achieving a biologically driven end. Early research (Tjepkema and Burris, 1976) found that nitrogenase activity was specifically associated with switchgrass in artificial prairies, although the inputs were low (3.6 kg N $ha^{-1}yr^{-1}$). Subsequent work in this area indicated that seedlings had much greater yields and higher N concentrations when inoculated with root fragments and rhizosphere soil collected from native prairies (Brejda et al., 1998). Subsequent efforts isolated and identified specific plant growth promoting rhizobacteria that may fix N, solubilize phosphorus (P), produce auxins and promote tillering (Ker et al., 2012, 2014). Other plant-microbe relationship studies have shown that specific bacterial endophyte strains may both increase early growth and benefit plant stress responses (Wang et al., 2015a,b). More recent data have shown that associative N fixation declines as soil fertility levels increase (Roley et al., 2018). Such plant-microbe relations are a new area of exploration, representing a largely unharnessed means of mitigating plant abiotic stress, improving plant productivity and increasing soil carbon sequestration (Robertson et al., 2017).

Response to P in production studies generally is limited beyond the establishment year (e.g. McKenna and Wolf, 1990) and sometimes not observed on low-P soils (e.g. Kering et al., 2012). Evidence over several studies would suggest that responses to P (and N) are mediated by mycorrhiza or root-associated bacteria (Fike et al., 2014), and better understanding of these associations and how best to take advantage of them may further the opportunity to reduce system inputs and increase resource-use sustainability. The literature presents little evidence that switchgrass responds to K and responses to pH adjustment may be less a function of absolute pH and more a change in nutrient availability (Fike et al., 2014).

4.4 Harvest management: timing and frequency

Long-term sustainability of a switchgrass stand is facilitated by harvest timing and intensity. Nitrogen needs and plant nutrient concentrations are relatively high early in the growing season, with biomass yields increasing to the time of anthesis; these then decline as the growing season progresses (Waramit et al., 2011). Although yields are lower when switchgrass is harvested after killing frost, this management strategy provides the plant sufficient time to translocate nutrients back into its extensive root system, reducing future nutrient input needs and benefitting long-term stand maintenance (Mitchell and Schmer, 2012). A late-season harvest strategy may also support greater feedstock quality for thermochemical conversion systems (e.g. through lower ash concentrations).

In some cases, an additional harvest during the growing season can be justified, although this may be a regionally specific practice. Some research indicates that with some upland cultivars (which are less productive than lowlands), a cutting during the growing season, coupled with a dormant season harvest, can increase yields over a single harvest by 30% or more in the upper Southeast (Fike et al., 2006a,b) but the opposite pattern was found on the Great Plains (Vogel et al., 2002). Under the right conditions, such a strategy could both increase yield and improve system logistics, but it must be considered relative to the additional costs of harvest, nutrients removed from the system and effects on the process system (Cundiff et al., 2009).

5 Economic, policy and social issues

Despite encompassing two legs of the sustainability stool, we give only cursory discussion of the economic, policy and social issues associated with development of a sustainable biofuels industry. These legs are critical and central to the progress of biofuels production efforts in the Unites States. Indeed, they have contributed to biofuel's current staying power in the face of current low energy prices and growth of fossil fuel exports, although some have argued that the policy approach may have hindered as much as helped (BIO, n.d.).

The justifications for bioenergy may be changed to fit the need, and at various points (and in some combination) over the last 40+ years, biofuels production for the United States has been viewed as a way to improve the country's national security position (by reducing reliance on oil supplies from unreliable or unsavory partners), of improving its domestic economy (through rural development) and of addressing climate change. From a global perspective the current goals for biofuels are similar: climate mitigations, increasing energy access and reducing air pollution (IEA, 2019b). But, in the case of transport fuel, they are 'Not on track' to meet current production goals (IEA, 2019a). However, policy may have a more recalcitrant nature than a political justification, and this can have consequences for system performance.

Markets and economic issues may be able to drive biofuels forward, but they do not stand independent of policy. Given the slow pace of development, many view policy as the key driver to meet the goals of increased biofuels production. Such policy might include 'application of...advanced biofuels quotas and financial de-risking measures' given 'modest' production of advanced biofuels and 'mixed performance' of commercial ethanol facilities (IEA, 2019a). However, critics of United States policy lambaste them for being 'based on hopes, not reality' (Foreman, 2018). In addition to issues of supply and attendant logistics challenges and concerns about ethanol suitability for gasoline engines, Foreman (2018) argued that current polices have put cellulosic ethanol at a disadvantage because it 'must...compete for space...limited by the blend wall' and that regulatory uncertainty has deterred investment. Moreover, policy approaches face opposition from entrenched industries and interest groups, both outside and within government. In the current context of liquid fuel production, those who support ethanol seek its increased use in liquid fuel blends to overcome an existing 'blend wall' and do so on the basis of its potential economic and environmental benefits. Those who are opposed, however, are concerned with the added costs and potential negative consequences for existing petroleum industries. After framing this debate, McMillan (2019) stated simply, 'It is hard to determine what is ultimately best for society'.

While Foreman (2018) argued that United States biofuel policies have supported an expensive solution in search of a problem that domestic oil production has already solved, it is important to ask whether the right problem truly has been addressed. Although new sources of (or at least greater access to) fossil energy have increased and kept the price of these liquid fuels low, the consequences of their sustained use would seem a far greater problem for sustainability based on current predictions of even moderate warming (Hoegh-Guldberg et al., 2018). Perhaps policies and rewards specifically aimed at climate mitigation (and full-fare accounting of C emissions) would do as much or more to create equal footing for different energy sources and better support the development of a sustainable biofuels industry. For example, a number of policy options that include renewable energy standards and carbon taxation were outlined by Smolinski and Cox (2016).

Of course, there are other larger social and policy issues that biofuels production may face given their potential impacts to local and global ecosystems. Production of switchgrass and other biofuels in the United States likely will be viewed favorably if they serve to provide wildlife habitat, restore soil health and improve water quality while displacing fossil energy. However, if their further development and expansion has negative consequences for ecosystems and societies within or outside this country – for example, through food vs. fuel competition or habitat loss from indirect land use change – they could hardly be sustainable. Such issues and ideas are behind the growing efforts to develop landscape design approaches to land use for multiple outcomes (e.g. see Comer, 2017; Dale et al., 2016).

6 Case study

The issues around policy and on what basis to move forward with biofuels stand in some contrast with our brief case study. Although much of the interest has been for delivering significant quantities of biomass for liquid fuels, the scenario presented here suggests that biomass markets might be advanced with simpler technologies applied at smaller scales which could in turn support and inform broader industry development.

A cooperative Farm to Fuel project based on burning switchgrass was embarked on in 2011. The essential entities involved in this collaborative effort are FDC Enterprises (FDCE), the Piedmont Geriatric Hospital (PGH), and Nottoway County (Virginia), with support from the Virginia Tobacco Region Revitalization Board (the Tobacco Commission). FDCE is the harvester, aggregator, preprocessor and deliverer of ground switchgrass. PGH, the recipient of the ground biomass, burns it for heat and steam. The Tobacco Commission has provided the infrastructure for bale storage and for the size reduction equipment used to process the bales while Nottoway County contributed the land needed for the facility site. In addition, 13 landowners contribute to the project, receiving payment for biomass produced for the effort.

The partnership arose as PGH, a state-run facility, was looking for a way to reduce fuel costs and dampen the fluctuations in fuel prices associated with the purchase of #2 fuel oil. The hospital faced large price swings which made budgeting challenging. Additional goals for hospital administrators were to reduce GHG emissions and improve the local farm economy and environment. FDCE, a successful commercial contractor based out of Ohio, USA, and which had planted thousands of acres of native grasses across the country for conservation plantings, saw this as an opportunity to expand their commercial establishment efforts in Virginia, USA, which was facilitated by county and Tobacco Commission support.

The initial plans for the project included having local operators harvest the biomass, but over time FDCE took on the role of harvester and processor to ensure the quality of the feedstocks. A long-term, 10-year contract is paid to FDCE for the delivery of ground biomass at $175/ton. This provides stability to all entities and is especially helpful to PGH, which would have been exposed to oil prices ranging from $0.41 to $1.02/l ($1.55 to $3.86/gal). Since 2011, PGH has burned over 10000 Mg (11364 tons) of biomass during 1125 days. This cheaper fuel has saved the state of Virginia a total of over $1.1M and displaced about 4.54 M liters (1.20 M gal) of #2 fuel oil. Estimated reductions in fossil GHG emissions approach 11900 Mg (13118 tons) when counting both the displaced fuel oil and added soil C in the 1200+ hectares (nearly 3000 acres) of switchgrass planted. This operation adds value to the surrounding land base in terms of greater water quality and increased wildlife habitat and provides payments of $247/hectare/year ($100/acre/year) to the landowner, which is several-fold greater than the typical land rent for the area. We note, however, that such rent values would not be competitive in much of the Corn Belt.

While the system uses simple combustion technology (i.e. it is not converting straw to liquid fuel), it does require a boiler that can handle biomass. Switchgrass is a more suitable feedstock than sawdust, the other leading available resource, because it has less than 15% moisture, compared with 40–50% moisture in the sawdust. In addition to the clean-ness of the burn, the energy and steam outputs from switchgrass have been higher than had been predicted based on literature values.

7 Conclusion

In its broadest sense, any bioenergy crop that is sustainable will provide suitable economic returns across the market spectrum, have net positive environmental impacts and be broadly socially acceptable. Switchgrass, a productive, perennial warm-season grass, native to much of North America, has potential to meet these metrics. The species has much potential for supplying abundant biomass for bioenergy, and the fact that it is native and broadly adapted makes it an excellent feedstock candidate. The species has potential to sequester carbon, and its use in perennial energy systems is expected to improve water quality, provide wildlife habitat and improve agroecosystem biodiversity. Switchgrass may not be the most productive species across all environments, but it is likely to be as or more productive in the Southeast and much of the Plains states, where it could have alternative uses, since it is a good fit for beef production systems common to those regions. It may be less competitive as a feedstock in the Midwest, because that region has high amounts of available stover, which is viewed as a capturable 'co-resource' along with the primary commodity (corn grain). Regardless of feedstock system, current challenges with supply logistics and feedstock conversion systems continue to slow industry expansion. Agronomic management will be important for optimizing switchgrass systems in terms of high productivity with limited inputs, and this will have implications for carbon sequestration and greenhouse gas emissions. Renewable liquid fuels production has not met production capacity envisioned by government policy, and investment has been constrained because of regulatory uncertainty and opposition from entrenched interests which has likely kept the energy 'playing field' unlevel for bioenergy. Accounting for carbon balance can be challenging, but doing so would likely serve bioenergy well relative to fossil fuel sources. Primary focus on biomass for liquid fuels may have been a hindrance, and an existing case study shows that switchgrass-based biomass production can be competitive with fossil energy sources when used in a small-scale combustion system.

8 Future trends

Continued effort at conversion will be a driver for biomass to liquid fuel systems. Greater effort to develop and apply remote sensing, positioning and information technologies may also create opportunities to deploy switchgrass and other energy crops, particularly in environmentally sensitive sites that currently are used (unprofitably) for commodity crops such as corn or soybeans. Efforts with landscape design offer promise in this regard, although such approaches likely will not be without controversy, as scientists, land owners and policy makers work to determine what should or should not be grown on the land base under such management models. Finally, all of these issues must be considered in the context of climate change, which may make energy cropping both more difficult (due to greater weather variability and extremes) and more urgent.

9 Where to look for further information

As noted earlier, stable, consistent policy is seen by many as one of the key requirements to enhance development of biofuel productions systems. In a best-case scenario, policy makers will look to scientific evidence gathered from federal agencies, universities and others to develop the most appropriate policies to advance the biofuel industry. To this end, the United States has developed numerous programs and centers devoted to conducting the highest priority research and collecting the most important data related to all three legs of the biofuel sustainability stool. These centers or initiatives are necessarily crosscutting by nature in order to best assess all aspects of sustainability. Some of the specific centers are described below:

- The US Department of Energy (US DOE) in 2017 announced the establishment or continuation of four such centers: (1) the Great Lakes Bioenergy Research Center; (2) The Center for Bioenergy Innovation; (3) the Joint BioEnergy Institute and (4) the Center for Advanced Bioenergy and Bioproducts Innovation. These centers received around US$40 million in FY2018 alone with plans to continue the centers for an additional four years (https://genomicscience.energy.gov/centers/). Each centers conducts transdisciplinary research on sustainability, feedstock development, deconstruction and conversion.
- In 2010, the US Department of Agriculture (USDA) designated four Regional Biomass Research Centers across the country (https://www.ars.usda.gov/natural-resources-and-sustainable-agricultural-systems/biorefining/docs/regional-biomass-research-centers/). These regional centers have the charge to 'develop the best feedstocks and sustainable feedstock production systems for specific agro-eco regions where advanced biofuels will likely be produced. In addition, each Regional Biomass Research Center fosters collaborative research within the complete bioenergy supply chain so as to accelerate the creation of commercial supply chains for the production of advanced biofuels'.
- In 2002, the Sun Grant Initiative (SGI) was authorized in the 2002 USDA Farm Bill and subsequently reauthorized in the 2008, 2014 and 2018 Farm Bills. The SGI is comprised of five Regional Sun Grant Centers located at land-grant universities with the mission to 'enhance national energy security; provide opportunities for rural economic development in America's traditional agricultural communities; promote environmentally sustainable and diversified production opportunities for agricultural and forestry resources; and encourage further bioenergy research collaboration between government agencies and land-grant colleges and universities'. These efforts are accomplished through regional multidisciplinary competitive grants programs. To date, the SGI has received funding not only from USDA, but also through the US DOE and the US Department of Transportation.

This is not an exhaustive list of centers or activities funded by the US federal agencies, but it highlights some of the key crosscutting initiatives across the country. The Department of Energy's Office of Energy Efficiency and Renewable Energy (EERE) also maintains a website describing the activities of various national laboratories and their functions relative to bioenergy research: https://www.energy.gov/eere/bioenergy/bioenergy-national-laboratories.

10 References

Ameen, A., Han, L. and Xie, G. H. 2019. Dynamics of soil moisture, pH, organic carbon, and nitrogen under switchgrass cropping in a semiarid sandy wasteland.

Communications in Soil Science and Plant Analysis 50(7), 922–33. doi:10.1080/00103624.2019.1594883.

Ashworth, A. J., Weiss, S. A., Keyser, P. D., Allen, F. L., Tyler, D. D., Taylor, A., Beamer, K. P., West, C. P. and Pote, D. H. 2016. Switchgrass composition and yield response to alternative soil amendments under intensified heat and drought conditions. *Agriculture, Ecosystems and Environment* 233, 415–24. doi:10.1016/j.agee.2016.09.041.

Beaty, E. R., Engel, J. L. and Powell, J. D. 1978. Tiller development and growth in switchgrass. *Journal of Range Management* 31(5), 361–5. doi:10.2307/3897360.

Biermacher, J. T., Haque, M., Mosali, J. and Rogers, J. K. 2017. Economic feasibility of using switchgrass pasture to produce beef cattle gain and bioenergy feedstock. *BioEnergy Research* 10(3), 740–9. doi:10.1007/s12155-017-9835-6.

BIO (Biotechnology Innovation Organization). n.d. Estimating another year of chilled investment in advanced biofuels due to RFS uncertainty. Available at: https://www.bio.org/sites/default/files/Estimating_Another_Year_of_Chilled_Investment.pdf (accessed on 9 May 2019).

Bonner, I. J., Cafferty, K. G., Muth Jr., D. J., Tomer, M. D., James, D. E., Porter, S. A. and Karlen, D. 2014. Opportunities for energy crop production based on subfield scale distribution of profitability. *Energies* 7(10), 6509–26. doi:10.3390/en7106509.

Bonner, I., McNunn, G., Muth Jr., D., Tyner, W., Leirer, J. and Dakins, M. 2016. Development of integrated bioenergy production systems using precision conservation and multicriteria decision analysis techniques. *Journal of Soil and Water Conservation* 71(3), 182–93. doi:10.2489/jswc.71.3.182.

Bow, J. R., Muir, J. P., Weindor, D. C., Rosiere, R. E. and Butler, T. J. 2008. Integration of cool-season annual legumes and dairy manure compost with switchgrass. *Crop Science* 48(4), 1621–8. doi:10.2135/cropsci2007.12.0697.

Brandes, E., McNunn, G. S., Schulte, L. A., Bonner, I. J., Muth, D. J., Babcock, B. A., Sharma, B. and Heaton, E. A. 2016. Subfield profitability analysis reveals an economic case for cropland diversification. *Environmental Research Letters* 11(1), 014009. doi:10.1088/1748-9326/11/1/014009.

Brandes, E., McNunn, G. S., Schulte, L. A., Muth, D. J., VanLoocke, A. and Heaton, E. A. 2018a. Targeted subfield switchgrass integration could improve the farm economy, water quality, and bioenergy feedstock production. *GCB Bioenergy* 10(3), 199–212. doi:10.1111/gcbb.12481.

Brandes, E., Plastina, A. and Heaton, E. A. 2018b. Where can switchgrass production be more profitable than corn and soybean? An integrated subfield assessment in Iowa, USA. *GCB Bioenergy* 10(7), 473–88. doi:10.1111/gcbb.12516.

Brejda, J. J. 2000. Fertilization of native warm-season grasses. Native warm-season grasses: research trends and issues. *Proceedings of the Native Warm-Season Grass Conference and Expo*, Des Moines, IA, USA, 12–13 September 1996, pp. 177–200.

Brejda, J. J., Moser, L. E. and Vogel, K. P. 1998. Evaluation of switchgrass rhizosphere microflora for enhancing seedling yield and nutrient uptake. *Agronomy Journal* 90(6), 753–8. doi:10.2134/agronj1998.00021962009000060006x.

Butler, T. J., Muir, J. P., Huo, C. and Guretzky, J. A. 2013. Switchgrass biomass and nitrogen yield over-seeded with cool-season forages in the southern Great Plains. *BioEnergy Research* 6(1), 44–52. doi:10.1007/s12155-012-9225-z.

Chatterjee, A., Long, D. S. and Pierce, F. J. 2018. Change in soil organic carbon after five years of continuous winter wheat or switchgrass. *Soil Science Society of America Journal* 82(2), 332–42. doi:10.2136/sssaj2017.08.0290.

Chavas, J. P. and Holt, M. T. 1990. Acreage decisions under risk: the case of corn and soybeans. *American Journal of Agricultural Economics* 72(3), 529-38. doi:10.2307/1243021.

Comer, K. 2017. Enabling sustainable landscape design for continual improvement of operating bioenergy supply systems. 2017 Project Peer Review. Availability and Sustainability Session. 9 March 2017. Denver, CO. US Department of Energy Bioenergy Technologies Office. Available at: https://www.energy.gov/sites/prod/files/2017/05/f34/analysis_and_sustainability_comer_clark_4.2.2.60.pdf.

Cundiff, J. S., Fike, J. H., Parrish, D. J. and Alwang, J. 2009. Logistic constraints in developing dedicated large-scale bioenergy systems in the Southeastern United States. *Journal of Environmental Engineering* 135(11), 1086–96. doi:10.1061/(ASCE)EE.1943-7870.0000123.

Dale, V. H., Kline, K. L., Buford, M. A., Volk, T. A., Tattersall Smith, C. and Stupak, I. 2016. Incorporating bioenergy into sustainable landscape designs. *Renewable and Sustainable Energy Reviews* 56, 1158–71. doi:10.1016/j.rser.2015.12.038.

Das, S., Teuffer, K., Stoof, C. R., Walter, M. F., Walter, M. T., Steenhuis, T. S. and Richards, B. K. 2018. Perennial grass bioenergy cropping on wet marginal land: impacts on soil properties, soil organic carbon, and biomass during initial establishment. *BioEnergy Research* 11(2), 262–76. doi:10.1007/s12155-018-9893-4.

DeGraff, M. A., Jastrow, J. D., Adkins, J., Johns, A. C., Morris, G. and Six, J. 2016. Impacts of root traits and genotypic diversity in switchgrass cropping systems on biogeochemical cycling of soil carbon and nitrogen. *American Geophysical Union Fall Meeting*, San Francisco, CA, abstract #B33E-0660.

Dumortier, J. 2016. Impact of agronomic uncertainty in biomass production and endogenous commodity prices on cellulosic biofuel feedstock composition. *GCB Bioenergy* 8(1), 35–50. doi:10.1111/gcbb.12238.

Fike, J. H., Parrish, D. J., Wolf, D. D., Balasko, J. A., Green Jr., J. T., Rasnake, M. and Reynolds, J. H. 2006a. Long-term yield potential of switchgrass-for-biofuel systems. *Biomass and Bioenergy* 30(3), 198–206. doi:10.1016/j.biombioe.2005.10.006.

Fike, J. H., Parrish, D. J., Wolf, D. D., Balasko, J. A., Green Jr., J. T., Rasnake, M. and Reynolds, J. H. 2006b. Switchgrass production for the upper southeastern USA: influence of cultivar and cutting frequency on biomass yields. *Biomass and Bioenergy* 30(3), 207–13. doi:10.1016/j.biombioe.2005.10.008.

Fike, J. H., Parrish, D. J., Alwang, J. and Cundiff, J. S. 2007. Challenges for deploying dedicated, large-scale, bioenergy systems in the USA. *CAB Reviews: Perspectives in Agriculture, Veterinary Science, Nutrition and Natural Resources* 2(64), 28. doi:10.1079/PAVSNNR20072064.

Fike, J. H., Butler, T. J. and Mitchell, R. 2014. The agronomy of switchgrass for biomass. In: Luo, H., Wu, Y. and Kole, C. (Eds), *Compendium of Bioenergy Plants: Switchgrass*. CRC Press, Boca Raton, FL, pp. 16–66.

Fike, J. H., Pease, J. A., Owens, V. N., Farris, R. L., Hansen, J. L., Heaton, E. A., Mayton, H. S., Mitchell, R. B. and Viands, D. R. 2017. Switchgrass nitrogen response and cost of production on diverse sites. *GCB Bioenergy* 9, 1526–42. doi:10.1111/gcbb.12444/full.

Follett, R. F., Vogel, K. P., Varvel, G. E., Mitchell, R. B. and Kimble, J. 2012. Soil carbon sequestration by switchgrass and no-till maize grown for bioenergy. *BioEnergy Research* 5(4), 866–75. doi:10.1007/s12155-012-9198-y.

Foreman, R. D. 2018. On cellulosic ethanol, hope still outpacing reality Energy Tomorrow Blog. American Petroleum Institute, Breaking Energy.Com. Energy Tomorrow Blog.

Available at: https://breakingenergy.com/2018/12/19/on-cellulosic-ethanol-hope-still-outpacing-reality/ (accessed on 07 May 2019).

Foster, J. L., Guretzky, J. A., Huo, C., Kering, M. K. and Butler, T. J. 2013. Effects of row spacing, seeding rates, and planting date on establishment and biomass of switchgrass. *Crop Science* 53(1), 309–15. doi:10.2135/cropsci2012.03.0171.

Frank, A. B., Berdahl, J. D., Hanson, J. D., Liebig, M. A. and Johnson, H. A. 2004. Biomass and carbon partitioning in switchgrass. *Crop Science* 44(4), 1391–6. doi:10.2135/cropsci2004.1391.

Garten Jr., C. T., Smith, J. L., Tyler, D. D., Amonette, J. E., Bailey, V. L., Brice, D. J., Castro, H. F., Graham, R. L., Gunderson, C. A., Izaurralde, R. C., Jardine, P. M., Jastrow, J. D., Kerley, M. K., Matamala, R., Mayes, M. A., Metting, F. B., Miller, R. M., Moran, K. K., Post III, W. M., Sands, R. D., Schadt, C. W., Phillips, J. R., Thomson, A. M., Vugteveen, T., West, T. O. and Wullschleger, S. D. 2010. Intra-annual changes in biomass, carbon, and nitrogen dynamics at 4-year old switchgrass field trials in west Tennessee, USA. *Agriculture, Ecosystems and Environment* 136(1–2), 177–84. doi:10.1016/j.agee.2009.12.019.

Garten Jr., C. T., Brice, D. J., Castro, H. F., Graham, R. L., Mayes, M. A., Phillips, J. R., Post, W. M., Schadt, C. W., Wullschleger, S. D., Tyler, D. D., Jardine, P. M., Jastrow, J. D., Matamala, R., Miller, R. M., Moran, K. K., Vugteveen, T. W., Izaurralde, R. C., Thomson, A. M., West, T. O., Amonette, J. E., Bailey, V. L., Metting, F. B. and Smith, J. L. 2011. Response of "Alamo" switchgrass tissue chemistry and biomass to nitrogen fertilization in West Tennessee, USA. *Agriculture, Ecosystems and Environment* 140(1–2), 289–97. doi:10.1016/j.agee.2010.12.016.

Guretzky, J. A., Biermacher, J. T., Cook, B. J., Kering, M. K. and Mosali, J. 2011. Switchgrass for forage and bioenergy: harvest and nitrogen rate effects on biomass yields and nutrient composition. *Plant and Soil* 339(1–2), 69–81. doi:10.1007/s11104-010-0376-4.

Hamilton, S. K., Hussain, M. Z., Bhardwaj, A. K., Basso, B. and Robertson, G. P. 2015. Comparative water use by maize, perennial crops, restored prairie, and poplar trees in the US Midwest. *Environmental Research Letters* 10(6). doi:10.1088/1748-9326/10/6/064015: 064015.

Heggenstaller, A. H., Moore, K. J., Liebman, M. and Anex, R. P. 2009. Nitrogen influences biomass and nutrient partitioning by perennial, warm-season grasses. *Agronomy Journal* 101(6), 1363–71. doi:10.2134/agronj2008.0225x.

Hendrickson, J. R., Schmer, M. R. and Sanderson, M. A. 2013. Water use efficiency by switchgrass compared to a native grass or a native grass alfalfa mixture. *BioEnergy Research* 6(2), 746–54. doi:10.1007/s12155-012-9290-3.

Hitchcock, A. S. 1935. Manual of the grasses of the United States. Miscellaneous Publication 200. United States Department of Agriculture, Washington DC.

Hoegh-Guldberg, O., Jacob, D., Taylor, M., Bindi, M., Brown, S., Camilloni, I., et al. 2018. Impacts of 1.5°C Global Warming on Natural and Human Systems. In: Masson-Delmotte, V., Zhai, P., Pörtner, H.-O., Roberts, D., Skea, J., Shukla, P. R., Pirani, A., Moufouma-Okia, W., Péan, C., Pidcock, R., Connors, S., Matthews, J. B. R., Chen, Y., Zhou, X., Gomis, M. I., Lonnoy, E., Maycock, T., Tignor, M. and Waterfield, T. (Eds), *Global Warming of 1.5°C. An IPCC Special Report on the Impacts of Global Warming of 1.5°C above Pre-industrial Levels and Related Global Greenhouse Gas Emission Pathways, in the Context of Strengthening the Global Response to the Threat of Climate Change, Sustainable Development, and Efforts to Eradicate Poverty.*

Available at: https://www.ipcc.ch/site/assets/uploads/sites/2/2019/02/SR15_Chapter3_Low:Res.pdf (accessed on 9 May 2019).

Hultquist, S. J., Vogel, K. P., Lee, D. J., Arumuganathan, K. and Kaeppler, S. 1997. DNA content and chloroplast DNA polymorphisms among switchgrasses from remnant midwestern prairies. *Crop Science* 37(2), 595–8. doi:10.2135/cropsci1997.0011183X003700020047x.

IEA (International Energy Agency). 2019a. Biofuels for transport. Tracking clean energy progress. Available at: https://www.iea.org/tcep/transport/biofuels/ (accessed on 07 May 2019).

IEA (International Energy Agency). 2019b. Sustainable development goals. A cleaner and more inclusive energy future. Available at: https://www.iea.org/weo/weomodel/sds/ (accessed on 07 May 2019).

Johnston, C. A. 2014. Agricultural expansion: land use shell game in the U.S. Northern Plains. *Landscape Ecology* 29(1), 81–95. doi:10.1007/s10980-013-9947-0.

Ker, K., Seguin, P., Driscoll, B., Fyles, J. and Smith, D. 2012. Switchgrass establishment and seeding year production can be improved by inoculation with rhizosphere endophytes. *Biomass and Bioenergy* 47(12), 295–301. doi:10.1016/j.biombioe.2012.09.031.

Ker, K., Seguin, P., Driscoll, B., Fyles, J. and Smith, D. 2014. Evidence for enhanced N availability during switchgrass establishment and seeding year production following inoculation with rhizosphere endophytes. *Archives of Agronomy and Soil Science* 60(110), 1553–63. doi:10.1080/03650340.2014.898840.

Kering, M. K., Biermacher, J. T., Butler, T. J., Mosali, J. and Guretzky, J. A. 2012. Biomass yield and nutrient responses of switchgrass to phosphorus application. *BioEnergy Research* 5(1), 71–8. doi:10.1007/s12155-011-9174-y.

Kibet, L. C., Blanco-Canqui, H., Mitchell, R. B. and Schacht, W. H. 2016. Root biomass and soil carbon response to growing perennial grasses for bioenergy. *Energy, Sustainability and Society* 6(1), 8.

Lai, L., Kumar, S., Osborne, S. and Owens, V. N. 2018a. Switchgrass impact on selected soil parameters, including soil organic carbon, within six years of establishment. *Catena* 163, 288–96. doi:10.1016/j.catena.2017.12.030.

Lai, L., Oh Hong, C., Kumar, S., Osborne, S. L., Lehman, R. M. and Owens, V. N. 2018b. Soil nitrogen dynamics in switchgrass seeded to a marginal cropland in South Dakota. *GCB Bioenergy* 10(1), 28–38. doi:10.1111/gcbb.12475.

Lee, K. H., Isenhart, T. M., Schultz, R. C. and Mickelson, S. K. 1998. Nutrient and sediment removal by switchgrass and cool-season grass filter strips in Central Iowa, USA. *Agroforestry Systems* 44(2/3), 121–32. doi:10.1023/A:1006201302242.

Lee, D. K., Owens, V. N. and Doolittle, J. J. 2007. Switchgrass and soil carbon sequestration response to ammonium nitrate, manure, and harvest frequency on conservation reserve program land. *Agronomy Journal* 99(2), 462–8. doi:10.2134/agronj2006.0152.

Lee, D. K., Aberle, E., Anderson, E. K., Anderson, W., Baldwin, B. S., Baltensperger, D., Barrett, M., Blumenthal, J., Bonos, S., Bouton, J., Bransby, D. I., Brummer, C., Burks, P. S., Chen, C., Daly, C., Egenolf, J., Farris, R. L., Fike, J. H., Gaussoin, R., Gill, J. R., Gravois, K., Halbleib, M. D., Hale, A., Hanna, W., Harmoney, K., Heaton, E. A., Heiniger, R. W., Hoffman, L., Hong, C. O., Kakani, G., Kallenbach, R., Macoon, B., Medley, J. C., Missaoui, A., Mitchell, R., Moore, K. J., Morrison, J. I., Odvody, G. N., Richwine, J. D., Ogoshi, R., Parrish, J. R., Quinn, L., Richard, E., Rooney, W. L., Rushing, J. B.,

Schnell, R., Sousek, M., Staggenborg, S. A., Tew, T., Uehara, G., Viands, D. R., Voigt, T., Williams, D., Williams, L., Wilson, L. T., Wycislo, A., Yang, Y. and Owens, V. 2018. Biomass production of herbaceous energy crops in the United States: field trial results and yield potential maps from the multiyear regional feedstock partnership. *GCB Bioenergy* 10(10), 698–716. doi:10.1111/gcbb.12493.

Lemus, R., Charles Brummer, E., Lee Burras, C., Moore, K. J., Barker, M. F. and Molstad, N. E. 2008. Effects of nitrogen fertilization on biomass yield and quality in large fields of established switchgrass in southern Iowa, USA. *Biomass and Bioenergy* 32(12), 1187–94. doi:10.1016/j.biombioe.2008.02.016.

Li, R. and Chen, J. 2018. Planning the next-generation biofuel crops based on soil-water constraints. *Biomass and Bioenergy* 115, 19-26. doi:10.1016/j.biombioe.2018.04.003.

Liu, X. A., Fike, J. H., Galbraith, J. M. and Fike, W. B. 2014. Switchgrass response to cutting frequency and biosolids amendment: biomass yield, feedstock quality, and theoretical ethanol yield. *BioEnergy Research* 7(4), 1191–200. doi:10.1007/s12155-014-9454-4.

Liu, X. A., Fike, J. H., Galbraith, J. M. and Fike, W. B. 2015. Biosolids amendment and harvest frequency affect nitrogen use dynamics of switchgrass grown for biofuel production. *BioEnergy Research* 8(2), 560–9. doi:10.1007/s12155-014-9543-4.

Luo, C., Brink, D. L. and Blanch, H. W. 2002. Identification of potential fermentation inhibitors in conversion of hybrid poplar hydrolyzate to ethanol. *Biomass and Bioenergy* 22(2), 125–38. doi:10.1016/S0961-9534(01)00061-7.

Ma, Z., Wood, C. W. and Bransby, D. I. 2000. Soil management impacts on soil carbon sequestration by switchgrass. *Biomass and Bioenergy* 18(6), 469–77. doi:10.1016/S0961-9534(00)00013-1.

Ma, Z., Wood, C. W. and Bransby, D. I. 2001. Impact of row spacing, nitrogen rate, and time on carbon partitioning of switchgrass. *Biomass and Bioenergy* 20(6), 413-9. doi:10.1016/S0961-9534(01)00008-3.

Madakadze, I. C., Stewart, K. A., Peterson, P. R., Coulman, B. E. and Smith, D. L. 1999. Cutting frequency and nitrogen fertilization effects on yield and nitrogen concentration of switchgrass in a short season area. *Crop Science* 39(2), 552-7. doi:10.2135/cropsci1999.0011183X003900020041x.

Mbonimpa, E. G., Hong, C. O., Owens, V. N., Lehman, R. M., Osborne, S. L., Schumacher, T. E., Clay, D. E. and Kumar, S. 2015. Nitrogen fertilizer and landscape position impacts on CO_2 and CH_4 fluxes from a landscape seeded to switchgrass. *GCB Bioenergy* 7(4), 836–49. doi:10.1111/gcbb.12187.

McKenna, J. R. and Wolf, D. D. 1990. No-till switchgrass establishment as affected by limestone, phosphorus, and carbofuran. *Journal of Production Agriculture* 3(4), 475–9. doi:10.2134/jpa1990.0475.

McMillan, D. A. 2019. U.S. ethanol future tied to bumper crop of D.C. politics. *Producer.com*. Available at: https://www.producer.com/2019/02/u-s-ethanol-future-tied-to-bumper-crop-of-d-c-politics/ (accessed on 06 May 2019).

Mersie, W., Seybold, C. A., McNamee, C. and Huang, J. 1999. Effectiveness of switchgrass filter strips in removing dissolved atrazine and metolachlor from runoff. *Journal of Environmental Quality* 28(3), 816–21. doi:10.2134/jeq1999.00472425002800030012x.

Mitchell, R. and Schmer, M. 2012. Switchgrass harvest and storage. In: Monti, A. (Ed.), *Switchgrass. Green Energy and Technology*. Springer, London, pp. 113-27.

Mitchell, R., Vogel, K. P. and Uden, D. R. 2012. The feasibility of switchgrass for biofuel production. *Biofuels* 3(1), 47–59. doi:10.4155/bfs.11.153.

Mitchell, R. B., Schmer, M. R., Anderson, W. F., Jin, V., Balkcom, K. S., Kiniry, J., Corrin, A. and White, P. 2016. Dedicated energy crops and crop residues for bioenergy feedstocks in the Central and Eastern USA. *Bioenergy Research* 9, 384–398. doi:10.1007/s12155-016-9734-2.

Nikièma, P., Rothstein, D. E., Min, D.-H. and Kapp, C. J. 2011. Nitrogen fertilization of switchgrass increases biomass yield and improves net greenhouse gas balance in northern Michigan, U.S.A. *Biomass and Bioenergy* 35(10), 4356–67. doi:10.1016/j.biombioe.2011.08.006.

Noe, R. R., Nachman, E. R., Heavenrich, H. R., Keeler, B. L., Hernández, D. L. and Hill, J. D. 2016. Assessing uncertainty in the profitability of prairie biomass production with ecosystem service compensation. *Ecosystem Services* 21, 103–8. doi:10.1016/j.ecoser.2016.05.004.

Oates, L. G., Duncan, D. S., Gelfand, I., Millar, N., Robertson, G. P. and Jackson, R. D. 2016. Nitrous oxide emissions during establishment of eight alternative cellulosic bioenergy cropping systems in the North Central United States. *GCB Bioenergy* 8(3), 539–49. doi:10.1111/gcbb.12268.

Oliveira, J. A., West, C. P., Afif, E. and Palencia, P. 2017. Comparison of Miscanthus and switchgrass cultivars for biomass yield, soil nutrients, and nutrient removal in northwest Spain. *Agronomy Journal* 109(1), 122–30. doi:10.2134/agronj2016.07.0440.

Parrish, D. J. and Fike, J. H. 2005. The biology and agronomy of switchgrass for biofuels. *Critical Reviews in Plant Sciences* 24(5–6), 423–59. doi:10.1080/07352680500316433.

Quinn, L. D., Allen, D. J. and Stewart, J. R. 2010. Invasiveness potential of *Miscanthus sinensis*: implications for bioenergy production in the United States. *Global Change Biology Bioenergy* 2(6), 310–20. doi:10.1111/j.1757-1707.2010.01062.x.

Robertson, G. P., Hamilton, S. K., Barham, B. L., Dale, B. E., Izaurralde, R. C., Jackson, R. D., Landis, D. A., Swinton, S. M., Thelen, K. D. and Tiedje, J. M. 2017. Cellulosic biofuel contributions to a sustainable energy future: choices and outcomes. *Science* 356(6345). doi:10.1126/science.aal2324.

Roley, S. S., Duncan, D. S., Liang, D., Garoutte, A., Jackson, R. D., Tiedje, J. M. and Robertson, G. P. 2018. Associative nitrogen fixation (ANF) in switchgrass (*Panicum virgatum*) across a nitrogen input gradient. *PLoS ONE* 13(6), e0197320. doi:10.1371/journal.pone.0197320.

Roth, A. M., Sample, D. W., Ribic, C. A., Paine, L., Undersander, D. J. and Bartelt, G. A. 2005. Grassland bird response to harvesting switchgrass as a biomass energy crop. *Biomass and Bioenergy* 28(5), 490–8. doi:10.1016/j.biombioe.2004.11.001.

Sanderson, M. A., Jones, R. M., McFarland, M. J., Stroup, J., Reed, R. L. and Muir, J. P. 2001. Nutrient movement and removal in a switchgrass biomass-filter strip system treated with dairy manure. *Journal of Environmental Quality* 30(1), 210–6. doi:10.2134/jeq2001.301210x.

Schmer, M. R., Vogel, K. P., Mitchell, R. B., Moser, L. E., Eskridge, K. M. and Perrin, R. K. 2006. Establishment stand thresholds for switchgrass grown as a bioenergy crop. *Crop Science* 46(1), 157–61. doi:10.2135/cropsci2005.0264.

Schmer, M. R., Vogel, K. P., Mitchell, R. B. and Perrin, R. K. 2008. Net energy of cellulosic ethanol from switchgrass. *Proceedings of the National Academy of Sciences of the United States of America* 105(2), 464–9. doi:10.1073/pnas.0704767105.

Searchinger, T., Heimlich, R., Houghton, R. A., Dong, F., Elobeid, A., Fabiosa, J., Tokgoz, S., Hayes, D. and Yu, T. -H. 2008. Use of U.S. croplands for biofuels increases greenhouse gases through emissions from land-use change. *Science* 319(5867), 1238–40. doi:10.1126/science.1151861.

Shinners, K. J., Binversie, B. N., Muck, R. E. and Weimer, P. J. 2007. Comparison of wet and dry corn stover harvest and storage. *Biomass and Bioenergy* 31(4), 211–21. doi:10.1016/j.biombioe.2006.04.007.

Smith, P. 2012. Agricultural greenhouse gas mitigation potential globally, in Europe and in the UK: what have we learnt in the last 20 years? *Global Change Biology* 18(1), 35–43. doi:10.1111/j.1365-2486.2011.02517.x.

Smith, K. A. and Searchinger, T. D. 2012. Crop-based biofuels and associated environmental concerns. *GCB Bioenergy* 4(5), 479–84. doi:10.1111/j.1757-1707.2012.01182.x.

Smolinski, S. and Cox, S. 2016. Policies to enable bioenergy deployment. Key considerations and good practices. National Renewable Energy Laboratory, Clean Energy Solutions Center, p. 19. NREL/TP-7A40-66322. Available at: https://www.nrel.gov/docs/fy16osti/66322.pdf (accessed on 16 May 2019).

Soldavini, S. and Tyner, W. E. 2018. Determining switchgrass breakeven prices in a landscape design system. *BioEnergy Research* 11(1), 191–208. doi:10.1007/s12155-017-9888-6.

Springer, T. L. 2017. Effect of nitrogen fertilization and residual nitrogen on biomass yield of switchgrass. *BioEnergy Research* 10(3), 648–56. doi:10.1007/s12155-017-9827-6.

Springer, T. L., Aiken, G. E. and McNew, R. W. 2001. Combining ability of binary mixtures of native, warm-season grasses and legumes. *Crop Science* 41(3), 818–23. doi:10.2135/cropsci2001.413818x.

Tilman, D., Hill, J. and Lehman, C. 2006a. Carbon-negative biofuels from low-input high-diversity grassland biomass. *Science* 314(5805), 1598–600. doi:10.1126/science.1133306.

Tilman, D., Reich, P. B. and Knops, J. M. H. 2006b. Biodiversity and ecosystem stability in a decade-long grassland experiment. *Nature* 441(7093), 629–32. doi:10.1038/nature04742.

Tjepkema, J. D. and Burris, R. H. 1976. Nitrogenase activity associated with some Wisconsin prairie grasses. *Plant and Soil* 45(1), 81–94. doi:10.1007/BF00011131.

Uden, D. R., Mitchell, R. B., Allen, C. R., Guan, Q. and McCoy, T. D. 2013. The feasibility of producing adequate feedstock for year-round cellulosic ethanol production in an intensive agricultural fuelshed. *BioEnergy Research* 6(3), 930–8. doi:10.1007/s12155-013-9311-x.

USEIA (United States Energy Information Administration). 2018. EPA finalizes Renewable Fuel Standard for 2019, reflecting cellulosic biofuel shortfalls. Today in Energy. Available at: https://www.eia.gov/todayinenergy/detail.php?id=37712# (accessed on 06 May 2019).

Valdez, Z. P., Hockaday, W. C., Masiello, C. A., Gallagher, M. E. and Robertson, G. P. 2017. Soil carbon and nitrogen responses to nitrogen fertilizer and harvesting rates in switchgrass cropping systems. *BioEnergy Research* 10(2), 456–64. doi:10.1007/s12155-016-9810-7.

VanLoocke, A., Twine, T. E., Zeri, M. and Bernacchi, C. J. 2012. A regional comparison of water use efficiency for Miscanthus, switchgrass and maize. *Agricultural and Forest Meteorology* 164, 82–95. doi:10.1016/j.agrformet.2012.05.016.

Vogel, K. P., Brejda, J. J., Walters, D. T. and Buxton, D. R. 2002. Switchgrass biomass production in the Midwest USA: harvest and nitrogen management. *Agronomy Journal* 94(3), 413–20. doi:10.2134/agronj2002.0413.

Wang, B. J. R., Seiler, J. R. and Mei, C. 2015a. *Burkholderia phytofirmans* strain PsJN advanced development and altered leaf level physiology of switchgrass. *Biomass and Bioenergy* 83, 493–500. doi:10.1016/j.biombioe.2015.10.029.

Wang, B., Seiler, J. R. and Mei, C. 2015b. A microbial endophyte enhanced growth of switchgrass under two drought cycles improving leaf level physiology and leaf development. *Environmental and Experimental Botany* 122, 100–8. doi:10.1016/j.envexpbot.2015.09.004.

Waramit, N., Moore, K. J. and Heggenstaller, A. H. 2011. Composition of native warm-season grasses for bioenergy production in response to nitrogen fertilization rate and harvest date. *Agronomy Journal* 103(3), 655–62. doi:10.2134/agronj2010.0374.

Warwick, K., Allen, F. L., Keyser, P. D., Ashworth, A. J., Bates, G. E., Tyler, D. D., Lambdin, P. L. and Harper, C. A. 2016. Biomass and integrated forage/biomass yields of switchgrass as affected by intercropped cool- and warm-season legumes. *Journal of Soil and Water Conservation* 71(1), 21–8. doi:10.2489/jswc.71.1.21.

Woodbury, P. B., Kemanian, A. R., Jacobson, M. and Langholtz, M. 2018. Improving water quality in the Chesapeake Bay using payments for ecosystem services for perennial biomass for bioenergy and biofuel production. *Biomass and Bioenergy* 114, 132–42. doi:10.1016/j.biombioe.2017.01.024.

Wright, C. K. and Wimberly, M. C. 2013. Recent land use change in the Western Corn Belt threatens grasslands and wetlands. *Proceedings of the National Academy of Sciences of the United States of America* 110(10), 4134–9. doi:10.1073/pnas.1215404110.

Zhang, Y., Zalapa, J., Jakubowski, A. R., Price, D. L., Acharya, A., Wei, Y. L., Brummer, E. C., Kaeppler, S. M. and Casler, M. D. 2011. Natural hybrids and gene flow between upland and lowland switchgrass. *Crop Science* 51(6), 2626–41. doi:10.2135/cropsci2011.02.0104.

Zilverberg, C. J., Johnson, W. C., Owens, V., Boe, A., Schumacher, T., Reitsma, K., Hong, C. O., Novotny, C., Volke, M. and Werner, B. 2014. Biomass yield from planted mixtures and monocultures of native prairie vegetation across a heterogeneous farm landscape. *Agriculture, Ecosystems and Environment* 186, 148–59. doi:10.1016/j.agee.2014.01.027.

Zilverberg, C. J., Teoh, K., Boe, A., Johnson, W. C. and Owens, V. 2015. Strategic use of native species on environmental gradients increases diversity and biomass relative to switchgrass monocultures. *Agriculture, Ecosystems and Environment* 215, 110–21. doi:10.1016/j.agee.2015.09.006.

Chapter 9

Sustainable production of willow for biofuel use

M. Weih, P.-A. Hansson, J. A. Ohlsson, M. Sandgren, A. Schnürer and A.-C. Rönnberg-Wästljung, Swedish University of Agricultural Sciences, Sweden

1 Introduction

Willows (genus *Salix*, family Salicaceae) are fast-growing deciduous trees and shrubs occurring mostly in temperate and arctic zones of the northern hemisphere (Dickmann and Kuzovkina, 2014). Traditionally, various species and cultivars of willows have been used in many countries of the world for centuries, for example, for basket-making, fencing, and small-scale husbandry. More recently, intensively managed plantations of willows and other fast-growing trees on agricultural land are gaining increasing interest in many countries of the world, mainly due to their efficient and sustainable land use in combination with an increasing demand for biofuel sources (Karp and Shield, 2008). The largest willow plantations for industrial, biofuel, and environmental purposes are currently found in China, Argentina, North America, and some European countries (FAO, 2016) (Fig. 1). In some of these countries, the shoots from

http://dx.doi.org/10.19103/AS.2019.0027.17

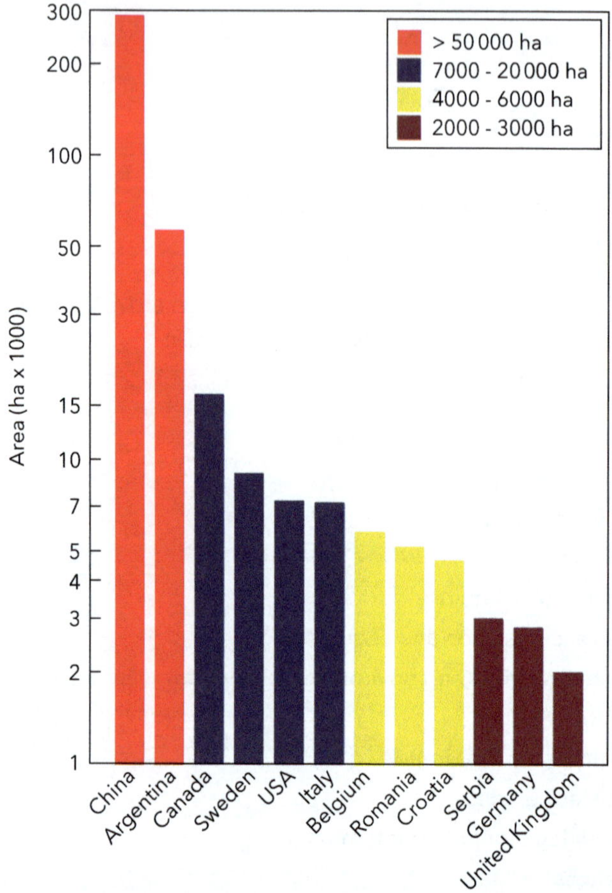

Figure 1 Areas of planted willow (*Salix* spp.) worldwide in 2015 (FAO, 2016). Only countries with willow plantation areas ≥2000 ha are shown. Note the $\log_e$ scale of the *y*-axis.

willows have been used commercially as biofuel in the form of wood chips for direct combustion in heat and power plants since the late 1980s (Kuzovkina et al., 2008) (Fig. 2). There is now also an increasing interest of using biomass from willows as raw material for conversion into biofuels such as biogas and bioethanol (Karp et al., 2011; Phitsuwan et al., 2013), a development that requires an enhanced focus in research and development on feedstock quality when willow is to be used as raw material for different supply chains.

Willow production for biofuel is mostly based on an intensive production system known as short rotation coppice (SRC) system (Larsson et al., 2007). The SRC cultivation practice is based on intensive site preparation prior to planting, including weed control, mechanical planting, in many cases mineral nutrient

Figure 2 Commercial plantations of willow for biofuel are a common view in the agricultural landscape of South and Central Sweden. Photos: M. Weih (left) and J. Svennås-Gillner (right), SLU, Sweden.

fertilizer inputs at the beginning of each cutting cycle, and multiple harvests on the basis of 2- to 5-year cutting cycles (Fig. 3) (Weih, 2013; Verwijst et al., 2013). The plantations consist of densely grown (from cuttings), high-yielding varieties of willow cultivated on agricultural land at densities of 10 000–20 000 plants ha^{-1}. The shoots are typically harvested during winter and processed to wood chips on-site by using specialized machinery (Fig. 4). The rootstock or stools remain in the ground after harvest with new shoots resprouting the following spring. A plantation could be viable for more than 20 years before reestablishment becomes necessary.

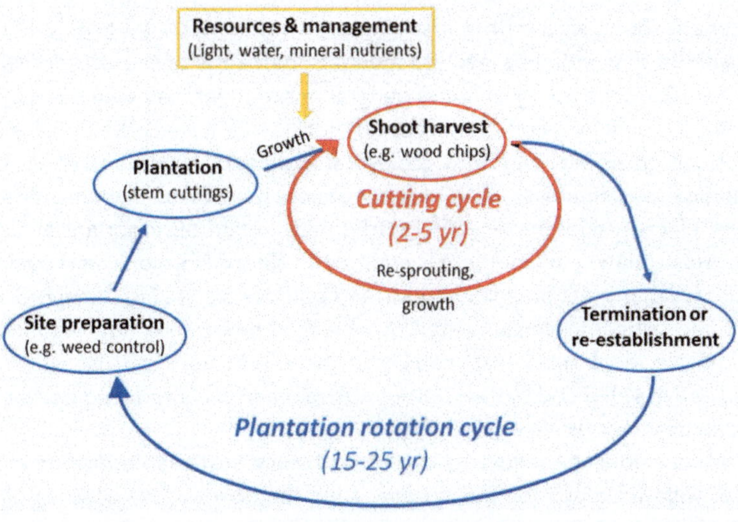

Figure 3 Overview of the actions and processes in a typical willow (*Salix*) short rotation coppice plantation grown for biofuel. Source: redrawn and modified after Weih (2013).

Figure 4 Willow plantations for direct combustion of biomass in heat and power plants are harvested during winter with specialized machinery that both cuts the rods and processes them directly into wood chips, which are supplied to the end user. Photo: N.-E. Nordh, SLU, Sweden.

The cultivation of plantations for biofuel use is often motivated in terms of beneficial effects on the environment and sustainable development (Karp and Shield, 2008). A major challenge is striking the balance between long-term ecological sustainability and reaching short-term crop productivity goals. In terms of sustainable development, the overall framework is here the document known as the Brundtland report (WCED, 1987), according to which sustainable development involves meeting the needs of the present without compromising the ability of future generations to meet their own needs. In this chapter we only focus on the primary dimension of sustainable development, that is, safeguarding long-term ecological sustainability (Holden et al., 2014). Ecological sustainability in biofuel production involves minimal chemical inputs, efficient nutrient recycling, and enhancement of important ecological processes related to nutrient acquisition, decomposition, and cropping security (e.g. protection against pests and pathogens), that is, those processes that often appear to be stimulated by increased biodiversity. In a resource-use perspective, ecological sustainability in the production and use of biofuels could aim at enhanced biofuel output with maintained or reduced depletion of natural resources (Weih et al., 2014).

We consider the following key issues for the sustainable production and use of willow for biofuel: the quality of the feedstock to be used as biofuel, feedstock productivity, cropping security, biodiversity, nutrient uptake and use, carbon accumulation and sequestration, and ecosystem services and environmental impact. Crop improvement programs in terms of selection and breeding mostly target feedstock productivity and quality traits at the individual plant scale, whereas several of the above key issues only can be evaluated at stand or landscape scales. This chapter provides an overview of the main challenges and key issues in the sustainable production and use of willow as raw material for biofuel production in temperate regions, and links crop improvement issues acting at individual plant scale to ecological sustainability issues frequently operating at higher scales.

2 Feedstock quality for biofuel use

In addition to the traditional use of willow biomass as biofuel for the production of heat and electricity by direct combustion (Ledin, 1996), there is recently a growing interest for the conversion of willow biomass into transportation fuels, with a strong focus on biogas and bioethanol (Sassner et al., 2008b; Serapiglia et al., 2013b; Pawar et al., 2018; Horn et al., 2011; Estevez et al., 2014). The biogas and bioethanol are produced by exposing the biomass to various microbiological processes. In bioethanol production, the cellulosic carbohydrates in the biomass are converted into fermentable monomers through an initial pretreatment, typically chemical or thermal, followed by enzymatic digestion. The monomers are further converted by yeast to ethanol, either simultaneously with enzymatic digestion or in separate steps (Sassner et al., 2008b). In biogas production, the biomass is degraded by a consortium of microbes under anaerobic conditions, known as anaerobic digestion. This process can, in theory, proceed without any pretreatment; however, a pretreatment breaking up the lignocellulosic structure improves accessibility and can increase both biogas yields and production rates (Horn et al., 2011; Estevez et al., 2012; Alexandropoulou et al., 2017).

For bioethanol production, the most important parameters are arguably the cellulose content of the biomass and secondary cell wall resistance (commonly referred to as biomass recalcitrance). For biogas, a high cellulose content and/or low lignin content of the willow biomass is desirable to achieve high biogas yields (Pawar et al., 2018). In general, quantities of major chemical components in willow shoot biomass vary greatly depending on the environment in which the feedstock is grown, feedstock genetics (species and variety), and also the method applied for their determination (Table 1). For example in one study, cellulose was found to be more variable than other biomass components in willow, but most of the variability was due to the environment and not the willow genotype (Fabio et al., 2017). Also biomass recalcitrance is highly variable in willow (Serapiglia et al., 2013b; Ray et al., 2012). Recently, several high-throughput screening facilities for rapid assessment of biomass recalcitrance have been constructed (Decker et al., 2009). These facilities enable the

Table 1 Major chemical components (%) of willow shoot biomass reported in the literature. Cellulose and hemicellulose values may be reported as glucan and xylan, respectively, in the reference but are treated interchangeably here

Cellulose	Hemicellulose	Lignin	Ash	Method	Reference
55.9 (0.5)	14.0 (1.0)	13.8 (0.4)	1.3 (0.1)	Wet chemistry	Szczukowski et al. (2002)[a]
41.5	15.0	25.2	1.9	Wet chemistry	Sassner et al. (2008a)
39.7 (1.8)	22.5 (1.6)	22.7 (1.6)	N/A	Wet chemistry	Serapiglia et al. (2009)
47.6 (3.1)	28.3 (3.4)	24.7 (1.4)	N/A	Wet chemistry	Sandak and Sandak (2011)
42.5 (0.4)	26.1 (0.5)	23.0 (0.3)	N/A	Wet chemistry	Stolarski et al. (2011)
34.8–41.8	12.2–15.2	23.9–28.8	N/A	Wet chemistry	Ray et al. (2012)
44.1 (0.2)	21.8 (0.1)	20.4 (0.3)	2.1 (0.1)	Wet chemistry	Stolarski et al. (2013)
41.6 (3.3)	14.0 (1.2)	28.0 (2.0)	1.9 (0.3)	Wet chemistry	Serapiglia et al. (2015)
44.4 (1.7)	31.0 (1.2)	25.5 (0.7)	1.3 (0.2)	Wet chemistry	Krzyzaniak et al. (2014)
39–41	32–33	22–23	N/A	HR-TGA	Serapiglia et al. (2012)[b]
38.7–44.5	30.1–33.6	20.9–24.1	0.5–4.5	HR-TGA	Serapiglia et al. (2013a)
38.4–45.3	31.1–34.9	20.3–23.2	N/A	HR-TGA	Serapiglia et al. (2013b)
41.3–44.0	18.3–19.4	25.0–25.9	1.6–2.7	HR-TGA	Fabio et al. (2017)[c]
37.7 (1.8)	33.4 (2.3)	25.6 (1.2)	N/A	NIR spectroscopy	Sandak et al. (2017)

[a] Data refers to 3-year-old shoots.
[b] Values estimated from figures.
[c] Range refers to genotypic means.
Values are reported either as ranges or as means with standard deviations in parentheses, and in one case only a single value was reported. N/A means the value was not reported.
Abbreviations: HR-TGA, high-resolution thermogravimetric analysis; NIR, near infrared.

screening of large numbers of biomass samples for recalcitrance characteristics by breeders and researchers with an interest in elucidating the genetics and functional mechanisms underlying recalcitrance; a recent example is the screening of recalcitrance characteristics in poplars, which are taxonomically closely related to willows (Bhagia et al., 2016).

Independent of the biomass conversion process in consideration, the chemical composition of the biomass strongly influences the final biofuel yield. However, the most desirable chemical composition of the feedstock strongly depends on the supply chain targeted. Besides the biomass composition influencing the amount of output of the targeted product (e.g., transportation fuel or energy from direct combustion) per unit of biomass, the most important factor affecting biofuel yield is the productivity of the feedstock.

3 Feedstock productivity

Production rates (dry matter, DM) of more than 20 t ha^{-1} $year^{-1}$ were reported for willow plantations in Sweden and Canada after 3–5 years of growth in small-scale fertilized experimental plantations (Labrecque and Teodorescu, 2003; Christersson, 1986). In commercial willow plantations, biomass production rates in the range of 5–11 t DM ha^{-1} $year^{-1}$ or lower have been reported, with great variation between growers (Mola-Yudego and Aronsson, 2008; Dimitriou and Mola-Yudego, 2017; Aylott et al., 2008). Greater productivity rates are expected in the near future also for the commercial plantations, due to the more frequent use of improved willow varieties and better management of the plantations (Mola-Yudego, 2011).

In general, various plant-breeding methods are available to create more productive varieties of willow with improved quality traits for biofuel use. Owing to their shorter breeding histories, willows are less domesticated than most traditional agricultural crops; but due to a number of factors including short generation times, easy hybridization, clonal propagation, and large natural genetic variation, willow breeders have made great progress during the last decades (Karp et al., 2011). In addition, willows have a relatively small genome size and are closely related to *Populus*, a model genus for tree genetics and physiology (Ghelardini et al., 2014; Hanley et al., 2006; Berlin et al., 2010); and the genome of purple willow (*Salix purpurea*) has recently been sequenced (http://phytozome.jgi.doe.gov). Taken together, these factors and genomic tools can aid in developing a rich resource for genetic improvements of willows in the future.

As for most perennial crops, a major challenge in *Salix* breeding is the production of plant material with high productivity over many years, which stresses the need for specific breeding goals targeting traits important for cropping security (e.g. resistance to abiotic and biotic stresses) apart from productivity-related traits. Breeding programs for *Salix* were launched in many countries mainly of northern temperate regions (Kuzovkina et al., 2008; Stanton et al., 2014; Smart and Cameron, 2008), using both traditional crossings and breeding methods but increasingly also molecular methods and marker-assisted breeding. Willows can be used as raw material (biomass) for different biofuel supply chains, such as the production of wood chips for combustion or the production of gaseous or liquid fuels (e.g. biogas and bioethanol). Specific breeding objectives for biomass willows target not only productivity, but also growth habit (to adapt the crop to commercial harvesting machinery), water-use efficiency and drought tolerance, nitrogen-use efficiency, frost and disease resistance, and biofuel quality characteristics related to, for example, wood chip quality (Karp et al., 2011; Ronnberg-Wastljung et al., 2005; Weih et al., 2006; Berlin et al., 2014a; Fabio et al., 2017; Samils et al., 2011; Pawar et al., 2018).

Apart from the genetics of the plant material, the production system characteristics and management strongly affect both quantity and quality of the biomass produced in willow plantations (Verwijst et al., 2013; Adler et al., 2008). For example, the planting and weed control strategies have strong effects on the stand establishment, development, and biomass productivity (Albertsson et al., 2016; Welc et al., 2017; Edelfeldt et al., 2015; Bergkvist and Ledin, 1998; Nordh, 2005). Irrigation is currently not applied in most plantations of willow used for biofuel production, although the availability of water can be critical for achieving high productivity rates in some regions (Lindroth and Bath, 1999). High temperatures during the growing season are not critical for willow plantations, as long as the stands have good water supply (Bonosi et al., 2013). Nutrient fertilization increases willow productivity in many sites, but the effect of nutrient fertilization on productivity is strongly dependent on the plant material and site conditions (Aronsson et al., 2014; Weih and Nordh, 2005; Fabio and Smart, 2018b). Observational studies in differently composed tree stands suggest that also the composition of the plantation in terms of pure stands, consisting of single species or cultivars, or mixed stands can affect the biomass productivity (Liang et al., 2016; Verheyen et al., 2016). Currently, most commercial willow plantations grown for biomass production consist of single species or cultivars. So far, no clear evidence has been found for a strong positive effect of stand diversity on productivity in young willow plantations (Dillen et al., 2016; Hoeber et al., 2018), but it is possible that a (positive) diversity effect on productivity will evolve as plantations grow older. In addition, the specific species or cultivar composition in a willow plantation has been reported to significantly affect the biomass productivity and other growth factors in mixed willow stands, which suggests the possibility for designing specific willow mixtures to promote productivity as well as other characteristics beneficial for the ecological sustainability of willow plantations grown for biomass production (Hoeber et al., 2018; Baum et al., 2018).

4 Cropping security

Durable resistance toward pathogens and insects, tolerance toward abiotic factors such as frost and drought, and efficient nutrient acquisition and use are especially important in perennial crops such as willow SRC grown for biofuel production purposes, because these crops are supposed to grow for many years. One way to enhance crop protection and cropping security is to breed for increased resistance or tolerance toward the relevant abiotic and biotic factors. In willows grown for biofuel production, the ecophysiological mechanisms and genetic background of nutrient and water use as well as the resistance or tolerance toward frost, drought, and pathogens have been investigated (Tsarouhas et al., 2004; Pucholt et al., 2015; Samils et al., 2011; Weih et al., 2011, 2006; Pei et al., 2008; Bonosi et al., 2010; Ronnberg-Wastljung et al., 2005; Beyer et al., 2018), and many of the corresponding crop traits are frequently among the most prominent targets in willow breeding programs (Kuzovkina et al., 2008). Another way to increase crop protection and performance especially in perennial production systems is to utilize ecological (biological) processes and functions that support stress tolerance, resource-use efficiency and pest control, and thereby enhance the long-term sustainability of the production system (Weih et al., 2008; Rooney et al., 2009; Grossman et al., 2018), including the possibilities of biological control and integrated pest management (IPM) (Bjorkman et al., 2004; Moritz et al., 2017; Charles et al., 2014). Compared with the corresponding traditional management actions needed, utilization of these ecological processes usually requires more time to show effects, which provides opportunities particularly in perennial production systems with their considerably longer-time perspective compared to annual systems. The utilization of ecological processes to reduce the necessity of traditional management actions is therefore expected to require longer time periods (Fig. 5), but offers a viable alternative to enhance the long-term sustainability in perennial systems such as willow SRC. A major challenge in the utilization of ecological processes for improving long-term sustainability in these systems is to understand the complexity of the involved mechanisms. In willow grown for biofuel, approaches are being explored to utilize ecological information for improving crop protection toward harmful leaf beetles frequently damaging willow SRC (Bjorkman et al., 2004; Stenberg et al., 2010; Moritz et al., 2017), manipulating willow resource use, and enhancing pest resistance through the selection of appropriate soil microbe (e.g., mycorrhiza)–plant combinations (Fig. 6) (Rooney et al., 2009; Fransson et al., 2013; Baum et al., 2009b, 2018), and reducing damage to trees by pests and pathogens through the culture of more diverse (mixed) plantations (McCracken and Dawson, 1994, 1998; Muller et al., 2018). In addition, ecological research has addressed the induction of insect resistance in willows (Hoglund and Larsson, 2005; Hoglund et al., 2005),

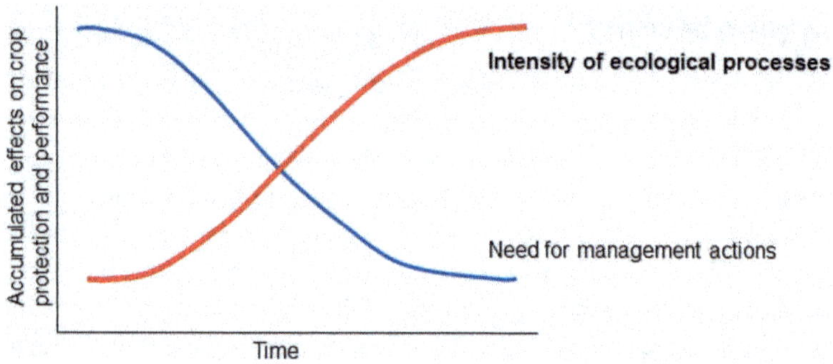

Figure 5 Relationship between time and the intensity of ecological (biological) processes supporting crop protection and performance (e.g. protection against pests and pathogens, enhancement of microbial-driven nutrient acquisition) and the mirror image, indicating the need for management actions (e.g. pesticides, nutrient fertilizers) to achieve the same effects on crop protection and performance. Source: redrawn and modified after Weih et al. (2008).

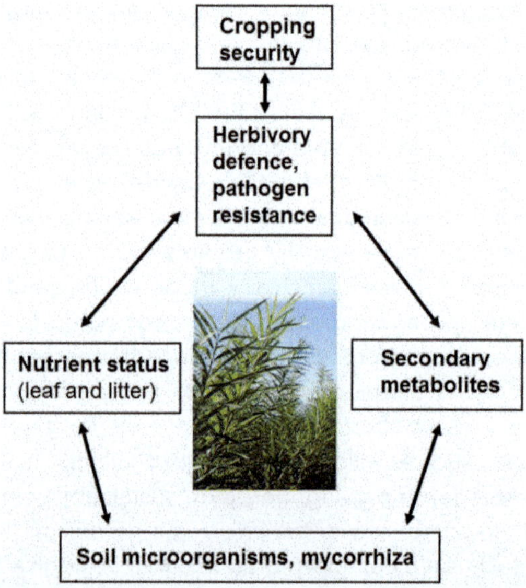

Figure 6 Potential effects of soil organisms and mycorrhizal fungi on cropping security in willow plantations. Mycorrhizal fungi are an important integral component of the plant-soil system, forming symbiotic associations with willow roots and having a fundamental role in plant nutrition and protection against pathogens through effects on the foliar concentrations of, for example, secondary metabolites. Source: modified after Rooney et al. (2009). Photo: J. Svennås-Gillner, SLU, Sweden.

the factors affecting palatability of willow leaves for herbivores (Albrectsen et al., 2007, 2004; Glynn et al., 2004, 2007; Baum et al., 2009b), and willow leaf beetle behave with particular focus on the common management practice of willow plantations and the possibilities for biological control including IPM and plant breeding (Charles et al., 2014; Bjorkman et al., 2004; Dalin et al., 2004; Bjorkman and Ahrne, 2005; Bjorkman and Eklund, 2006; Stenberg et al., 2010). However, utilizing the knowledge on the biology and ecology of the organisms involved to improve cropping security of willow SRC in a predictable way remains a great challenge, as the relevant biological interactions are often highly context-dependent and greatly affected by abiotic factors, for example, temperature (Puentes et al., 2015; Fransson et al., 2013).

5 Biodiversity

On the one hand, the previous sections discussed some potentially beneficial effects of more diverse willow plantations (i.e. species or variety mixtures) on productivity and cropping security. On the other hand, studies on sustainability aspects of biomass production systems often point out negative effects of these production systems on biodiversity (landscape level) as a key concern (Desiree et al. 2014). In general, plantations of trees can have positive or negative effects on biodiversity at the landscape level, depending on location, planting design (e.g., single-tree or SRC), and management of the plantation–but also previous land use and the kinds of organisms considered (Hartley, 2002). Plantations of willow SRC grown on agricultural land have been reported to improve plant diversity at landscape level, in particular if the plantations are established instead of cultures of cereals and spruce or fallow ground in a homogeneous agricultural landscape (Baum et al., 2009c, 2012). The willow plantations can provide benefits to bird and small mammal populations and enhance their value for biodiversity, especially if multiple age classes of willows are maintained in the landscape simultaneously (Berg, 2002; Campbell et al., 2012). Also, high number and diversity of beneficial insects has been found in willow plantations when compared to conventional agricultural fields (cereals, sugar beet, and potato) (Jaworska, 2010). Comprehensive information regarding the effects of willow SRC on the biodiversity of many animals is still lacking, but it has been concluded that (1) the diversity of breeding birds is higher than in agricultural cropland, but lower than in forest ecosystems (Berg, 2002; Campbell et al., 2012); (2) ground beetle diversity is higher in arable fields than in willow SRC (Jaworska, 2010); (3) willow SRC plantations accommodate both a greater diversity and higher abundance in most animal groups compared to poplar plantations (Schulz et al., 2009); and (4) the cultivation of willow SRC can lead to an increase in animal diversity in homogeneous agricultural landscapes, but to adverse effects in landscapes with high conservational value (Schulz et al., 2009). Little information is available on the soil fungal and bacterial responses to the conversion of conventional agricultural land to willow SRC, reporting enrichment of fungal communities (including mycorrhizal fungi) but indifferent responses of bacterial communities (Xue et al., 2016). In contrast to most conventional agricultural crops, willow roots can be colonized by ectomycorrhizal fungi, leading to the introduction of these fungi into arable soils and to considerable changes in the soil microbial colonization and activity when conventional agricultural land is converted to willow SRC. The abundance and diversity of mycorrhizal fungi associated with willows was found to be greatly affected by the previous land use, management (e.g. nutrient fertilization), the plant material, and the age of the willow plants (Hrynkiewicz et al., 2010, 2012; Baum et al., 2002, 2009a). Compared to conventional agricultural land, the non-tillage management and the high litter supply in willow SRC can increase the abundance of earthworms but decrease the abundance of certain ground-living beetles (Carabidae) (Baum et al., 2009a). In summary, the effects of planting willow SRC after conventional agricultural crops on landscape biodiversity are frequently positive, especially in open agricultural landscapes and when maintaining high structural diversity (e.g. multiple age classes), but effects are strongly context-dependent.

6 Nutrient uptake and use

Reduced resource depletion is an important element in sustainable land use for biofuel production purposes (Higman et al., 2005; Ra et al., 2012), and the availability of mineral nutrients and particularly nitrogen (N) is a critical factor in willow production for biofuel, because the manufacture of commercial fertilizers consumes much fossil-based energy and thereby contributes to global warming. Biomass productivity per unit plant-internal amount of N is considerably higher in willows compared to conventional agricultural crops such as wheat (Weih et al., 2018). Increased productivity at maintained nutrient accumulation in the crop would be one way to improve crop nutrient use, and some of the key factors and mechanisms determining the productivity per unit plant-internal N under different environmental conditions (e.g., drought) and using different plant material have been identified for willows used for biomass production (Weih et al., 2011; Weih and Nordh, 2002; Weih, 2009). Another way of increasing productivity is the use of nutrient fertilizers, and willows used for biomass production greatly vary in their growth response to different nutrient supply (Weih and Nordh, 2005; Fabio and Smart, 2018a). Thus, enhanced biomass production through increased nutrient uptake rates is a prominent target in improvement programs for biomass crops, but is not in line with the sustainability aim of reduced resource depletion, because higher productivity is usually associated with increased N absorption from the soil (Ra et al., 2012). In general, nutrient fertilization greatly enhances crop yields, but decreases the productivity per unit of plant-internal amount of N (Weih et al., 2018), reflecting the general pattern of more than proportional resource demand when crop yields are increased through fertilization (Tilman et al., 2002). In an ecological sustainability context, a major breeding focus for willow cultivars to be grown for biofuel production could be enhanced energy output per N resource depletion, and the identification of heritable traits affecting it (Weih et al., 2014). This ratio is expected to be influenced by various crop characteristics, for example, the productivity per unit plant-internal N and quality traits of the harvested biomass. The productivity per unit plant-internal N is expected to be enhanced through the traits optimizing canopy architecture and photosynthesis, for which genetic variation and breeding potential has been documented in willows (Weih and Ronnberg-Wastljung, 2007; Fabio and Smart, 2018a). Additional breeding opportunities relevant for enhanced energy output per N resource depletion arise from the genetic variation in heating value and biomass quality traits documented for willows (Fabio et al., 2017; Pawar et al., 2018). Compared to a traditional focus of crop breeding research on biomass yields and qualities, a resource-oriented approach adopts a more holistic approach including resource (here nutrients) aspects that can be helpful in the evaluation of biofuel crops in an ecological sustainability context. Such an approach could also consider the energy produced in relation to the fate of the targeted resource. For example, the nutrient resources to generate feedstock for biogas production can easily be reused for the cultivation of new feedstock, resulting in enhanced sustainability of the production system compared to a system without the immediate reuse of the nutrient resources.

7 Carbon accumulation and sequestration

Considering the frequently environmental motivation of willow grown for biofuel production, an important target for the further development of bioenergy crops should be increased carbon transfer from atmosphere to soil and enhanced carbon accumulation in soil. Increased carbon accumulation has been documented in willow plantations grown on formerly arable soils, while indifferent or positive effects on carbon accumulation were found when willow SRC was planted on former grassland sites (Kahle et al., 2005; Harris et al., 2015, 2017; Dimitriou et al., 2012b). Also, carbon balance models targeting the carbon fluxes between soil, biomass, and atmosphere in willow SRC indicated that these plantations act as carbon sinks and therefore have a mitigating effect on climate change, especially when plant material and management conditions are applied that support high biomass yields (Hammar et al., 2014). An important factor affecting the total carbon fluxes and the soil carbon accumulation in a willow stand is the previous land use (Grogan and Matthews, 2002). Also, the choice of the plant material, for example, willow genotype can affect below-ground biomass allocation in the plant and thereby soil carbon accumulation, which opens possibilities for plant breeding toward increased carbon accumulation under willow plantations grown for biofuel production (Cunniff et al., 2015; Gregory et al., 2018). Apart from the quantity of accumulated soil carbon, its specific composition is also important and determines the longevity of the carbon pool in the soil and thus the sustainability of any carbon accumulation seen in the soil under biomass crops (Baum et al., 2013). The formation of root mycorrhizal fungi can also enhance the amount of carbon allocated below-ground under willow (Jones et al., 1991), and these root-associated fungi can affect several host plant traits relevant for soil carbon accumulation both directly through the roots and indirectly through the influences on leaf and litter chemistry (Rooney et al., 2009). However, the interaction between the relevant willow plant traits and the fungi appears difficult to predict due to the complex nature of the corresponding interactions (Fransson et al., 2013).

8 Ecosystem services and environmental impact

Ecosystem services are the goods and services provided by an ecosystem to society, including watershed services, nutrient cycling, waste management, carbon storage and effects on climate change, scenic landscapes, and biodiversity and wildlife habitat. Many of these services were already discussed in the previous sections of this chapter regarding willows used for biofuel production, and are also reviewed elsewhere (Isebrands et al., 2014). This section highlights some environmental applications that are relevant in the context of the sustainable use of willows used for biofuel production, and summarizes the reported knowledge on the environmental impact of willow SRC including some results from life cycle assessment (LCA).

Productive willow plantations take up large quantities of water and nutrients, which has initiated various applications in which these plantations are used as recipients for municipal wastewater and industrial sludge along with the simultaneous production of biomass, sometimes referred to as multifunctional biomass plantations (Aronsson and Perttu, 2001; Labrecque and Teodorescu, 2003). A major challenge occurs with those applications in regions with extended dormant periods during winter, where only small quantities of nutrients are taken up by the willow plants; in those regions, wastewater can be stored in ponds or lagoons during the winter period for subsequent application to the plantations during the growing season (Dimitriou and Aronsson, 2005).

A large-scale shift from, for example, conventional agricultural crops to willow SRC is likely to impact various environmental issues, although there are many uncertainties with respect to the specific impacts in a given context, because willow SRC still is a new production system for most regions in which it could be grown in the future. Thus, results so far indicate many environmental benefits, but also negative impacts on the environment due to willow SRC implementation, and the effects on the environment strongly depend on the existing or previous land use, the plantation scale, the plant material used, and the management practices applied (Weih and Dimitriou, 2012; Langeveld et al., 2012; Bacenetti et al., 2016). Apart from the effects of willow SRC on soil carbon accumulation and biodiversity summarized previously, willow SRC has been reported to decrease leaching of nitrate to the groundwater, but increase phosphorus leaching compared to conventional agricultural crops (Dimitriou et al., 2012a; Schmidt-Walter and Lamersdorf, 2012), and decrease the soil concentration of cadmium (Cd), a hazardous trace element for human health in the food chain (Dimitriou et al., 2012b).

LCA revealed a considerably higher production efficiency and environmental performance of willow biomass used for direct combustion in combined heat and power plants (CHP) compared to fossil fuel-based power plants (Buonocore et al., 2012). Compared to other feedstock used for biofuel production, willow has a high energy output to input ratio and a low carbon footprint (Parajuli et al., 2017). In willow SRC used for biofuel, the most important factors affecting the energy efficiency are the application of nutrient fertilizers (affecting yield level), the harvest, and the transport of the harvested product (Hammar et al., 2017). The application of nutrient fertilizers to willow plantations enhances the productivity, but also the negative environmental impacts related to, for example, eutrophication and toxicity potentials. Taken together, it has been shown that the higher biomass productivity achieved with nutrient fertilization results in a better overall environmental profile in terms of net energy yield (i.e. net quantity of energy in MJ generated per land area) and global warming potential compared to unfertilized plantations (Gonzalez-Garcia et al., 2012). However, there is a trade-off between achieving a high net energy yield and achieving a high energy ratio (Nordborg et al., 2018), and a common notion in many LCA and environmental impact studies is the strong context-dependency of the environmental impacts associated with the plantation of willow SRC for biofuel.

9 Case study

In Sweden, willows have been grown for biofuel purpose for several decades (Christersson et al., 1993), and Sweden has highly ambitious climate goals aiming at zero net greenhouse gas emissions by 2045 (e.g. https://www.newscientist.com/article/2138008). Moving from a fossil-based to a bio-based society with reduced greenhouse gas emissions requires an increased use of multiple sources of renewable energy, and the sustainable use of willow for biofuel production provides a viable alternative for contributing to the transition from a fossil-based to a bio-based society. For willows to play a significant role in this development, new willow varieties are needed that can be used for the production of different biofuels and contribute to sustainable development in terms of, for example, enhanced productivity and cropping security, increased resource-use efficiency, and a large carbon sequestration potential.

In perennial plants such as willows grown in seasonal environments, the timing of the critical stages for growth initiation in spring (e.g. bud-burst and advancement of leaf unfolding) and growth termination in autumn (e.g. height growth cessation and leaf abscission), that is, leaf phenology, is an important driver for biomass productivity. Spring phenology determines the start of the growth period, but earlier bud-burst may increasingly predispose plants to greater risk of frost damage. Autumn phenology is closely linked to frost-hardening, and a correlation between frost resistance and height growth cessation has been reported in willow (Ogren, 1999). However, many willows continue to be photosynthetically active and accumulate biomass after height growth cessation and bud set (Bollmark et al., 1999). Thus, a delayed growth cessation and/or autumn leaf fall could go along with increased shoot production, but light and temperature conditions are rapidly deteriorating toward late autumn especially in high-latitude environments, and there should be a critical period in autumn after which delayed growth cessation and/or leaf fall is not any longer associated with enhanced productivity. Also for biomass willows, the bud and leaf phenologies have been pointed out as promising targets for the breeding (Lennartsson and Ogren, 2004; Ronnberg-Wastljung, 2001). However, there is still uncertainty in how the variation in phenology affects the biomass production, and also about the genetic background of those phenology characteristics that most strongly affect biomass production. Due to these uncertainties, various commercial willow varieties have been examined to investigate the effects of willow cultivar and environment on spring and autumn phenologies as well as the relationships between phenology and biomass production; the genetic background of the corresponding phenology traits was analyzed in a large breeding population. The duration of the leafy period, that is, the period between bud-burst and leaf abscission, and bud-burst date were found to strongly affect the productivity of the willows; the timing of bud-burst and leaf abscission was shown to be more important for willow biomass production than the timing of height growth cessation (Weih, 2009) (Fig. 7). Subsequent studies investigated the genetic background of the corresponding phenology characteristics in a willow breeding population by assessing genetic and phenotypic variation in bud-burst, growth cessation, and leaf abscission in different years and environments, and identified various genetic markers for these traits that will speed up the breeding of biomass willows in the future (Ghelardini et al., 2014; Berlin et al., 2017). In a similar approach, critical characteristics for willow resource (nutrients and water) economy were first analyzed at the phenotype level in a breeding population (Weih et al., 2011), and the genetic background of the same characteristics was subsequently evaluated and the corresponding genetic markers were identified

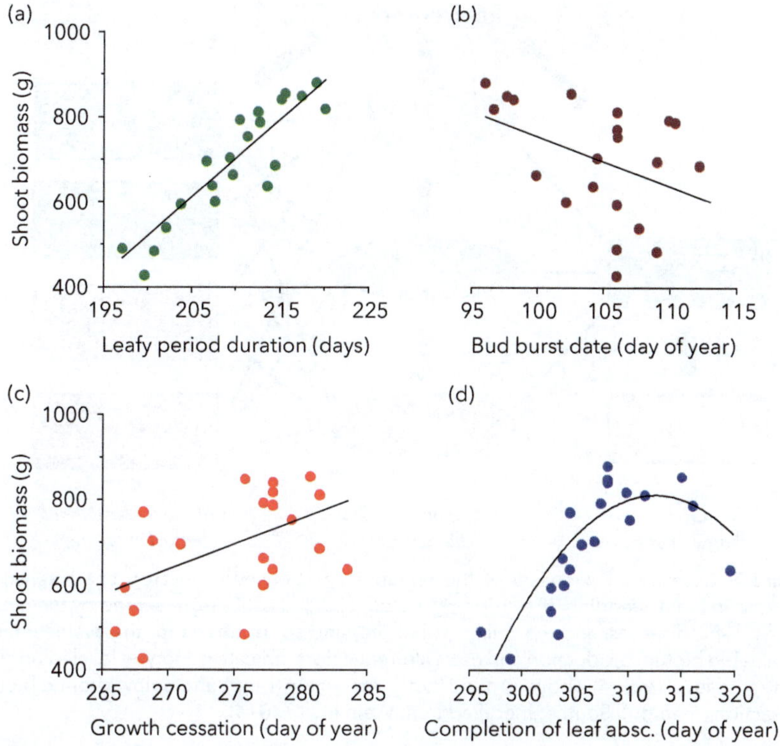

Figure 7 The shoot biomass production of willow as related to the duration of period in which the trees had leaves (a), bud-burst date (b), timing of growth cessation (c), and leaf abscission (d). The timing of bud-burst and leaf abscission are better correlated to biomass production than the timing of height growth cessation. Source: redrawn from Weih (2009).

to support rapid breeding for sustainable willow production and use (Berlin et al., 2014a). Apart from productivity characteristics focusing on the amount of biomass produced, recent research targets increasingly also biomass quality and different routes for biofuel use from willows, for example, the identification of genetic markers for different wood chemical traits, information that is important for willow breeding toward different uses of its biomass (Pawar et al., 2018). The knowledge and tools that have been built up within the framework of the above research are excellent resources for a continued plant breeding of willow and in addition also make it possible to develop willow into a model plant for studying breeding strategies in woody plant species.

When evaluating the sustainable use of willow for biofuel production, the breeding issues discussed so far need to be connected to ecological functions and ecological sustainability. In general, plant breeding is strictly trait-based and usually targets relatively short-term events occurring in more or less

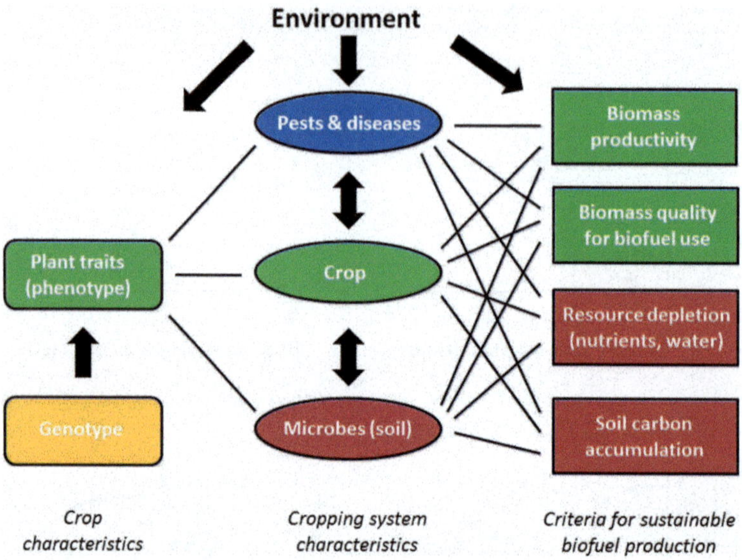

Figure 8 Conceptual overview of the relationships between the crop characteristics targeted in plant breeding; the cropping system characteristics represented by the crop, its pests and diseases, and the soil microbial life; and some criteria for the evaluation of sustainable biofuel production and use. Different colors indicate genotype level (orange), plant/crop level (green), above-ground biotic interaction (blue), and below-ground biotic interaction (maroon). Source: modified from Weih et al. (2014).

simple systems, whereas any ecological sustainability evaluation requires the integration of longer-term ecological processes occurring in more complex systems (Weih et al., 2008). Therefore, a trait-to-ecosystem perspective should be integrated into the breeding research for the sustainable use of plants as feedstock for biofuel production. In such a perspective, the cascading effects of plant characteristics targeted in breeding research on the ecosystem processes particularly important in the evaluation of the ecological sustainability of biofuel crops should be considered (Weih et al., 2014) (Fig. 8). Adopting such an approach, some willow genotypes have been found to positively affect community productivity, resource (nitrogen) efficiency, and soil carbon accumulation while others seem to have negative effects on the same functions (Hoeber et al., 2017, 2018; Gregory et al., 2018), but the corresponding specific plant traits behind those effects have yet to be identified.

10 Summary and future trends

Intensively managed willow plantations are gaining interest, due mainly to their efficient and sustainable land use in combination with an increasing demand for biofuel resources. In many countries, willows have been used commercially as biofuel in the form of wood chips for direct combustion in heat and power plants, but there is an increasing interest in using biomass from willows as raw material for conversion into biofuels such as biogas and bioethanol. The establishment of willow plantations for biofuel production purposes is often motivated in terms of beneficial effects on the environment and sustainable development. A major challenge is striking the balance between long-term sustainability and reaching short-term crop productivity goals. Plant breeding is an efficient tool to create more productive varieties with improved quality traits for biofuel use, and major breeding progress has been achieved in willows during the last decades using both traditional breeding methods and increasingly also molecular methods and marker-assisted breeding. Major challenge in willow breeding is the production of plant material with high productivity over many years, with focus on specific traits important for cropping security (e.g. resistance to abiotic and biotic stresses) apart from purely productivity-related traits including resource-use (water and nutrients) efficiency and biofuel quality. Besides genetics and breeding, the production system characteristics and management strongly affect the quantity and quality of the biomass produced. Productivity is affected by the composition of the plantation in terms of pure stands consisting of single species or cultivars vs. mixed stands, involving the possibility for designing specific willow mixtures to promote productivity as well as other characteristics beneficial for the ecological sustainability of willow biofuel plantations. Durable resistance toward pathogens and insects, tolerance toward abiotic factors such as frost and drought, and efficient nutrient acquisition and use are especially important in willows grown for biofuel production, because these perennial crops are supposed to grow for many years. An attractive alternative to achieve durable resistance is the utilization of ecological processes and functions that support stress tolerance, resource-use efficiency, and pest control and thereby enhance the long-term ecological sustainability of the production system; a major challenge in the utilization of ecological processes is to understand the complexity of the involved mechanisms. Studies on ecological sustainability aspects of biofuel production systems sometimes point out negative effects on landscape-level biodiversity. For willows, the effects on landscape biodiversity are frequently reported positive, although comprehensive information regarding the effects of willow plantations on the biodiversity of many organisms is still lacking. The reduced depletion of resources such as nutrient fertilizer is an important element in sustainable land use for biofuel production, and a major breeding focus could be enhanced energy output per nutrient depletion, and the identification of heritable traits affecting it. Increased carbon transfer from atmosphere to soil and enhanced soil carbon accumulation are other important goals in the sustainable use of willow for biofuel production, since these factors have been shown to be especially important for the total climate effects in a life cycle perspective. Willow plantations have been reported to frequently act as carbon sinks, but soil carbon accumulation is affected by various factors including site history, choice of plant material, and soil microbial life. Willow plantations are also used as recipients for municipal wastewater and industrial sludge along with the simultaneous production of biomass and can thus provide multiple ecosystem services, with positive environmental impacts. A large-scale shift from, for example, conventional agricultural crops to willow plantations is likely to impact various environmental issues, although there are many uncertainties with respect to the specific impacts in a given context. Compared to other feedstock used for biofuel production, willow has a high energy output to input ratio and a low carbon footprint, and research results indicate many environmental benefits, but also negative impacts on the environment.

Intensively grown willow plantations have the potential to become key players in the sustainable production of woody biomass to be used as raw material for various biofuel supply chains, and willows are thus interesting as feedstock not only for direct combustion in heat and power plants, but also in various biorefinery conversion processes (Sassner et al., 2008b; Horn et al., 2011; Serapiglia et al., 2013b). Development and breeding of willow plant material better suited for the use as biofuel crops compared to the currently available material requires an improved understanding of the structure and chemistry of willow wood and knowledge about the genetic variation in, and control over, the relevant wood traits. The relevant wood traits include the characteristics facilitating, for example, improved lignocellulose disruption and subsequent enzymatic saccharification, and their interactions with different microbiological and thermochemical biofuel production systems. Molecular breeding approaches using marker-assisted selection in plant breeding are promising for wood characteristics, as the relevant traits are time consuming and costly to measure. Thus, the development of genetic markers for key genes controlling wood traits would greatly facilitate the early selection and efficient breeding of willows for biofuel use, and thus increase the potential genetic gain. Almost half a century of pre-breeding research and commercial breeding of willow in Sweden and elsewhere gives access to a wide range of well-defined willow plant material to be used for future pre-breeding research, and there are several established research populations harboring genetic resources for future breeding programs. Especially in the context of the sustainability goals important for the use of willows for biofuel, holistic breeding approaches considering the ecological processes and functions supporting long-term cropping security (i.e. resistance and/or tolerance to pests and diseases, reduced depletion of water and nutrient resources) probably will be important research subjects in the future, although the complexity of the involved biological and ecological processes will be a major challenge for the predictable utilization of ecological interactions in terms of, for example, the biological control of pests including IPM (Charles et al., 2014; Weih et al., 2008). Especially in light of the future development toward the use of willow for diverse biofuel outputs, more research is needed on the effects of carbon-rich waste products from various biorefinery processes returned to the soil on the below-ground processes driving carbon cycling and sequestration; the latter is one of the ecosystem services of great public concern in relation to climate change and maintenance of soil fertility. For example, the effects of plant material (e.g. variety) identity, and plantation diversity (e.g., defined variety mixtures) on biofuel productivity and quality, cropping security of the plantation, and below-ground carbon and nutrient cycling need to be evaluated in the future to ensure a sustainable use of willow for different biofuel production chains. In this context, LCA has emerged as a powerful tool for quantifying and improving the environmental impacts of biofuel production in a sustainability perspective (Ahlgren et al., 2012), and supported by recent progress in methodology developments including dynamic modeling of carbon dynamics and global temperature increase using response functions (Ericsson et al., 2013; Hammar et al., 2014). A future trend is thus to combine, in an integrated approach, the research on willow genetics and breeding with investigations on biological and chemical wood conversion processes, ecological processes, and biological control including IPM, carbon and nutrient cycling, and environmental impact by using, for example, LCA methodology (Fig. 9).

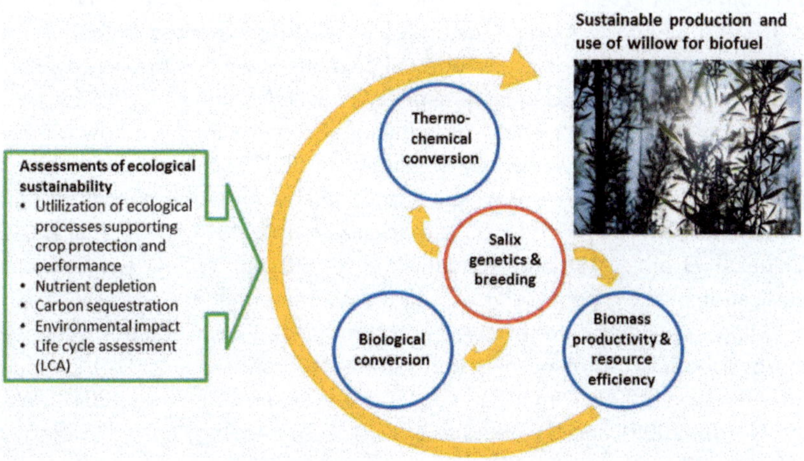

Figure 9 Outline of a research approach integrating willow genetics and breeding with investigations on biological and chemical wood conversion processes and the assessment of ecological sustainability criteria related to the utilization of ecological processes for, for example, biological control, carbon and nutrient cycling, and environmental impact by using, for example, LCA methodology. Photo: V. Wrange, SLU, Sweden.

11 Acknowledgements

The compilation of this chapter was partly funded by the The Swedish Research Council for Environment, Agricultural Sciences and Spatial Planning (Formas, project no. 942-2016-31).

12 Where to look for further information

Updated information on all aspects of willows grown for biofuel production and other purposes is found at the web page of the International Poplar Commission (IPC) at: www.fao.org/forestry/ipc. Regular conferences addressing current aspects of willow (and poplar) production and use for various purposes are organized by the IPC (information at the above website) and by the International Union of Forest Research Organizations (IUFRO), unit 2.08.04–Poplars and Willows; see further information at: www.iufro.org/science/divisions/division-2/20000/20800/20804. Several reviews have been published regarding the sustainable production and use of willow (and other plants) for biofuel, for example, Volk et al. (2004, 2016), Verwijst et al. (2013), Karp and Shield (2008), Amichev et al. (2014), and Weih (2004). Overviews on various willow taxonomy, genetics, domestication, and breeding aspects are given by Gullberg (1993), Larsson (1997), Kuzovkina et al. (2008), Dickmann and Kuzovkina (2014), Stanton et al. (2014), Smart and Cameron (2008), Karp et al. (2011), Hanley and Karp (2014), Berlin et al. (2014b), Hallingback et al. (2016), and Ronnberg-Wastljung and Gullberg (1999), while management aspects are reviewed by Verwijst et al. (2013) and Stanturf and van Oosten (2014) and documented in manuals for growers published for different regional contexts at various publicly available web pages. Basic wood quality aspects of willows are reported by Ledin (1996), while wood properties, processing, and utilization of willow biomass for energetic use is reviewed by Balatinecz et al. (2014) and various biofuel quality aspects for different willow supply chains are subjects of the more specialized papers by Jirjis (2005), Serapiglia et al. (2013b), Pawar et al. (2018), Whittaker et al. (2018), and Krzyzaniak et al. (2016a). Willow environmental applications, ecosystem services, and environmental impact assessments including biodiversity are reviewed and summarized in several book chapters and/or special issues of journals, for example, Isebrands et al. (2014), Dimitriou and Aronsson (2005), Dimitriou et al. (2009), Weih and Dimitriou (2012), Weih and Nordh (2009), and Bacenetti et al. (2016). The ecology and physiology of willows, including the responses to abiotic stresses, have been reviewed by Mitchell et al. (1992), Richardson et al. (2014), and Marron et al. (2014), whereas specific reviews on willow diseases and pests including IPM where published by Ostry et al. (2014) and Charles et al. (2014). The mathematical modeling of biomass production in willow plantations grown under different environmental conditions has been addressed (Wang et al., 2015; Sannervik et al., 2006; Cerasuolo et al., 2016; Philippot, 1996), as well as different methodologies of LCA to simulate environmental and climate impacts for different willow biofuel supply chains (Heller et al., 2003; Gonzalez-Garcia et al., 2012; Buonocore et al., 2012; Budsberg et al., 2012; Krzyzaniak et al., 2016b; Ericsson et al., 2017; Hammar et al., 2017).

13 References

Adler, A., Dimitriou, I., Aronsson, P., Verwijst, T. and Weih, M. 2008. Wood fuel quality of two *Salix viminalis* stands fertilised with sludge, ash and sludge-ash mixtures. *Biomass and Bioenergy* 32(10), 914–25. doi:10.1016/j.biombioe.2008.01.013.

Ahlgren, S., Roos, E., Lucia, L. D., Sundberg, C. and Hansson, P. A. 2012. EU sustainability criteria for biofuels: uncertainties in GHG emissions from cultivation. *Biofuels* 3(4), 399–411. doi:10.4155/bfs.12.33.

Albertsson, J., Verwijst, T., Rosenqvist, H., Hansson, D., Bertholdsson, N. O. and Ahman, I. 2016. Effects of mechanical weed control or cover crop on the growth and economic viability of two short-rotation willow cultivars. *Biomass and Bioenergy* 91, 296–305. doi:10.1016/j.biombioe.2016.05.030.

Albrectsen, B. R., Gardfjell, H., M. Orians, C., Murray, B. and S. Fritz, R. 2004. Slugs, willow seedlings and nutrient fertilization: intrinsic vigor inversely affects palatability. *Oikos* 105(2), 268–78. doi:10.1111/j.0030-1299.2004.12892.x.

Albrectsen, B. R., Gutierrez, L., Fritz, R. S., Fritz, R. D. and Orians, C. M. 2007. Does the differential seedling mortality caused by slugs alter the foliar traits and subsequent susceptibility of hybrid willows to a generalist herbivore? *Ecological Entomology* 32, 070130195410003-???. doi:10.1111/j.1365-2311.2006.00860.x.

Alexandropoulou, M., Antonopoulou, G., Ntaikou, I. and Lyberatos, G. 2017. Fungal pretreatment of willow sawdust with *Abortiporus biennis* for anaerobic digestion: impact of an external nitrogen source. *Sustainability* 9(1), 14. doi:10.3390/su9010130.

Amichev, B. Y., Hangs, R. D., Konecsni, S. M., Stadnyk, C. N., Volk, T. A., Belanger, N., Vujanovic, V., Schoenau, J. J., Moukoumi, J. and Van Rees, K. C. J. 2014. Willow short-rotation production systems in Canada and northern United States: a review. *Soil Science Society of America Journal* 78(S1), S168–82. doi:10.2136/sssaj2013.08.0368nafsc.

Aronsson, P. and Perttu, K. 2001. Willow vegetation filters for wastewater treatment and soil remediation combined with biomass production. *The Forestry Chronicle* 77(2), 293–9. doi:10.5558/tfc77293-2.

Aronsson, P., Rosenqvist, H. and Dimitriou, I. 2014. Impact of nitrogen fertilization to short-rotation willow coppice plantations grown in Sweden on yield and economy. *Bioenergy Research* 7(3), 993–1001. doi:10.1007/s12155-014-9435-7.

Aylott, M. J., Casella, E., Tubby, I., Street, N. R., Smith, P. and Taylor, G. 2008. Yield and spatial supply of bioenergy poplar and willow short-rotation coppice in the UK. *The New Phytologist* 178(2), 358–70. doi:10.1111/j.1469-8137.2008.02396.x.

Bacenetti, J., Bergante, S., Facciotto, G. and Fiala, M. 2016. Woody biofuel production from short rotation coppice in Italy: environmental-impact assessment of different species and crop management. *Biomass and Bioenergy* 94, 209–19. doi:10.1016/j.biombioe.2016.09.002.

Balatinecz, J., Mertens, P., De Boever, L., Hua, Y. K., Jin, J. W. and Van Acker, J. 2014. Properties, processing and utilization. In: Isebrands, J. G. and Richardson, J. (Eds), *Poplars and Willows: Trees for Society and the Environment*. CABI Publishing, Wallingford.

Baum, C., Weih, M., Verwijst, T. and Makeschin, F. 2002. The effects of nitrogen fertilization and soil properties on mycorrhizal formation of *Salix viminalis*. *Forest Ecology and Management* 160(1–3), 35–43. doi:10.1016/S0378-1127(01)00470-4.

Baum, C., Leinweber, P., Weih, M., Lamersdorf, N. and Dimitriou, I. 2009a. Effects of short rotation coppice with willows and poplar on soil ecology. *Landbauforschung Volkenrode* 59, 183–96.

Baum, C., Toljander, Y. K., Eckhardt, K.-U. and Weih, M. 2009b. The significance of host-fungus combinations in ectomycorrhizal symbioses for the chemical quality of willow foliage. *Plant and Soil* 323(1–2), 213–24. doi:10.1007/s11104-009-9928-x.

Baum, S., Weih, M., Busch, G., Kroiher, F. and Bolte, A. 2009c. The impact of Short Rotation Coppice plantations on phytodiversity. *Landbauforschung Volkenrode* 59, 163–70.

Baum, S., Bolte, A. and Weih, M. 2012. High value of short rotation coppice plantations for phytodiversity in rural landscapes. *Global Change Biology Bioenergy* 4(6), 728–38. doi:10.1111/j.1757-1707.2012.01162.x.

Baum, C., Eckhardt, K. U., Hahn, J., Weih, M., Dimitriou, I. and Leinweber, P. 2013. Impact of poplar on soil organic matter quality and microbial communities in arable soils. *Plant, Soil and Environment* 59(3), 95–100. doi:10.17221/548/2012-PSE.

Baum, C., Hrynkiewicz, K., Szymanska, S., Vitow, N., Hoeber, S., Fransson, P. M. A. and Weih, M. 2018. Mixture of *Salix* genotypes promotes root colonization with dark septate endophytes and changes P cycling in the mycorrhizosphere. *Frontiers in Microbiology* 9, 1012. doi:10.3389/fmicb.2018.01012.

Berg, Å 2002. Breeding birds in short-rotation coppices on farmland in central Sweden - the importance of *Salix* height and adjacent habitats. *Agriculture, Ecosystems and Environment* 90(3), 265–76. doi:10.1016/S0167-8809(01)00212-2.

Bergkvist, P. and Ledin, S. 1998. Stem biomass yields at different planting designs and spacings in willow coppice systems. *Biomass and Bioenergy* 14(2), 149–56. doi:10.1016/S0961-9534(97)10021-6.

Berlin, S., Lagercrantz, U., Von Arnold, S., Ost, T. and Ronnberg-Wastljung, A. C. 2010. High-density linkage mapping and evolution of paralogs and orthologs in *Salix* and *Populus*. *BMC Genomics* 11, 129. doi:10.1186/1471-2164-11-129.

Berlin, S., Ghelardini, L., Bonosi, L., Weih, M. and Ronnberg-Wastljung, A. C. 2014a. QTL mapping of biomass and nitrogen economy traits in willows (*Salix* spp.) grown under contrasting water and nutrient conditions. *Molecular Breeding* 34(4), 1987–2003. doi:10.1007/s11032-014-0157-5.

Berlin, S., Trybush, S. O., Fogelqvist, J., Gyllenstrand, N., Hallingback, H. R., Ahman, I., Nordh, N. E., Shield, I., Powers, S. J., Weih, M., et al. 2014b. Genetic diversity, population structure and phenotypic variation in European *Salix viminalis* L. (Salicaceae). *Tree Genetics and Genomes* 10(6), 1595–610. doi:10.1007/s11295-014-0782-5.

Berlin, S., Hallingback, H. R., Beyer, F., Nordh, N. E., Weih, M. and Ronnberg-Wastljung, A. C. 2017. Genetics of phenotypic plasticity and biomass traits in hybrid willows across contrasting environments and years. *Annals of Botany* 120(1), 87–100. doi:10.1093/aob/mcx029.

Beyer, F., Jäck, O., Manzoni, S. and Weih, M. 2018. Relationship between foliar δ13C and sapwood area indicates different water use patterns across 236 *Salix* genotypes. *Trees-Structure and Function* 32(6), 1737–50, doi:10.1007/s00468-018-1747-3.

Bhagia, S., Muchero, W., Kumar, R., Tuskan, G. A. and Wyman, C. E. 2016. Natural genetic variability reduces recalcitrance in poplar. *Biotechnology for Biofuels* 9, 106. doi:10.1186/s13068-016-0521-2.

Bjorkman, C. and Ahrne, K. 2005. Influence of leaf trichome density on the efficiency of two polyphagous insect predators. *Entomologia Experimentalis et Applicata* 115(1), 179–86. doi:10.1111/j.1570-7458.2005.00284.x.

Bjorkman, C. and Eklund, K. 2006. Factors affecting willow leaf beetles (*Phratora vulgatissima*) when selecting overwintering sites. *Agricultural and Forest Entomology* 8(2), 97–101. doi:10.1111/j.1461-9555.2006.00288.x.

Bjorkman, C., Bommarco, R., Eklund, K. and Hoglund, S. 2004. Harvesting disrupts biological control of herbivores in a short-rotation coppice system. *Ecological Applications* 14(6), 1624–33. doi:10.1890/03-5341.

Bollmark, L., Sennerbyforsse, L. and Ericsson, T. 1999. Seasonal dynamics and effects of nitrogen supply rate on nitrogen and carbohydrate reserves in cutting-derived *Salix viminalis* plants. *Canadian Journal of Forest Research* 29(1), 85–94. doi:10.1139/x98-183.

Bonosi, L., Ghelardini, L. and Weih, M. 2010. Growth responses of 15 Salix genotypes to temporary water stress are different from the responses to permanent water shortage. *Trees-Structure and Function* 24(5), 843–54. doi:10.1007/s00468-010-0454-5.

Bonosi, L., Ghelardini, L. and Weih, M. 2013. Towards making willows potential bio-resources in the South: northern *Salix* hybrids can cope with warm and dry climate when irrigated. *Biomass and Bioenergy* 51, 136–44. doi:10.1016/j.biombioe.2013.01.009.

Budsberg, E., Rastogi, M., Puettmann, M. E., Caputo, J., Balogh, S., Volk, T. A., Gustafson, R. and Johnson, L. 2012. Life-cycle assessment for the production of bioethanol from willow biomass crops via biochemical conversion. *Forest Products Journal* 62(4), 305–13. doi:10.13073/FPJ-D-12-00022.1.

Buonocore, E., Franzese, P. P. and Ulgiati, S. 2012. Assessing the environmental performance and sustainability of bioenergy production in Sweden: a life cycle assessment perspective. *Energy* 37(1), 69–78. doi:10.1016/j.energy.2011.07.032.

Campbell, S. P., Frair, J. L., Gibbs, J. P. and Volk, T. A. 2012. Use of short-rotation coppice willow crops by birds and small mammals in central New York. *Biomass and Bioenergy* 47, 342–53. doi:10.1016/j.biombioe.2012.09.026.

Cerasuolo, M., Richter, G. M., Richard, B., Cunniff, J., Girbau, S., Shield, I., Purdy, S. and Karp, A. 2016. Development of a sink-source interaction model for the growth of short-rotation coppice willow and in silico exploration of genotype x environment effects. *Journal of Experimental Botany* 67(3), 961–77. doi:10.1093/jxb/erv507.

Charles, J. G., Nef, L., Allegro, G., Collins, C. M., Delplanque, A., Gimenez, R., Hoglund, S., Jiafu, H., Larsson, S., Luo, Y., et al. 2014. Insect and other pests of poplars and willows. In: Isebrands, J. G. and Richardson, J. (Eds), *Poplars and Willows: Trees for Society and the Environment*. CABI Publishing, Wallingford.

Christersson, L. 1986. High technology biomass production by *Salix* clones on a sandy soil in southern Sweden. *Tree Physiology* 2(1_2_3), 261–72. doi:10.1093/treephys/2.1-2-3.261.

Christersson, L., Sennerby-Forsse, L. and Zsuffa, L. 1993. The role and significance of woody biomass plantations in Swedish agriculture. *The Forestry Chronicle* 69(6), 687–93. doi:10.5558/tfc69687-6.

Cunniff, J., Purdy, S. J., Barraclough, T. J. P., Castle, M., Maddison, A. L., Jones, L. E., Shield, I. F., Gregory, A. S. and Karp, A. 2015. High yielding biomass genotypes of willow (*Salix* spp.) show differences in below ground biomass allocation. *Biomass and Bioenergy* 80, 114–27. doi:10.1016/j.biombioe.2015.04.020.

Dalin, P., Bjorkman, C. and Eklund, K. 2004. Leaf beetle grazing does not induce willow trichome defence in the coppicing willow *Salix viminalis*. *Agricultural and Forest Entomology* 6(2), 105–9. doi:10.1111/j.1461-9555.2004.00211.x.

Decker, S. R., Brunecky, R., Tucker, M. P., Himmel, M. E. and Selig, M. J. 2009. High-throughput screening techniques for biomass conversion. *Bioenergy Research* 2(4), 179–92. doi:10.1007/s12155-009-9051-0.

Desiree, J. I., Pita, A. V., Floor, V. D. H. and Andre, P. C. F. 2014. Biodiversity impacts of bioenergy crop production: a state-of-the-art review. *Global Change Biology Bioenergy* 6, 183–209.

Dickmann, D. I. and Kuzovkina, J. 2014. Poplars and willows of the world, with emphasis on silviculturally important species. In: Isebrands, J. G. and Richardson, J. (Eds), *Poplars and Willows: Trees for Society and the Environment*. CABI Publishing, Wallingford.

Dillen, M., Vanhellemont, M., Verdonckt, P., Maes, W. H., Steppe, K. and Verheyen, K. 2016. Productivity, stand dynamics and the selection effect in a mixed willow clone short rotation coppice plantation. *Biomass and Bioenergy* 87, 46–54. doi:10.1016/j.biombioe.2016.02.013.

Dimitriou, I. and Aronsson, P. 2005. Willows for energy and phytoremediation in Sweden. *Unasylva (English Education)* 56, 47–50.

Dimitriou, I. and Mola-Yudego, B. 2017. Poplar and willow plantations on agricultural land in Sweden: area, yield, groundwater quality and soil organic carbon. *Forest Ecology and Management* 383, 99–107. doi:10.1016/j.foreco.2016.08.022.

Dimitriou, I., Baum, C., Baum, S., Busch, G., Schulz, U., Kohn, J., Lamersdorf, N., Leinweber, P., Aronsson, P., Weih, M., et al. 2009. The impact of Short Rotation Coppice (SRC) cultivation on the environment. *Landbauforschung Volkenrode* 59, 159–62.

Dimitriou, I., Mola-Yudego, B. and Aronsson, P. 2012a. Impact of willow short rotation coppice on water quality. *Bioenergy Research* 5(3), 537–45. doi:10.1007/s12155-012-9211-5.

Dimitriou, I., Mola-Yudego, B., Aronsson, P. and Eriksson, J. 2012b. Changes in organic carbon and trace elements in the soil of willow short-rotation coppice plantations. *Bioenergy Research* 5(3), 563–72. doi:10.1007/s12155-012-9215-1.

Edelfeldt, S., Lundkvist, A., Forkman, J. and Verwijst, T. 2015. Effects of cutting length, orientation and planting depth on early willow shoot establishment. *Bioenergy Research* 8(2), 796–806. doi:10.1007/s12155-014-9560-3.

Ericsson, N., Porso, C., Ahlgren, S., Nordberg, Å, Sundberg, C. and Hansson, P. A. 2013. Time-dependent climate impact of a bioenergy system – methodology development and application to Swedish conditions. *Global Change Biology Bioenergy* 5(5), 580–90. doi:10.1111/gcbb.12031.

Ericsson, N., Sundberg, C., Nordberg, Å, Ahlgren, S. and Hansson, P. A. 2017. Time-dependent climate impact and energy efficiency of combined heat and power production from short-rotation coppice willow using pyrolysis or direct combustion. *Global Change Biology Bioenergy* 9(5), 876–90. doi:10.1111/gcbb.12415.

Estevez, M. M., Linjordet, R. and Morken, J. 2012. Effects of steam explosion and co-digestion in the methane production from *Salix* by mesophilic batch assays. *Bioresource Technology* 104, 749–56. doi:10.1016/j.biortech.2011.11.017.

Estevez, M. M., Sapci, Z., Linjordet, R., Schnurer, A. and Morken, J. 2014. Semi-continuous anaerobic co-digestion of cow manure and steam-exploded *Salix* with recirculation of liquid digestate. *Journal of Environmental Management* 136, 9–15. doi:10.1016/j.jenvman.2014.01.028.

Fabio, E. S. and Smart, L. B. 2018a. Differential growth response to fertilization of ten elite shrub willow (*Salix* spp.) bioenergy cultivars. *Trees-Structure and Function* 32(4), 1061–72. doi:10.1007/s00468-018-1695-y.

Fabio, E. S. and Smart, L. B. 2018b. Effects of nitrogen fertilization in shrub willow short rotation coppice production – a quantitative review. *Global Change Biology Bioenergy* 10(8), 548–64. doi:10.1111/gcbb.12507.

Fabio, E. S., Volk, T. A., Miller, R. O., Serapiglia, M. J., Kemanian, A. R., Montes, F., Kuzovkina, Y. A., Kling, G. J. and Smart, L. B. 2017. Contributions of environment and genotype to variation in shrub willow biomass composition. *Industrial Crops and Products* 108, 149–61. doi:10.1016/j.indcrop.2017.06.030.

FAO. 2016. *Poplars and Other Fast-Growing Trees – Renewable Resources for Future Green Economies. Synthesis of Country Progress Reports.* Forestry Policy and Resources Division, Food and Agriculture Organization, Rome, Italy.

Fransson, P. M. A., Toljander, Y. K., Baum, C. and Weih, M. 2013. Host plant–ectomycorrhizal fungus combination drives resource allocation in willow: evidence for complex species interaction from a simple experiment. *Ecoscience* 20(2), 112–21. doi:10.2980/20-2-3576.

Ghelardini, L., Berlin, S., Weih, M., Lagercrantz, U., Gyllenstrand, N. and Ronnberg-Wastljung, A. C. Y. 2014. Genetic architecture of spring and autumn phenology in Salix. *BMC Plant Biology* 14, 31. doi:10.1186/1471-2229-14-31.

Glynn, C., Ronnberg-Wastljung, A., Julkunen-Tiitto, R. and Weih, M. 2004. Willow genotype, but not drought treatment, affects foliar phenolic concentrations and leaf-beetle resistance. *Entomologia Experimentalis et Applicata* 113(1), 1–14. doi:10.1111/j.0013-8703.2004.00199.x.

Glynn, C., Herms, D. A., Orians, C. M., Hansen, R. C. and Larsson, S. 2007. Testing the growth-differentiation balance hypothesis: dynamic responses of willows to nutrient availability. *The New Phytologist* 176(3), 623–34. doi:10.1111/j.1469-8137.2007.02203.x.

Gonzalez-Garcia, S., Mola-Yudego, B., Dimitriou, I., Aronsson, P. and Murphy, R. 2012. Environmental assessment of energy production based on long term commercial willow plantations in Sweden. *The Science of the Total Environment* 421–422, 210–9. doi:10.1016/j.scitotenv.2012.01.041.

Gregory, A. S., Dungait, J. A. J., Shield, I. F., Macalpine, W. J., Cunniff, J., Durenkamp, M., White, R. P., Joynes, A. and Richter, G. M. 2018. Species and genotype effects of bioenergy crops on root production, carbon and nitrogen in temperate agricultural soil. *Bioenergy Research* 11(2), 382–97. doi:10.1007/s12155-018-9903-6.

Grogan, P. and Matthews, R. 2002. A modelling analysis of the potential for soil carbon sequestration under short rotation coppice willow bioenergy plantations. *Soil Use and Management* 18(3), 175–83. doi:10.1111/j.1475-2743.2002.tb00237.x.

Grossman, J. J., Vanhellemont, M., Barsoum, N., Bauhus, J., Bruelheide, H., Castagneyrol, B., Cavender-Bares, J., Eisenhauer, N., Ferlian, O., Gravel, D., et al. 2018. Synthesis and future research directions linking tree diversity to growth, survival, and damage in a global network of tree diversity experiments. *Environmental and Experimental Botany* 152, 68–89. doi:10.1016/j.envexpbot.2017.12.015.

Gullberg, U. 1993. Towards making willows pilot species for coppicing production. *The Forestry Chronicle* 69(6), 721–6. doi:10.5558/tfc69721-6.

Hallingback, H. R., Fogelqvist, J., Powers, S. J., Turrion-Gomez, J., Rossiter, R., Amey, J., Martin, T., Weih, M., Gyllenstrand, N., Karp, A., et al. 2016. Association mapping in *Salix viminalis* L. (Salicaceae) – identification of candidate genes associated with growth and phenology. *Global Change Biology. Bioenergy* 8(3), 670–85. doi:10.1111/gcbb.12280.

Hammar, T., Ericsson, N., Sundberg, C. and Hansson, P. A. 2014. Climate impact of willow grown for bioenergy in Sweden. *Bioenergy Research* 7(4), 1529–40. doi:10.1007/s12155-014-9490-0.

Hammar, T., Hansson, P. A. and Sundberg, C. 2017. Climate impact assessment of willow energy from a landscape perspective: a Swedish case study. *Global Change Biology Bioenergy* 9(5), 973–85. doi:10.1111/gcbb.12399.

Hanley, S. J. and Karp, A. 2014. Genetic strategies for dissecting complex traits in biomass willows (*Salix* spp.). *Tree Physiology* 34(11), 1167–80. doi:10.1093/treephys/tpt089.

Hanley, S. J., Mallott, M. D. and Karp, A. 2006. Alignment of a *Salix* linkage map to the *Populus* genomic sequence reveals macrosynteny between willow and poplar genomes. *Tree Genetics and Genomes* 3(1), 35–48. doi:10.1007/s11295-006-0049-x.

Harris, Z. M., Spake, R. and Taylor, G. 2015. Land use change to bioenergy: a meta-analysis of soil carbon and GHG emissions. *Biomass and Bioenergy* 82, 27–39. doi:10.1016/j.biombioe.2015.05.008.

Harris, Z. M., Alberti, G., Viger, M., Jenkins, J. R., Rowe, R., Mcnamara, N. P. and Taylor, G. 2017. Land-use change to bioenergy: grassland to short rotation coppice willow has an improved carbon balance. *Global Change Biology Bioenergy* 9(2), 469–84. doi:10.1111/gcbb.12347.

Hartley, M. J. 2002. Rationale and methods for conserving biodiversity in plantation forests. *Forest Ecology and Management* 155(1–3), 81–95. doi:10.1016/S0378-1127(01)00549-7.

Heller, M. C., Keoleian, G. A. and Volk, T. A. 2003. Life cycle assessment of a willow bioenergy cropping system. *Biomass and Bioenergy* 25(2), 147–65. doi:10.1016/S0961-9534(02)00190-3.

Higman, S., Mayers, J., Bass, S., Judd, N. and Nussbaum, R. 2005. *The Sustainable Forestry Handbook*. Earthscan Publications, London, UK.

Hoeber, S., Fransson, P., Prieto-Ruiz, I., Manzoni, S. and Weih, M. 2017. Two *Salix* genotypes differ in productivity and nitrogen economy when grown in monoculture and mixture. *Frontiers in Plant Science* 8, 231. doi:10.3389/fpls.2017.00231.

Hoeber, S., Arranz, C., Nordh, N. E., Baum, C., Low, M., Nock, C., Scherer-Lorenzen, M. and Weih, M. 2018. Genotype identity has a more important influence than genotype diversity on shoot biomass productivity in willow short-rotation coppices. *Global Change Biology Bioenergy* 10(8), 534–47. doi:10.1111/gcbb.12521.

Hoglund, S. and Larsson, S. 2005. Abiotic induction of susceptibility in insect-resistant willow. *Entomologia Experimentalis et Applicata* 115(1), 89–96. doi:10.1111/j.1570-7458.2005.00260.x.

Hoglund, S., Larsson, S. and Wingsle, G. 2005. Both hypersensitive and non-hypersensitive responses are associated with resistance in *Salix viminalis* against the gall midge

Dasineura marginemtorquens. *Journal of Experimental Botany* 56(422), 3215–22. doi:10.1093/jxb/eri318.

Holden, E., Linnerud, K. and Banister, D. 2014. Sustainable development: our common future revisited. *Global Environmental Change-Human and Policy Dimensions* 26, 130–9. doi:10.1016/j.gloenvcha.2014.04.006.

Horn, S. J., Estevez, M. M., Nielsen, H. K., Linjordet, R. and Eijsink, V. G. H. 2011. Biogas production and saccharification of *Salix* pretreated at different steam explosion conditions. *Bioresource Technology* 102(17), 7932–6. doi:10.1016/j.biortech.2011.06.042.

Hrynkiewicz, K., Baum, C., Leinweber, P., Weih, M. and Dimitriou, I. 2010. The significance of rotation periods for mycorrhiza formation in Short Rotation Coppice. *Forest Ecology and Management* 260(11), 1943–9. doi:10.1016/j.foreco.2010.08.020.

Hrynkiewicz, K., Toljander, Y. K., Baum, C., Fransson, P. M. A., Taylor, A. F. S. and Weih, M. 2012. Correspondence of ectomycorrhizal diversity and colonisation of willows (*Salix* spp.) grown in short rotation coppice on arable sites and adjacent natural stands. *Mycorrhiza* 22(8), 603–13. doi:10.1007/s00572-012-0437-z.

Isebrands, J. G., Aronsson, P., Carlson, M., Ceulemans, R., Coleman, M., Dickinson, N., Dimitriou, J., Doty, S., Gardiner, E., Heinsoo, K., et al. 2014. Environmental applications of poplars and willows. In: Isebrands, J. G. and Richardson, J. (Eds), *Poplars and Willows: Trees for Society and the Environment*. CABI Publishing, Wallingford.

Jaworska, J. 2010. The biodiversity of flying beneficial insects on a short rotation willow. *Progress in Plant Protection* 50, 1486–90.

Jirjis, R. 2005. Effects of particle size and pile height on storage and fuel quality of comminuted *Salix viminalis*. *Biomass and Bioenergy* 28(2), 193–201. doi:10.1016/j.biombioe.2004.08.014.

Jones, M. D., Durall, D. M. and Tinker, P. B. 1991. Fluxes of carbon and phosphorus between symbionts in willow ectomycorrhizas and their changes with time. *New Phytologist* 119(1), 99–106. doi:10.1111/j.1469-8137.1991.tb01012.x.

Kahle, P., Baum, C. and Boelcke, B. 2005. Effect of afforestation on soil properties and mycorrhizal formation. *Pedosphere* 15, 754–60.

Karp, A. and Shield, I. 2008. Bioenergy from plants and the sustainable yield challenge. *The New Phytologist* 179(1), 15–32. doi:10.1111/j.1469-8137.2008.02432.x.

Karp, A., Hanley, S. J., Trybush, S. O., Macalpine, W., Pei, M. and Shield, I. 2011. Genetic improvement of willow for bioenergy and biofuels. *Journal of Integrative Plant Biology* 53(2), 151–65. doi:10.1111/j.1744-7909.2010.01015.x.

Krzyzaniak, M., Stolarski, M. J., Waliszewska, B., Szczukowski, S., Tworkowski, J., Zaluski, D. and Snieg, M. 2014. Willow biomass as feedstock for an integrated multi-product biorefinery. *Industrial Crops and Products* 58, 230–7. doi:10.1016/j.indcrop.2014.04.033.

Krzyzaniak, M., Stolarski, M. J., Niksa, D., Tworkowski, J. and Szczukowski, S. 2016a. Effect of storage methods on willow chips quality. *Biomass and Bioenergy* 92, 61–9. doi:10.1016/j.biombioe.2016.06.007.

Krzyzaniak, M., Stolarski, M. J., Szczukowski, S. and Tworkowski, J. 2016b. Life cycle assessment of new willow cultivars grown as feedstock for integrated biorefineries. *Bioenergy Research* 9(1), 224–38. doi:10.1007/s12155-015-9681-3.

Kuzovkina, Y. A., Weih, M., Abalos Romero, M., Charles, J., Hurst, S., Mcivor, I., Karp, A., Trybush, S., Labrecque, M. and Teodorescu, T. I. 2008. *Salix*: botany and global horticulture. *Horticultural Reviews* 34, 447–89.

Labrecque, M. and Teodorescu, T. I. 2003. High biomass yield achieved by *Salix* clones in SRIC following two 3-year coppice rotations on abandoned farmland in southern Quebec, Canada. *Biomass and Bioenergy* 25(2), 135–46. doi:10.1016/S0961-9534(02)00192-7.

Langeveld, H., Quist-Wessel, F., Dimitriou, I., Aronsson, P., Baum, C., Schulz, U., Bolte, A., Baum, S., Kohn, J., Weih, M., et al. 2012. Assessing environmental impacts of short rotation coppice (SRC) expansion: model definition and preliminary results. *Bioenergy Research* 5(3), 621–35. doi:10.1007/s12155-012-9235-x.

Larsson, S. 1997. Commercial breeding of willow for short rotation coppice. *Aspects of Applied Biology* 49, 215–8.

Larsson, S., Nordh, N. E., Farrell, J. and Tweddle, P. 2007. Manual for SRC willow growers. Lantmännen Agroenergi AB, Örebro, Sweden. Available at: www.agroenergi.se.

Ledin, S. 1996. Willow wood properties, production and economy. *Biomass and Bioenergy* 11(2–3), 75–83. doi:10.1016/0961-9534(96)00022-0.

Lennartsson, M. and Ogren, E. 2004. Clonal variation in temperature requirements for budburst and dehardening in *Salix* species used for biomass production. *Scandinavian Journal of Forest Research* 19(4), 295–302. doi:10.1080/02827580410030145.

Liang, J. J., Crowther, T. W., Picard, N., Wiser, S., Zhou, M., Alberti, G., Schulze, E. D., McGuire, A. D., Bozzato, F., Pretzsch, H., et al. 2016. Positive biodiversity-productivity relationship predominant in global forests. *Science* 354(6309), 12. doi:10.1126/science.aaf8957.

Lindroth, A. and Bath, A. 1999. Assessment of regional willow coppice yield in Sweden on basis of water availability. *Forest Ecology and Management* 121(1–2), 57–65. doi:10.1016/S0378-1127(98)00556-8.

Marron, N., Gielen, B., Brignolas, F., Gao, J., Johnson, J. D., Karnosky, D. F., Polle, A., Scarascia-Mugnozza, G., Schroeder, W. R. and Ceulemans, R. 2014. Abiotic stresses. In: Isebrands, J. G. and Richardson, J. (Eds), *Poplars and Willows: Trees for Society and the Environment*. CABI Publishing, Wallingford.

McCracken, A. R. and Dawson, W. M. 1994. Experiences in the use of mixed-clonal stands of *Salix* as a method of reducing the impact of rust diseases. *Norwegian Journal of Agricultural Sciences (Suppl.)* (18), 101–9.

McCracken, A. R. and Dawson, W. M. 1998. Interaction of willow (*Salix*) clones growing in mixtures. *Annals of Applied Biology* 132, 54–5.

Mitchell, C. P., Ford-Robertson, J. B., Hinckley, T. and Sennerby, F. L. (Eds). 1992. *Ecophysiology of Short Rotation Forest Crops*. Elsevier Science Publishers Ltd, Barking, UK.

Mola-Yudego, B. 2011. Trends and productivity improvements from commercial willow plantations in Sweden during the period 1986–2000. *Biomass and Bioenergy* 35(1), 446–53. doi:10.1016/j.biombioe.2010.09.004.

Mola-Yudego, B. and Aronsson, P. 2008. Yield models for commercial willow biomass plantations in Sweden. *Biomass and Bioenergy* 32(9), 829–37. doi:10.1016/j.biombioe.2008.01.002.

Moritz, K. K., Bjorkman, C., Parachnowitsch, A. L. and Stenberg, J. A. 2017. Plant sex effects on insect herbivores and biological control in a Short Rotation Coppice willow. *Biological Control* 115, 30–6. doi:10.1016/j.biocontrol.2017.09.006.

Muller, M., Klein, A. M., Scherer-Lorenzen, M., Nock, C. A. and Staab, M. 2018. Tree genetic diversity increases arthropod diversity in willow short rotation coppice. *Biomass and Bioenergy* 108, 338–44. doi:10.1016/j.biombioe.2017.12.001.

Nordborg, M., Berndes, G., Dimitriou, I., Henriksson, A., Mola-Yudego, B. and Rosenqvist, H. 2018. Energy analysis of willow production for bioenergy in Sweden. *Renewable and Sustainable Energy Reviews* 93, 473–82. doi:10.1016/j.rser.2018.05.045.

Nordh, N. E. 2005. Long term changes in stand structure and biomass production in short rotation willow coppice. Doctoral thesis. Swedish University of Agricultural Sciences.

Ogren, E. 1999. Fall frost resistance in willows used for biomass production. II. Predictive relationships with sugar concentration and dry matter content. *Tree Physiology* 19(11), 755–60. doi:10.1093/treephys/19.11.755.

Ostry, M., Ramstedt, M., Newcombe, G. and Steenackers, M. 2014. Diseases of poplars and willows. In: Isebrands, J. G. and Richardson, J. (Eds), *Poplars and Willows: Trees for Society and the Environment*. CABI Publishing, Wallingford.

Parajuli, R., Knudsen, M. T., Djomo, S. N., Corona, A., Birkved, M. and Dalgaard, T. 2017. Environmental life cycle assessment of producing willow, alfalfa and straw from spring barley as feedstocks for bioenergy or biorefinery systems. *The Science of the Total Environment* 586, 226–40. doi:10.1016/j.scitotenv.2017.01.207.

Pawar, P. M. A., Schnurer, A., Mellerowicz, E. J. and Ronnberg-Wastljung, A. C. 2018. QTL mapping of wood FT-IR chemotypes shows promise for improving biofuel potential in short rotation coppice willow (*Salix* spp.). *Bioenergy Research* 11(2), 351–63. doi:10.1007/s12155-018-9901-8.

Pei, M. H., Lindegaard, K., Ruiz, C. and Bayon, C. 2008. Rust resistance of some varieties and recently bred genotypes of biomass willows. *Biomass and Bioenergy* 32(5), 453–9. doi:10.1016/j.biombioe.2007.11.002.

Philippot, S. 1996. Simulation models of short-rotation forestry production and coppice biology. *Biomass and Bioenergy* 11(2–3), 85–93. doi:10.1016/0961-9534(96)00008-6.

Phitsuwan, P., Sakka, K. and Ratanakhanokchai, K. 2013. Improvement of lignocellulosic biomass in planta: a review of feedstocks, biomass recalcitrance, and strategic manipulation of ideal plants designed for ethanol production and processability. *Biomass and Bioenergy* 58, 390–405. doi:10.1016/j.biombioe.2013.08.027.

Pucholt, P., Sjodin, P., Weih, M., Ronnberg-Wastljung, A. C. and Berlin, S. 2015. Genome-wide transcriptional and physiological responses to drought stress in leaves and roots of two willow genotypes. *BMC Plant Biology* 15, 244-. doi:10.1186/s12870-015-0630-2.

Puentes, A., Torp, M., Weih, M. and Bjorkman, C. 2015. Direct effects of elevated temperature on a tri-trophic system: *Salix*, leaf beetles and predatory bugs. *Arthropod-Plant Interactions* 9(6), 567–75. doi:10.1007/s11829-015-9401-0.

Ra, K., Shiotsu, F., Abe, J. and Morita, S. 2012. Biomass yield and nitrogen use efficiency of cellulosic energy crops for ethanol production. *Biomass and Bioenergy* 37, 330–4. doi:10.1016/j.biombioe.2011.12.047.

Ray, M. J., Brereton, N. J. B., Shield, I., Karp, A. and Murphy, R. J. 2012. Variation in cell wall composition and accessibility in relation to biofuel potential of short rotation coppice willows. *Bioenergy Research* 5(3), 685–98. doi:10.1007/s12155-011-9177-8.

Richardson, J., Isebrands, J. G. and Ball, J. B. 2014. Ecology and physiology of poplars and willows. In: Isebrands, J. G. and Richardson, J. (Eds), *Poplars and Willows: Trees for Society and the Environment*. CABI Publishing, Wallingford.

Ronnberg-Wastljung, A. C. 2001. Genetic structure of growth and phenological traits in *Salix viminalis*. *Canadian Journal of Forest Research* 31(2), 276–82. doi:10.1139/x00-175.

Ronnberg-Wastljung, A. C. and Gullberg, U. 1999. Genetics of breeding characters with possible effects on biomass production in *Salix viminalis* (L.). *Theoretical and Applied Genetics* 98(3-4), 531-40. doi:10.1007/s001220051101.

Ronnberg-Wastljung, A. C., Glynn, C. and Weih, M. 2005. QTL analyses of drought tolerance and growth for a *Salix dasyclados* × *Salix viminalis* hybrid in contrasting water regimes. *Theoretical and Applied Genetics* 110(3), 537-49. doi:10.1007/s00122-004-1866-7.

Rooney, D. C., Killham, K., Bending, G. D., Baggs, E., Weih, M. and Hodge, A. 2009. Mycorrhizas and biomass crops: opportunities for future sustainable development. *Trends in Plant Science* 14(10), 542-9. doi:10.1016/j.tplants.2009.08.004.

Samils, B., Ronnberg-Wastljung, A. C. and Stenlid, J. 2011. QTL mapping of resistance to leaf rust in *Salix*. *Tree Genetics and Genomes* 7(6), 1219-35. doi:10.1007/s11295-011-0408-0.

Sandak, J. and Sandak, A. 2011. Fourier transform near infrared assessment of biomass composition of shrub willow clones (*Salix* sp.) for optimal bio-conversion processing. *Journal of Near Infrared Spectroscopy* 19(5), 309-18. doi:10.1255/jnirs.950.

Sandak, A., Sandak, J., Waliszewska, B., Zborowska, M. and Mleczek, M. 2017. Selection of optimal conversion path for willow biomass assisted by near infrared spectroscopy. *iForest - Biogeosciences and Forestry* 10(2), 506-14. doi:10.3832/ifor1987-010.

Sannervik, A. N., Eckersten, H., Verwijst, T., Kowalik, P. and Nordh, N. E. 2006. Simulation of willow productivity based on radiation use efficiency, shoot mortality and shoot age. *European Journal of Agronomy* 24(2), 156-64. doi:10.1016/j.eja.2005.07.007.

Sassner, P., Galbe, M. and Zacchi, G. 2008a. Techno-economic evaluation of bioethanol production from three different lignocellulosic materials. *Biomass and Bioenergy* 32(5), 422-30. doi:10.1016/j.biombioe.2007.10.014.

Sassner, P., Martensson, C. G., Galbe, M. and Zacchi, G. 2008b. Steam pretreatment of H2SO4-impregnated *Salix* for the production of bioethanol. *Bioresource Technology* 99(1), 137-45. doi:10.1016/j.biortech.2006.11.039.

Schmidt-Walter, P. and Lamersdorf, N. P. 2012. Biomass production with willow and poplar short rotation coppices on sensitive areas-the impact on nitrate leaching and groundwater recharge in a drinking water catchment near Hanover, Germany. *Bioenergy Research* 5(3), 546-62. doi:10.1007/s12155-012-9237-8.

Schulz, U., Brauner, O. and Gruss, H. 2009. Animal diversity on short-rotation coppices - a review. *Landbauforschung Volkenrode* 59, 171-81.

Serapiglia, M. J., Cameron, K. D., Stipanovic, A. J. and Smart, L. B. 2009. Analysis of biomass composition using high-resolution thermogravimetric analysis and percent bark content for the selection of shrub willow bioenergy crop varieties. *Bioenergy Research* 2(1-2), 1-9. doi:10.1007/s12155-008-9028-4.

Serapiglia, M. J., Cameron, K. D., Stipanovic, A. J. and Smart, L. B. 2012. Correlations of expression of cell wall biosynthesis genes with variation in biomass composition in shrub willow (*Salix* spp.) biomass crops. *Tree Genetics and Genomes* 8(4), 775-88. doi:10.1007/s11295-011-0462-7.

Serapiglia, M. J., Cameron, K. D., Stipanovic, A. J., Abrahamson, L. P., Volk, T. A. and Smart, L. B. 2013a. Yield and woody biomass traits of novel shrub willow hybrids at two contrasting sites. *Bioenergy Research* 6(2), 533-46. doi:10.1007/s12155-012-9272-5.

Serapiglia, M. J., Humiston, M. C., Xu, H. W., Hogsett, D. A., De Orduna, R. M., Stipanovic, A. J. and Smart, L. B. 2013b. Enzymatic saccharification of shrub willow genotypes with differing biomass composition for biofuel production. *Frontiers in Plant Science* 4, 57. doi:10.3389/fpls.2013.00057.

Serapiglia, M. J., Gouker, F. E., Hart, J. F., Unda, F., Mansfield, S. D., Stipanovic, A. J. and Smart, L. B. 2015. Ploidy level affects important biomass traits of novel shrub Willow (*Salix*) hybrids. *Bioenergy Research* 8(1), 259–69. doi:10.1007/s12155-014-9521-x.

Smart, L. B. and Cameron, K. D. 2008. Genetic improvement of willow (*Salix* spp.) as a dedicated bioenergy crop. In: Vermerris, W. (Ed.), *Genetic Improvement of Bioenergy Crops*. Springer, New York.

Stanton, B. J., Serapiglia, M. J. and Smart, L. B. 2014. The domestication and conservation of *Populus* and *Salix* genetic resources. In: Isebrands, J. G. and Richardson, J. (Eds), *Poplars and Willows: Trees for Society and the Environment*. CABI Publishing, Wallingford.

Stanturf, J. A. and Van Oosten, C. 2014. Operational poplar and willow culture. In: Isebrands, J. G. and Richardson, J. (Eds), *Poplars and Willows: Trees for Society and the Environment*. CABI Publishing, Wallingford.

Stenberg, J. A., Lehrman, A. and Bjorkman, C. 2010. Uncoupling direct and indirect plant defences: novel opportunities for improving crop security in willow plantations. *Agriculture, Ecosystems and Environment* 139(4), 528–33. doi:10.1016/j.agee.2010.09.013.

Stolarski, M. J., Szczukowski, S., Tworkowski, J., Wroblewska, H. and Krzyzaniak, M. 2011. Short rotation willow coppice biomass as an industrial and energy feedstock. *Industrial Crops and Products* 33(1), 217–23. doi:10.1016/j.indcrop.2010.10.013.

Stolarski, M. J., Szczukowski, S., Tworkowski, J. and Klasa, A. 2013. Yield, energy parameters and chemical composition of short-rotation willow biomass. *Industrial Crops and Products* 46, 60–5. doi:10.1016/j.indcrop.2013.01.012.

Szczukowski, S., Tworkowski, J., Klasa, A. and Stolarski, M. 2002. Productivity and chemical composition of wood tissues of short rotation willow coppice cultivated on arable land. *Rostlinna Vyroba* 48, 413–7.

Tilman, D., Cassman, K. G., Matson, P. A., Naylor, R. and Polasky, S. 2002. Agricultural sustainability and intensive production practices. *Nature* 418(6898), 671–7. doi:10.1038/nature01014.

Tsarouhas, V., Gullberg, U. and Lagercrantz, U. 2004. Mapping of quantitative trait loci (QTLs) affecting autumn freezing resistance and phenology in *Salix*. *Theoretical and Applied Genetics* 108(7), 1335–42. doi:10.1007/s00122-003-1544-1.

Volk, T. A., Verwijst, T., Tharakan, P. J., Abrahamson, L. P. and White, E. H. 2004. Growing fuel: a sustainability assessment of willow biomass crops. *Frontiers in Ecology and the Environment* 2(8), 411–8. doi:10.1890/1540-9295(2004)002[0411:GFASAO]2.0.CO;2.

Volk, T. A., Heavey, J. P. and Eisenbies, M. H. 2016. Advances in shrub-willow crops for bioenergy, renewable products, and environmental benefits. *Food and Energy Security* 5(2), 97–106. doi:10.1002/fes3.82.

Wang, D., Jaiswal, D., Lebauer, D. S., Wertin, T. M., Bollero, G. A., Leakey, A. D. B. and Long, S. P. 2015. A physiological and biophysical model of coppice willow (*Salix* spp.) production yields for the contiguous USA in current and future climate scenarios. *Plant, Cell and Environment* 38(9), 1850–65. doi:10.1111/pce.12556.

WCED. 1987. *Our Common Future*. Oxford University Press, Oxford.

Weih, M. 2004. Intensive short rotation forestry in boreal climates: present and future perspectives. *Canadian Journal of Forest Research-Revue Canadienne De Recherche Forestiere* 34(7), 1369–78. doi:10.1139/x04-090.

Weih, M. 2009. Genetic and environmental variation in spring and autumn phenology of biomass willows (*Salix* spp.): effects on shoot growth and nitrogen economy. *Tree Physiology* 29(12), 1479–90. doi:10.1093/treephys/tpp081.

Weih, M. 2013. Willow. In: Singh, B. P. (Ed.), *Biofuel Crops: Production, Physiology and Genetics*. CABI Publishing, Wallingford, UK.

Weih, M. and Dimitriou, I. 2012. Environmental impacts of short rotation coppice (SRC) grown for biomass on agricultural land. *BioEnergy Research* 5(3), 535–6. doi:10.1007/s12155-012-9230-2.

Weih, M. and Nordh, N. E. 2002. Characterising willows for biomass and phytoremediation: growth, nitrogen and water use of 14 willow clones under different irrigation and fertilisation regimes. *Biomass and Bioenergy* 23(6), 397–413. doi:10.1016/S0961-9534(02)00067-3.

Weih, M. and Nordh, N. E. 2005. Determinants of biomass production in hybrid willows and prediction of field performance from pot studies. *Tree Physiology* 25(9), 1197–206. doi:10.1093/treephys/25.9.1197.

Weih, M. and Nordh, N. E. 2009. *Biomass Production With Fast-Growing Trees on Agricultural Land in Cool-Temperate Regions: Possibilities, Limitations, Challenges*. Nova Science Publishers, Inc, Hauppauge.

Weih, M. and Ronnberg-Wastljung, A. C. 2007. Shoot biomass growth is related to the vertical leaf nitrogen gradient in Salix canopies. *Tree Physiology* 27(11), 1551–9. doi:10.1093/treephys/27.11.1551.

Weih, M., Ronnberg-Wastljung, A. C. and Glynn, C. 2006. Genetic basis of phenotypic correlations among growth traits in hybrid willow (*Salix dasyclados* × *S-viminalis*) grown under two water regimes. *The New Phytologist* 170(3), 467–77. doi:10.1111/j.1469-8137.2006.01685.x.

Weih, M., Didon, U. M. E., Ronnberg-Wastljung, A.-C. and Bjorkman, C. 2008. Integrated agricultural research and crop breeding: allelopathic weed control in cereals and long-term productivity in perennial bion ass crops. *Agricultural Systems* 97(3), 99–107. doi:10.1016/j.agsy.2008.02.009.

Weih, M., Bonosi, L., Ghelardini, L. and Ronnberg-Wastljung, A. C. 2011. Optimizing nitrogen economy under drought: increased leaf nitrogen is an acclimation to water stress in willow (*Salix* spp.). *Annals of Botany* 108(7), 1347–53. doi:10.1093/aob/mcr227.

Weih, M., Hoeber, S., Beyer, F. and Fransson, P. 2014. Traits to ecosystems: the ecological sustainability challenge when developing future energy crops. *Frontiers in Energy Research* 2, 1–5. doi:10.3389/fenrg.2014.00017.

Weih, M., Hamner, K. and Pourazari, F. 2018. Analyzing plant nutrient uptake and utilization efficiencies: comparison between crops and approaches. *Plant and Soil* 430(1–2), 7–21. doi:10.1007/s11104-018-3738-y.

Welc, M., Lundkvist, A. and Verwijst, T. 2017. Effects of cutting phenology (non-dormant versus dormant) on early growth performance of three willow clones grown under different weed treatments and planting dates. *Bioenergy Research* 10(4), 1094–104. doi:10.1007/s12155-017-9871-2.

Verheyen, K., Vanhellemont, M., Auge, H., Baeten, L., Baraloto, C., Barsoum, N., Bilodeau-Gauthier, S., Bruelheide, H., Castagneyrol, B., Godbold, D., et al. 2016. Contributions of a global network of tree diversity experiments to sustainable forest plantations. *Ambio* 45(1), 29–41. doi:10.1007/s13280-015-0685-1.

Verwijst, T., Lundkvist, A., Edelfeldt, S. and Albertsson, J. 2013. Development of sustainable willow short rotation forestry in northern Europe. In: Matovic, M. D. (Ed.), *Biomass Now*. IntechOpen. doi:10.5772/55072. Available at: https://www.intechopen.com/books/biomass-now-sustainable-growth-and-use/development-of-sustainable-willow-short-rotation-forestry-in-northern-europe.

Whittaker, C., Yates, N. E., Powers, S. J., Misselbrook, T. and Shield, I. 2018. Dry matter losses and quality changes during short rotation coppice willow storage in chip or rod form. *Biomass and Bioenergy* 112, 29–36. doi:10.1016/j.biombioe.2018.02.005.

Xue, C., Penton, C. R., Zhang, B., Zhao, M., Rothstein, D. E., Mladenoff, D. J., Forrester, J. A., Shen, Q. and Tiedje, J. M. 2016. Soil fungal and bacterial responses to conversion of open land to short-rotation woody biomass crops. *Global Change Biology Bioenergy* 8(4), 723–36. doi:10.1111/gcbb.12303.

Chapter 10

Sustainable use of seaweed for biofuel

Jay Liu, Boris Brigljević and Peyman Fasahati, Pukyong National University, South Korea

1 Introduction

While fossil fuels still remain the lifeblood of modern societies, they are a finite resource facing the risk of depletion. Coupled with the large greenhouse gas emissions they produce, it is evident that new energy sources will soon have to replace fossil fuels. In the last two decades, biofuel research has intensified as a viable replacement for crude oil derivatives. In comparison to crude oil reserves, biological feedstocks for biofuel production are widespread and renewable. However, biofuel production has some disadvantages. Large government subsidies for the development of feedstock (crops) and production facilities gave rise to the so-called biofuel boom which peaked in 2005 and subsequently led to a rise in global food prices.[1]

These problems with the use of crops as a feedstock highlighted the potential use of non-edible alternatives such as aquatic biomass for biofuel production (i.e. microalgae and macroalgae), commonly known as third-generation biofuel feedstocks.[2-4] Aquatic biomass does not require land for cultivation, has a large potential (marine) cultivation area, and can also absorb carbon dioxide more effectively (6–8% photosynthetic efficiency) than any known terrestrial biomass (1.8–2.2% photosynthetic efficiency).[5] However, there are challenges in large-scale fuel production, including high moisture and

http://dx.doi.org/10.19103/AS.2019.0027.19

mineral content.[6] There are three general routes for the conversion of biomass: biological, chemical and thermochemical. Each has its own advantages and disadvantages, which vary in respect of factors such as the type of biomass, available resources and location.[7]

This chapter will first review seaweed's biological advantages and disadvantages as an energy crop and assess global seaweed production. The chapter will discuss seaweed cultivation techniques, together with harvesting and further processing in the supply chain. The chapter also provides an overview of conversion routes to produce biofuels and other value-added products from seaweed. Finally, the chapter provides case studies of theoretical industrial-scale processes involving three different conversion routes. These case studies illustrate the current economic feasibility, benefits and drawbacks of seaweed utilization as a biofuel. The chapter concludes by assessing the current state of seaweed biofuel technology and likely future trends.

2 Seaweed as a feedstock for third-generation biofuels

2.1 Classification and composition of seaweed

Seaweeds or macroalgae are a general denomination of approximately 9000 species of multicellular, eukaryotic aquatic plants.[8–10] Primary constituents of the majority of macroalgae are the blade or lamina, the stipe (analogous to the blade and vein of a leaf in terrestrial plants) and a holdfast which serves to anchor the seaweed to the substrate.[8,9,11]

In general, macroalgae are widely distributed all over the globe, mostly residing in coastal areas where the substrate (seafloor) and sunlight are available. Apart from the coast, they also occur in open water in floating form, which is seen as having particular potential for cultivation on a large scale.[11] They do not share a common genetic ancestry but are divided into three phenotype groups: green algae (Chlorophyta; 1200 species), red algae (Rhodophyta; 6000 species) and brown algae (Phaeophyceae; 2000 species) (Fig. 1). Out of 9000 known species, only 221 are commercially exploited worldwide in the food, fertilizer, cosmetic and chemical industries.[12,13] The

Figure 1 Three phenotype groups of seaweed; From left to right: green seaweed (*Codium fragile*), brown seaweed (*Saccorhiza polyschides*), red seaweed (*Calliblepharis jubata*) [Source: www.seaweed.ie; accessed on July 10, 2018; SUBJECT TO COPYRIGHT].

majority (76%) of cultivated macroalgal biomass comes from only five genera: *Laminaria, Undaria, Porphyra, Eucheuma* and *Gracilaria*.[11]

A distinctive feature (and potential disadvantage) of macroalgal biomass is the high water content of freshly harvested seaweed (75–90 wt. %), together with high mineral (ash) content (15–45% dry wt.).[14–16] Seaweed contains quite low amounts of lignin, which can be as low as 0.03 g/kg of dry wt.,[17] making it easier to process in this respect. The main components of seaweed are carbohydrates (25–60% dry wt.), which are the main components of cell walls. The majority of carbohydrates are polysaccharides (60–70% of total carbohydrate weight), which vary according to the seaweed phenotype group. For instance, mannan and ulvan are polysaccharides commonly found in green seaweeds.[8,17] Carrageenan and agar are usually found in red seaweed,[8,14] whereas brown seaweeds contain laminarin, mannitol and alginate.[15] Moderate quantities of starch and cellulose are also found in green and red seaweeds, and small quantities of cellulose in brown seaweeds. Monosaccharides (30–40% of total carbohydrate weight) are also present in seaweeds, most notably glucose, mannose, galactose, agarose and uronic acid. Carbohydrates are followed by proteins (7–30% dry wt.), and lastly relatively low amount of lipids (1–5 dry wt.). A summary of the major chemical constituents of seaweeds for each phenotype groups is shown in Table 1.

The chemical composition of seaweeds will vary through the year depending on factors such as temperature, salinity and nutrient availability. This seasonal variation in chemical composition is a crucial factor for harvesting and efficient biofuel processing. The variation of chemical composition depending on the month of the year can be observed in Figs. 2 and 3 (figures adapted from Adams et al.[6]). The ash content starts declining as the weather becomes warmer (May–June) and reaches its minimum at the peak of summer months (July–August) with weight ratios ranging from 13% to 16 wt. %. This is attributed to increased primary production (photosynthetic activity), as the sunlight is available at longer daily intervals, which, in turn, results in increased concentration of carbohydrates. A similar trend can also be observed by tracking key carbohydrates. In the case of brown seaweeds, these are laminarin and mannitol. In both cases there is a spike in concentration (as measured by mass) as the weather becomes warmer, resulting in an increased higher heating value (HHV). In geographical regions with seasonal changes, the optimal harvesting period for seaweed in terms of biofuel production is when the carbohydrate content is the highest and mineral content the lowest.

2.2 Current production trends

Today 43 countries worldwide actively harvest and utilize seaweed. In 2016 global production of seaweed reached 30 million wet tons, which is

Table 1 Summary of chemical composition of three groups of seaweeds; combined and/or averaged values from referenced works

Seaweed phenotype group	Green[18-23]	Brown[24-28]	Red[8,18,29-31]
Species example	*Ulva lactuca*	*Saccharina japonica*	*Gracilaria cacalia*
Major compound species; % dry wt.			
Carbohydrates	47.0	60.0	46.0
Proteins	13.6	12.0	29.4
Lipids	2.7	2.0	1.5
Proximate analysis; % dry wt.			
Moisture	8.0	2.8	11.7
Inorganic solids (Ash)	15.2	23.0	14.8
Volatile matter	70.2	70.9	54.5
Fixed carbon	6.5	3.3	19.0
Ultimate analysis; % dry wt.			
C	29.9	42.1	35.2
H	5.1	4.4	6.3
N	2.0	1.5	0.9
S	1.7	0.4	2.3
O	61.4	51.6	55.2
Mineral analysis; ppm (mg/kg of seaweed)			
K	24140	85150	59731
Na	23000	31110	14380
Ca	27800	23000	1425
Mg	27000	7570	2003
Fe	1027	290	1914
P	1300	1850	1944

approximately 4.5 million dry tons, or $35.8 billion.[32] Only 6% of the global amount produced in 2012 was from harvesting wild seaweed, with the great majority coming from seaweed farming. The sector has grown significantly since 1995 when production stood at 7.55 million wet tons with 52% from wild harvesting, as shown in Fig. 4.[9,32-34] The growth in seaweed cultivation is mainly concentrated in China, Indonesia, the Philippines and the Republic of Korea, which accounted for 96% of global seaweed production in 2016. China and Indonesia currently account for 47.9% and 38.7% of overall production, respectively, making them dominant players in the seaweed farming market.

Brown and red seaweed represent most of the cultivated species because of their rich carbohydrate content and are a feedstock for many value-added

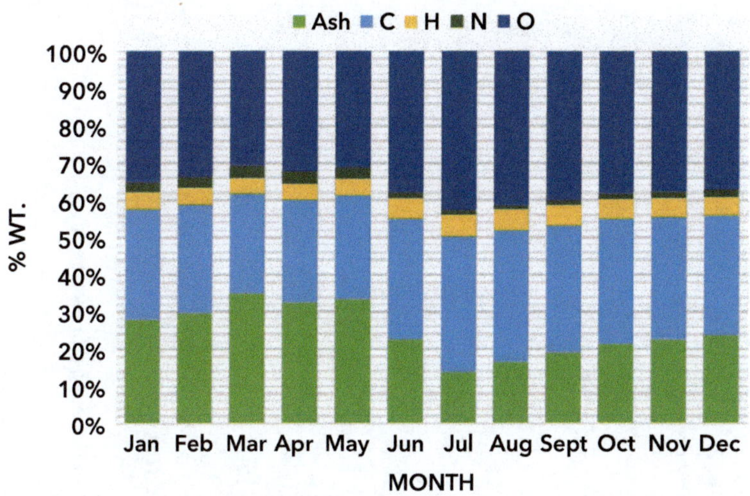

Figure 2 Seasonal variation of elemental composition of brown seaweed *Laminaria digitata*.

products, predominately in the pharmaceutical and cosmetic industries. Particularly important species are those of the *Eucheuma* and *Gracilaria* families for red seaweed, comprising 80% of the total red seaweed produced. In the case of brown seaweed, key carbohydrate-rich species are *Undaria pinnatifida* (jap. Wakame) and *Saccharina japonica*, which account for 19.7% and 78.2%, respectively, of total brown seaweed cultivation (Table 2).

2.3 Seaweed cultivation and supply chain

In terms of feedstock supply for downstream processing, there are three main sources of seaweed: collection of dead seaweed either from the shore or floating mass, harvesting wild seaweed species from the marine environment, and harvesting selected cultivated species from an aquaculture system. The primary focus in this chapter will be the latter source since the bulk of seaweed produced globally is in aquaculture systems.

Seaweed farming operations can generally be divided into three stages: production of propagules or seedlings, planting and harvesting. Propagule production is an essential tool for every large-scale, economically viable farming operation. Seaweeds are sensitive organisms which are easily damaged by environmental stress, which means material for vegetative propagation is not always readily available. Many seaweed species, particularly those from more temperate climates, tend to propagate through spore formation, rather than vegetative means. This facilitates the large-scale nursery production of healthy seaweed propagules. Where climatic conditions are relatively unchanging (e.g.

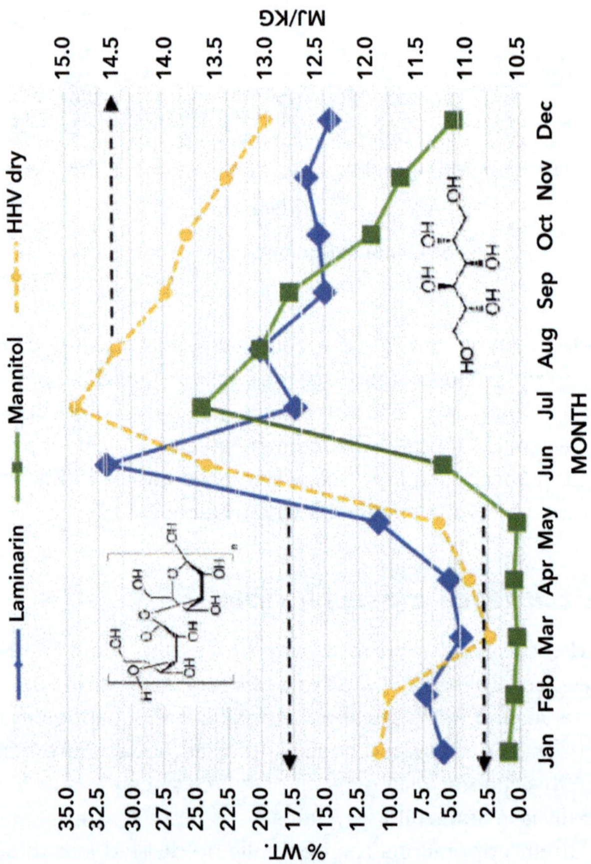

Figure 3 Seasonal variation of key carbohydrate proportions and higher heating value of dry *Laminaria digitata* seaweed.

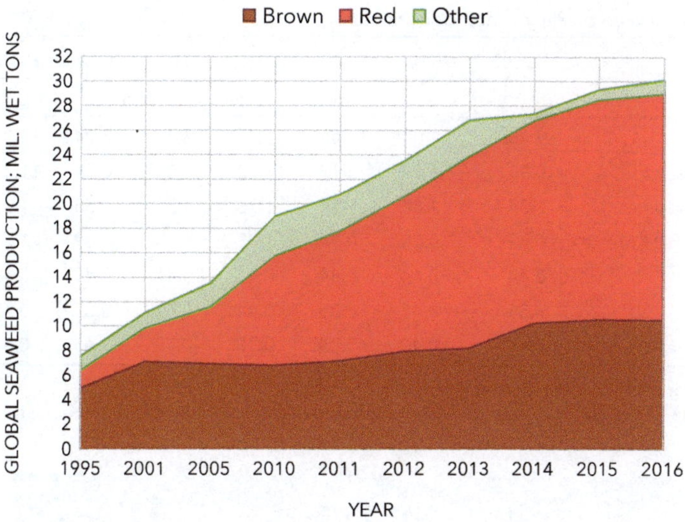

Figure 4 Global seaweed production; combined data from referenced sources for the period of 1995–2016.

in some tropical countries), simple vegetative propagation is possible, saving the cost of establishing a nursery.

After propagules are obtained, they are tied or attached (planted) directly on substrates (e.g. ropes), which are put in the open marine environment for the growing season (Fig. 5). Most cultivation techniques utilize ropes as substrates. Cultivation techniques can generally be divided into the following types:

1 line and net;
2 floating raft;
3 tank or pond (closed system); and
4 seafloor (rock-based farming).

Line and net cultivation techniques are perhaps the most widespread cultivation techniques in modern seaweed farming. Seaweed propagules are attached to long lines of rope (10–50 m) or nets of rope. Widths between ropes can vary between 0.5 and 1 m and more, with the widths affecting the density of the fully grown crop. Ropes are planted in a given depth of water, which can range from semi-submerged, shallow submerged, to deeply submerged (1–3 m or more). This technique allows both onshore and offshore cultivation, but it requires significant maintenance, often involving boats.

Floating raft cultivation comprises of a rigid frame upon which lines and ropes are attached. The frame is made of buoyant material, ensuring constant positioning of the seaweed culture at the surface. This technique allows

Table 2 Seaweed production trends by country

	2005	2010	2011	2012	2013	2014	2015	2016
Year	Million wet tons produced							
China	9.45	11.00	11.48	12.75	13.48	13.24	13.84	14.39
Indonesia	0.91	3.92	5.17	6.52	9.30	10.08	11.27	11.63
The Philippines	1.34	1.80	1.84	1.75	1.56	1.55	1.57	1.41
Republic of Korea	0.62	0.90	0.99	1.02	1.13	1.09	1.20	1.35
DPRK[a]	0.44	0.44	0.44	0.44	0.44	0.49	0.49	0.49
Japan	0.51	0.43	0.35	0.44	0.42	0.37	0.40	0.39
Others[b]	0.18	0.40	0.44	0.55	0.45	0.45	0.52	0.40

[a] Democratic People's Republic of Korea.
[b] Malaysia, Tanzania, Chile, Vietnam, Kiribati, Solomon Islands, India, Madagascar, Papua New Guinea.

Figure 5 Seaweed seedlings attached to a rope. http://www.mcpicarrageenan.com/images/cwd-line.jpg (July 25, 2018).

for farming sites far from the shore. However, it is also labor intensive and is complex to install. Figure 6 shows the principles of this technique (based on Peteiro et al.[35]).

Tank or pond cultivation involves closed and environmentally controlled systems for intensive production of delicate species. Another major application of closed seaweed cultivation systems is wastewater treatment, as seaweed culture thrives in a nutrient-abundant environment (aquaponics).

Seafloor cultivation implies direct or indirect (using an elastic rope) planting of seaweed to a rock substrate. Compared to other techniques, this is the simplest and the least labor intensive. However, farming sites require biosecurity measures to protect them from potential pests (Fig. 7).

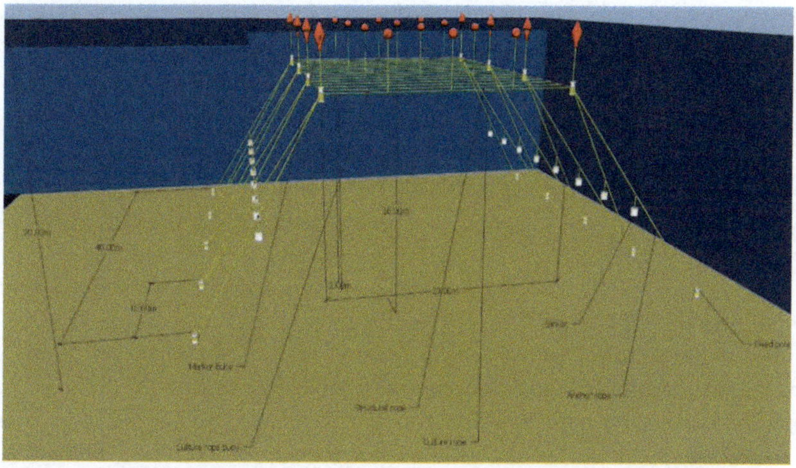

Figure 6 Open-sea cultivation schematic (floating raft with long-line rope).

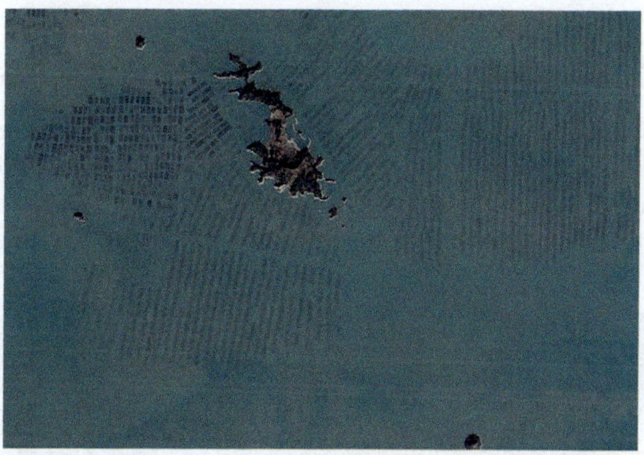

Figure 7 Large-scale seaweed farming in Republic of Korea (seen from space) [NASA Earth Observatory image by Jesse Allen, using Landsat data from the US Geological Survey].

The final step in seaweed farming is harvesting. This can be divided into the following broad categories:

- culture (wild versus cultivated); and
- technique (manual versus mechanical).

Wild harvesting or collecting seaweed from its natural environment comprises only about 4% of total global production and is not discussed here. The harvesting stage of seaweed cultivation is the main factor affecting the price of feedstock.

The type of seaweed harvesting technique will vary according to factors such as the species being farmed, the scale of operations and weather conditions. Manual harvesting is obviously the most labor-intensive option. It does, however, yield feedstock of the highest quality, since it provides an opportunity for the removal of possible contaminants (debris, fouling substances, parasites, etc.). This technique is mostly suitable for small-to-medium scale operations where the value of the end use of the feedstock can cover the cost of manual harvesting.

For large-scale farming operations, various mechanical harvesters are employed. These harvesters usually comprise of a boat or a barge of varying sizes. They include equipment such as a crane, conveyor, suction tubes, dredging cutters or rotating blades. Harvesters are generally custom designed for the species being cultivated. The first example of such equipment for harvesting *Laminaria digitata* was the 'scoubidou' boat developed in the late 1960s, with a hook-like gear that can turn on itself and uproot the seaweed (Fig. 8). Another example for the mechanical harvesting of *Laminaria hyperborera* is a boat with a rake-like tool on the end of a crane. The rake is dragged through a seaweed bed (like a comb through hair), where it uproots larger, heavier plants. It is capable of lifting 2 tons seaweed in 2 min (Fig. 9).

After harvesting, the seaweed has to be transported to a processing facility. Since these are predominately on shore, transport equipment will include a marine vehicle (for transport from cultivation location to shore), which is usually the harvester itself, and a land vehicle (for shore loading and movement to

Figure 8 Seaweed harvester with dual harvesting hooks (scoubidou) [© Julie Maguire, Indigo Rock Marine Research].

Figure 9 Scale model of a harvester barge and seaweed cultivation field.

the processing facility). To preserve the material, minimize shipping costs and prepare it for processing, a number of pretreatment steps are performed after harvesting.[11,34] The first operation is cleaning to remove foreign objects (e.g. mussels, snails, gravel). The next operation is size reduction, for example, by chopping and milling. This is beneficial for any kind of chemical processing since it increases the surface area of the material as well as makes transport easier. A final stage is dewatering and drying. When taken out of the sea, seaweed has about 85–90 wt. % in water. Dewatering is, therefore, important to reduce transportation weight. Partial dewatering sometimes occurs during harvesting (crane harvesters) if the material is pulled slowly out of the water, thus allowing the water to drain from the seaweed. In other cases, pressure filters or centrifuges will reduce the water content to about 30 wt. %. For completely dry material, conventional fuel-consuming dryers are used.[11] The development of large-scale standardized harvesting equipment can help increase throughput and thus control costs. Other options for reducing costs include the use of renewable energy technologies (e.g. solar-assisted convective dryers).

3 Conversion technologies for biofuel production from seaweed: biochemical conversion by anaerobic digestion

A biofuel can be defined as a combustible material in any phase which is derived from organic matter, in this case seaweed. This section provides an overview of the various conversion pathways (routes) for the production of biofuels from seaweed, starting with biochemical conversion routes. The first of these is anaerobic digestion (AD).

3.1 Anaerobic digestion to methane

The AD of biomass involves a biological decomposition of organic material to biogas by a mixture of microbes in the absence of oxygen.[36] The yield of the process is impacted by the type of digester selected, the pretreatment process, fermentation time and source of inoculum.[36-39] Anaerobic digesters can be categorized in various ways[40,41]:

- batch versus continuous process types;
- mesophilic (30–38°C) versus thermophilic (49–57°C) temperature operation
- high versus low solids loading; and
- single-stage versus multistage processes.

Experimental studies show that various types of macroalgae can be used to produce methane and biogas.[16,42] A comparison of the digestion characteristics of macroalgae with those of land-based biomass by Gunaseelan[43] showed that seaweed generally has a higher methane production capability at values between 0.31 and 0.48 m^3 CH_4 kg^{-1} volatile solid, which compares to 0.34–0.42 m^3 CH_4 kg^{-1} volatile solid for wood. Seaweeds are particularly attractive because of their lack of lignin, which reduces pretreatment costs. The effectiveness of AD of seaweed is affected by factors such as the type and composition of macroalgae, harvest season, geographic location, anaerobic bacteria culture and AD design.[44]

A review by Maneein et al.[45] on AD of seaweed biomass concluded that seaweed composition (in terms of salts, heavy metals, polyphenols and polysaccharides) is the major factor affecting biofuel yield. Rodriguez et al.[36] and Tedesco et al.[39] studied several pretreatment techniques to improve the digestibility of seaweed for increased methane yield. They showed that mechanical pretreatments results in a higher methane yield in comparison to other pretreatments such as ultrasound and microwave, thermal and chemical pretreatments. The methane production rate can also be improved by co-digesting seaweed with manure and waste-activated sludge.[44]

Miura et al.[37,38] evaluated marine sediments as a microbial source for AD of brown algae. Their study showed that the rate-limiting step in the process is acetolactic methanogenesis in brown algae fermentation, and methanogen production could be improved through batch-fed cultivation of microbes. In addition, AD under saline conditions allows wet brown algae containing about 80–90 wt. % water to be utilized. This makes it possible to bypass the biomass drying and washing steps prior to AD and to significantly reduce fresh water consumption in the biorefinery.

Anjaneyulu et al.[46] analyzed AD of seaweed using a mixed bacterial culture containing methanogenic bacteria and an alginate-degrading bacterial strain. They obtained specific rate constants during the lag, exponential and stationary (monomolecular) phases of bacterial growth. Their study indicated that about half of the total volume of biogas was generated during the exponential phase, irrespective of the concentration of seaweed in the digesters. Table 3 shows the average yield and operating conditions reported in the literature for AD of several types of brown algae.

AD of algal biomass is proven to have a higher carbon yield (g alcohol/g dry biomass) in comparison to fermentation processes[52-54] primarily because of lower release of CO_2 during AD and the fact that mixed anaerobic bacteria are capable of utilizing all non-lignin parts of biomass, including proteins, lipids and carbohydrates. However, AD processes take relatively longer to complete as they include four stages. These stages are:

- hydrolysis;
- acidogenesis;
- acetogenesis; and
- methanogenesis.

During hydrolysis, the polymeric structures of carbohydrates, proteins and lipids are broken down into simple sugars, amino acids and fatty acids, respectively. In the acidogenesis stage, monomer sugars are digested to produce volatile fatty acids (VFAs), ammonia, carbon dioxide and hydrogen sulfide. VFAs are further digested by acetogens to yield mainly acetic acid along with carbon dioxide and hydrogen. Finally, in the last methanogenesis stage, the methanogens convert the products obtained from the previous stages into methane, carbon dioxide and water.

Table 3 Average yield and operating conditions for anaerobic digestion of different types of brown algae

Brown algae type	CH_4 yield (m^3/kg VS)	Time (days)	Temperature (°C)	Reference
L. japonica	0.18	6~20	35	[38]
L. Digitata	0.14~0.17	21	37	[39]
L. japonica	0.297	17~28	55	[41]
L. japonica	0.25~0.28	14~46	35	[47]
L. japonica	0.24~0.30	9~30	35	[48]
L. saccharina	0.205~0.220	25	37	[49]
L. saccharina	0.230	24	35	[50]
Ascophyllum nodosum	0.110	24	35	[50]
L. hyperborea	0.280	24	35	[50]
Macrocystis pyrifera	0.277-0.310	18	38	[51]

Research has reported higher AD rates for carbohydrates of brown algae with the exception of alginate. Biological degradation of alginate is catalyzed by alginate lyases. Alginate lyase activity is inhibited by the polyphenols and divalent metal ions available in the cell wall of brown algae.[55,56] Moen et al.[55] reported that dissolved mannitol is entirely consumed within the first 10–12 h of fermentation while 13–50% of alginate remains insoluble. Østgaard et al.[57] investigated carbohydrate degradation and methane production during AD of *Laminaria saccharina*. Their results showed that mannitol and laminaran were reduced to less than 5% of the initial values within 24–48 h after inoculation, whereas 30% of the alginate content did not react, even after 30 days.

The final product of AD is biogas, which includes methane, CO_2, H_2S and water. Biogas can be used as a fuel to produce steam and power in a boiler/turbogenerator system[25,58] or be further purified and added to natural gas lines,[59] used in a fuel cell system to produce electricity[60] or purified and upgraded to higher-value fuels or chemicals.[61,62]

3.2 Anaerobic digestion to volatile fatty acids

Another product stream that could be produced through AD is VFAs such as acetic acid, propionic acid and butyric acid. These are made by adding inhibitors such as iodoform or bromoform to digesters to prevent methanogenesis stage. Pham et al.[62] studied the AD of brown algae *L. japonica*. They reported a VFA yield of 0.307–0.412 g VFA/g dry *L. japonica* after 5 days of fermentation at 35°C. Lee et al.[63] studied different bacteria cultures to optimize VFA production from *Gelidium amansii* red seaweed. Optimized bacterial strains were able to increase mixed acid production by 50.9% (g/g), reaching a productivity of 1.41 g/L/d in a semicontinuous AD with a biomass-loading rate of 2.77 g/L/d and hydraulic retention time of 30 days. While VFAs could be recovered, they could also be further processed as a carbon source for the production of higher-value chemicals or fuels.[64,65]

Figure 1 shows potential pathways and products that could be achieved using AD. One approach would be to recover and purify VFAs, and then catalytically hydrogenate them to create mixed alcohols such as methanol, propanol and butanol. This strategy could lead to a higher alcohol yield compared to fermentation to make ethanol. This is mainly because a lower amount of CO_2 is released during AD. Mixed anaerobic bacteria could use carbon in lipids and proteins in addition to carbohydrate for VFA production while, in the fermentation process, only carbohydrates of seaweed are hydrolyzed to sugars and later fermented to ethanol. The use of VFAs can be superior to fermentation in total yield (g fuel/g biomass) when the algal biomass contains proteins and lipids. However, for lignocellulose or starch-based biomass such as wood and corn that contains mostly carbohydrates, fermentation can be more appropriate, especially when the production of a specific alcohol such as ethanol is targeted.

3.3 Anaerobic digestion through MixAlco® process

A key challenge of the VFA approach is the high recovery costs of VFA from dilute AD effluents (5 wt. %), which accounts for about 40% of the production cost.[66,67] Therefore, alternative designs have been considered to increase VFA concentration and/or enhance recovery. The MixAlco® process is a biorefining technology that biochemically converts biomass through VFA production into valuable chemicals such as acetic acid, ketones and alcohols that can be further processed into renewable gasoline-type fuels.[68,69] The MixAlco® process includes a ketonization strategy (KS)[68–71] and an esterification strategy (ES).[72,73] In the KS, calcium carbonate is used as a buffer in the fermenters to produce calcium carboxylate salts. In the ES, ammonium bicarbonate (ABC) is added to the fermenters to produce ammonium carboxylate salts. The use of ABC as the buffering agent helps to condition the fermentation broth and provides nitrogen for the mixed culture, which enhances the fermentation process and total VFA yield.[74] However, ABC is expensive in comparison to $CaCO_3$ and requires high-energy input for production. In both strategies, VFA salts produced during AD are dewatered using a highly efficient multistage evaporator which concentrates the dilute stream of salts to 50% mass concentration.[28,68,75]

In the KS, concentrated salts are sent to a rotary drier to produce dry calcium carboxylate salts. Salts are subsequently degraded to ketone vapors and solid calcium carbonate at a temperature of 430°C using heat transferring media. A low pressure is maintained inside the reactor (30 mm Hg) using vacuum pumps to reduce the residence time and prevent ketones degradation.[70,71] A benefit of KS is that it joins carboxylates and, therefore, produces longer carbon chains which have higher energy density and could be more easily utilized as biofuel after further downstream processing. Ketone vapors are condensed and hydrogenated under high-pressure (55 bar) and isothermal (130°C) conditions to produce a range of secondary alcohols. Pham et al.[76] have described the optimal design for the hydrogenation of ketones as a series of three continually stirred tank reactors, which achieved a 98.4% conversion of the ketones into the alcohols. Alcohols can be easily recovered and separated using distillation columns. Alternatively, ketones can be catalytically upgraded to produce gasoline, diesel and jet fuels. Sankaranarayanapillai et al.[77] studied self-condensation of biomass-derived methyl ketones to create jet-diesel fuel precursors on a water-tolerant, solid-base catalyst. They showed that a dimerization yield above 60% for all methyl ketones containing 4–15 carbon atoms catalyzed by Si-Al-NHCH2CH2NH2 is achievable.

Ammonium carboxylate salts are highly soluble in water and decompose at elevated temperatures. In the ES, multistage evaporators can concentrate salts only to a certain limit, depending on operating conditions. To further recover the salts, concentrated ammonium carboxylates are reacted with a high-molecular-weight alcohol such as 1-heptanol in the presence of a solid acid catalyst to produce esters, water and ammonia. The reaction temperature for the esterification process depends on the type and molecular weight of the alcohol used. Reaction temperatures ranging from 100°C to 200°C have been reported in the literature, depending on alcohol molecular weight.[78,79] Since esterification is an endothermic reaction, higher temperatures can accelerate the forward reaction. In general, primary alcohols allow higher conversions to esters than secondary alcohols. In addition, high-boiling alcohols are reported to result in higher esterification rates and yields.[80]

A continuous removal of ammonia and water can speed up esterification, requiring a reactive distillation column (RDC). Esterification reactions occur in the reboiler of the RDC. The mixture of esters, water and ammonia exits from the top of the column and is partially condensed at 60°C. The partly condensed mixture is sent to a decanter to separate the ester-rich and water-rich liquids from the vapors. The vapors contain ammonia and water and are recycled to the ABC production unit. The water-rich phase is recycled to the RDC as reflux. A decanter at stage 10 of the RDC removes the water-rich phase and returns the ester-rich phase to stage 11. The water-rich phase contains a small amount of esters and is considered as wastewater from the process. The ester-rich phase is extracted from the last stage of the column, mixed with the recovered ester-rich phase from the decanter of the condenser.[28] In the hydrogenation unit, the esters are reacted with hydrogen to produce heptanol, ethanol, propanol and butanol. Fatty esters have also been hydrogenated to fatty alcohols in the presence of heterogeneous catalysts such as copper chromite, Raney nickel and Pd/ZnO at high temperatures (200–300°C) and high hydrogen pressures (200–300 atm).[81,82] A 97% conversion of esters at 200 psi and 200°C can be achieved in hydrogenation reactors.[82]

The esters received from the esterification unit and the recycled esters from the alcohol recovery unit are mixed with high-pressure hydrogen and are preheated, first by the products of the hydrogenation unit and later by the heat from a fired heater to reach the 200°C hydrogenation temperature. The fired heater uses the purge gasses of the hydrogenation unit and some of the unreacted esters recycled from the alcohol recovery unit. The resulting alcohols are cooled by preheating the feed to the hydrogenation reactors and cooling water. Two flash drums are used to depressurize the alcohols and recycle the unreacted hydrogen. A distillation column could be used to recover 1-heptanol and recycle it to the RDC. Use of heavier alcohols for esterification allows an easier and high-efficiency recovery in the distillation column. Light alcohol products from the hydrogenation unit could be further distilled to recover pure product streams.

4 Biochemical conversion by fermentation

Seaweeds contain a large amount of carbohydrates in their structure which makes them an ideal biomass for fermentation processes. Except for some types of red algae, seaweeds generally have no lignin in their chemical composition. Therefore, sugars can be released by simple operations such as milling or crushing. This bio-architectural feature provides a distinct advantage of seaweeds over lignocellulosic biomass and facilitates higher product yields while avoiding the need for energy-intensive biomass fractionation and hydrolysis processes prior to fermentation.[83] An analysis prepared for the US Department of Energy (DOE)[11] has shown that seaweed has an ethanol productivity twice that of sugarcane and five times higher than corn, achieving a cultivation yield of 59 dry metric tons/ha/year and an optimum ethanol yield of 0.254 kg ethanol/kg dry seaweed.

Among seaweed carbohydrates, mannitol and laminaran have a high-average dry mass fraction of 25% and 30%, respectively, and could be readily utilized by fermentation bacteria.[84] Generally, ethanol production from laminaran, cellulose and mannitol yields approximately 0.08–0.12 wt. % ethanol/wt dry brown algae.[11] Horn et al.[84,85] studied mannitol and laminaran fermentation to ethanol. They identified that *Zymobacter palmae* bacterium only converted mannitol to ethanol due to its lack of laminarase activity, whereas two yeast strains (*Kluyveromyces marxianus* and *Pachysolen tannophilus*) fermented laminaran but not mannitol. Only yeast *Pichia angophorae* has simultaneously utilized both mannitol and laminaran to produce ethanol. This yeast achieved the highest yield of 0.43 g ethanol/g substrate. Compared to glucose, mannitol is not readily fermented. Initially, it is oxidized to fructose, a reaction that generates NADH. Regeneration of NAD+ requires oxygen or transhydrogenase, which converts NADH to NADPH. Hence, ethanol production from mannitol is feasible only in the presence of electron shunts such as in micro-aerobic conditions.[84,86] In contrast to mannitol, each mole of alginate fermented to ethanol consumes two reducing equivalents. The catabolic pathway of alginate provides both an additional source of sugar and a counterbalance to the excess of reducing equivalents produced by mannitol catabolism, thus enabling simultaneous ethanol fermentation from all the sugar components in brown algae.[83]

It has been shown that higher enzymatic hydrolysis yields can be obtained by acid or alkaline pretreatment of biomass prior to fermentation. Acid or alkaline pretreatment at elevated temperatures is used to break down hemicelluloses into sugar monomers, reduce cellulose crystallinity and increase biomass porosity, consequently increasing the rate of enzymatic hydrolysis.[87] Many researchers have utilized acid thermal hydrolysis (ATH) followed by enzymatic hydrolysis for the pretreatment of brown algal biomass.[87-91] A literature review by Offei et al.[92] concluded that the major bottlenecks to enzymatic hydrolysis of seaweeds are the formation of inhibitors and polysaccharides selectivity. This is primarily because most of the currently developed enzymes are for starch-based and lignocellulosic biomass which have significantly different carbohydrates than seaweeds.

Genetic engineering is now being used to modify ethanologenic bacteria to facilitate the utilization of alginate for bioethanol production.[8,41] Takeda et al.[41] created novel ethanologenic strains from the alginate-assimilating bacterium *Sphingomonas* sp. A1. The ethanologenic recombinant strain A1 can incorporate intact polyuronic acid alginate, depolymerize it and convert it into ethanol in the cytosol. The bacteria are able to utilize 87% of the sodium alginate with 54% conversion efficiency. Similarly, Wargacki et al.[8] genetically modified *E. coli* bacteria that could yield 0.281 kg ethanol/kg dry *L. japonica* (which is equivalent to 80% of the maximum theoretical yield for the macroalgal carbohydrates). The new strain has demonstrated the ability to directly degrade, uptake and metabolize alginate without pretreatment, leaving laminarin, cellulose, mannitol and fucoidan to be hydrolyzed during the pretreatment and saccharification steps.[42,43]

Mannitol has high water solubility[44] and can be easily extracted (together with fucoidan) from the cell wall of brown algae using hot water.[15,16] Hal and Huijgen[45] have shown that mannitol can be efficiently extracted using hot water from fresh seaweed without prior pretreatment. They showed that, because of high osmotic stress in the seaweed, mannitol could migrate from the cells to the extraction water phase, without the need to break up the cells and/or the cell walls prior to extraction. Laminarin and cellulose, which constitute <20% of dry brown algae,[3] are the only carbohydrates to be pretreated by ATH. A detailed literature review has indicated that overall saccharification rates after a hot water wash (HWW) were slightly lower than those after the combined pretreatment (ATH).[8,31,34,39,40]

5 Thermochemical conversion routes

A thermochemical pathway implies the controlled decomposition of organic material (seaweed) into less complex products by the utilization of heat energy. Compared to biochemical and chemical conversion routes, thermochemical routes have two main advantages. First, thermochemical processes offer the most direct conversion which requires far fewer pretreatment steps.[7] Second, product composition can be modified by varying process parameters, making it easier to tailor products for particular markets.[7,93] Thermochemical (as

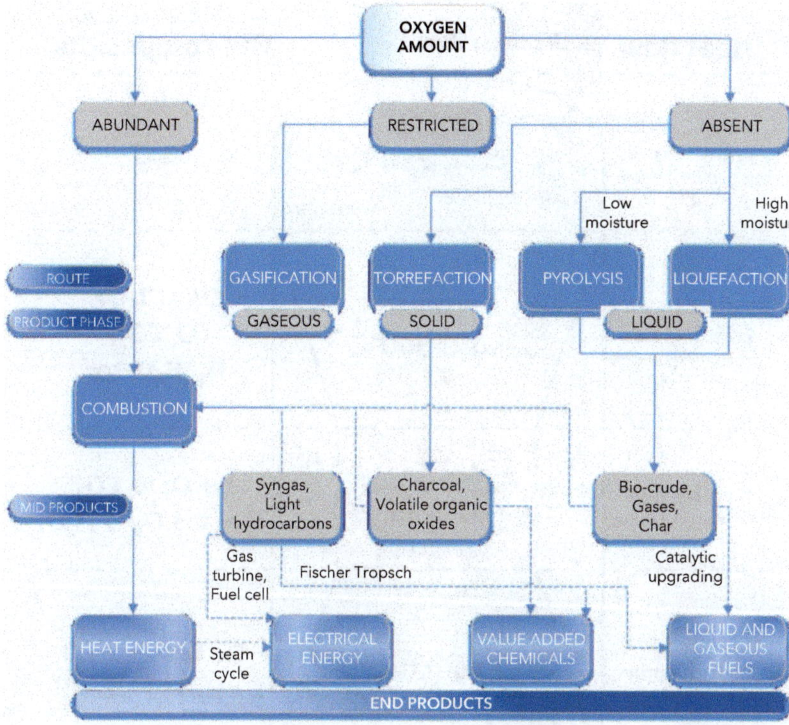

Figure 10 Thermochemical conversion routes and end products flowchart.

opposed to biochemical) processes can operate continuously, which is more economical and energy efficient than batch processes.[94,95] Thermochemical conversion methods can be classified according to air (oxygen) availability, as shown in Fig. 10.

5.1 Gasification

The process of gasification involves a thermal decomposition of organic material with the aim of maximizing the gaseous phase of the product. This is achieved by relatively high temperatures (800–1000°C) and long residence times. The basics of the process can be seen in Fig. 11. The air supply (or other oxidizing agents like steam) for the combustion section is sufficient to sustain the heat required by the endothermic parts of the process. As the biomass moves through the gasifier, it undergoes slow thermal decomposition, where it first losses moisture (drying zone), starts losing volatile compounds in the pyrolysis zone (primary cracking) and, finally, complex compounds are decomposed into simple molecules of primarily H_2, CO, and CH_4 or synthesis gas (syngas). The composition and heating quality of the syngas are particularly affected by

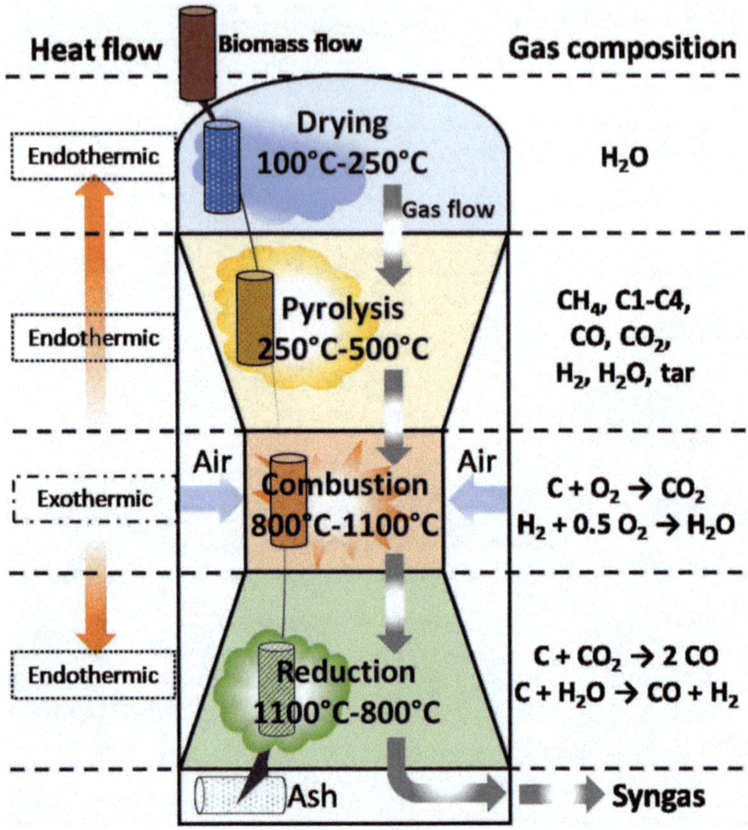

Figure 11 Gasification process schematic and distinct process zones.

the type of the feedstock, gasifying conditions and the type of gasifying agent. Steam produces the syngas of the best quality (15–20 MJ/m^3) as it minimizes the char yield and solvates the solid matter.[96,97]

Due to the naturally occurring high water content in macroalgae, this feedstock is a good candidate for this process. Two separate studies of five different gasified seaweeds have shown that the resulting gas contains very low amounts of char and higher amounts of hydrogen than gasses produced from some types of terrestrial biomass. The amount of CO was lower in hydrothermal gasification due to the higher WGS reaction which consumes CO.[98,99] The high concentration of minerals in the seaweeds (ash content) may have acted as gasification catalysts and promoted higher gas yields. Another study of supercritical water gasification of four brown macroalgae species has showed 30% higher yields of combustible gasses when catalyzed by ruthenium (Ru) compared to non-catalyzed processes.[100] A gasification process to convert the brown algae *Saccharina japonica* to biobutanol on an industrial scale (1200 tons per day) has been simulated.[101] Although estimated prices of butanol were relatively high (\$1.97–3.33/L) compared to the same product from lignocellulosic biomass, the CO_2e (CO_2 equivalent) avoided was \$620–2720/tCO_2e which was considerably higher than most (western countries accepted) \$50/tCO_2e. This means that, as a feedstock for industrial-scale gasification, seaweed is environmentally sustainable, though more needs to be done to improve its economic viability, especially in supply chain optimization.

5.2 Pyrolysis

Pyrolysis is a thermal decomposition process in which an organic substance is heated in a low or no oxygen environment to avoid combustion. Pyrolysis is generally achieved at temperature ranges of 300–600°C (depending on the feedstock) and at atmospheric pressure.[102] A general requirement for this process is dry feedstock, preferably with a large surface area (milled) for efficient heat transfer. Pyrolysis products are volatile organics in a vapor phase, non-condensable gasses such as CO_2, CO, H_2, C1–C4 hydrocarbons and carbon-rich solids (char). Bio-oil or biocrude is obtained by the condensation of volatile organic compounds. The main goal of the process is the maximization of liquid biofuel by the optimization of process parameters (such as temperature, heating rate, mineral content, residence time, choice of catalyst and reactor type). The process has been extensively researched for a number of feedstocks because it takes only 15% of the energy contained in the feedstock to sustain the pyrolytic process.[103]

There have been a number of studies on macroalgae pyrolysis.[26,27,104–106] The feed material undergoes three stages as the temperature increases: evaporation of moisture, primary devolatilization and residual decomposition. The thermogravimetric analysis of seaweed as it undergoes these stages is shown in Fig. 12 (adapted from[26]). The main challenge for pyrolysis of seaweed is its high mineral content, which hinders heat transfer, lowering heating values (especially biochar) and risking slag formation. Various pretreatments of macroalgae biomass increase the feasibility of this process, most notably the use of a water wash and/or acid wash.[107,108]

A system which provides the best liquid product yields, as well as requiring the least amount of input heat due to superior heat transfer, is the circulating fluidized bed reactor (Fig. 13).[109,110] Here, the dried feed material is put into direct contact with the hot heat transfer medium (usually a sand-based material like quartz or olivine) in the pyrolizer. The fluidization velocity is tuned for optimal heat transfer. Pyrolysis vapors are separated from solids (sand and char) and are condensed into biocrude liquid in a quenching tower. Non-condensable gasses are used for fluidization and for combustion

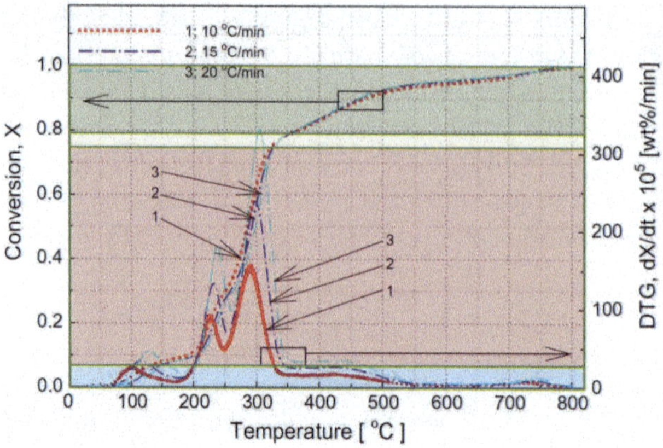

Component	Value, %wt. dry
Moisture	6.9
Volatile Matter	68.79
Fixed Carbon	4.1
Ash	20.21
C	32.89
H	6.17
N	0.93
O	60.01
HHV (MJ/kg)	12.11

Figure 12 Thermogravimetric, ultimate and proximate analyses of *Saccharina japonica* macroalgae.

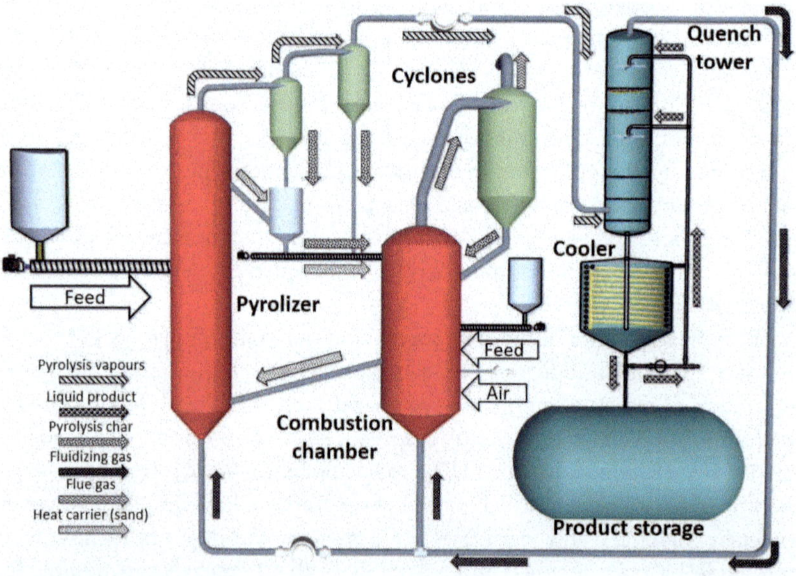

Figure 13 Circulating fluidized bed pyrolysis system schematic.

in the combustion chamber (along with char and extra fuel). Biocrude from macroalgae pyrolysis is a mixture of water and various organic compounds, mostly oxygenated hydrocarbons such as carboxylic acids, ketones, furans, phenols, sugars and heteromolecules containing N and S. Due to the relatively high water and oxygen content, this biocrude cannot be used as a fuel but has to undergo catalytic hydrogenation and deoxygenation to increase its heating value.

5.3 Liquefaction

Hydrothermal liquefaction (HTL) is one of the promising techniques that has attracted attention in recent years as the process does not require drying the feed, which is an energy-intensive, and therefore expensive, process. HTL converts solid feedstock into fuels rather than converting them into gasses or alcohols. The moist condition of the feed is an advantage in the HTL process as water plays a vital role in breaking down megamolecules.[111] Hydrothermal processing conditions range from 250°C to 350°C, with operating pressures from 10 to 20 MPa. Water plays an essential role in HTL (Fig. 14).

It is important to note that, with increasing temperature, the properties of water change significantly, making it more interactive as it splits its ions and thus breaks down the biomass. Recent research has focused on changing water reactivity to improve the process.[112] HTL of two different microalgae has been reported at a temperature range of 348–353°C.[113] Zhou et al.[114] studied the HTL of the macroalgae *Enteromorpha prolifera* at operating temperatures of 220–320°C to produce bio-oil. HTL of four different macroalgae has been studied by Anastasakis,[115] which suggests HTL is superior to fermentation and AD. Studying HTL of three different macroalgae at 250°C, Leticia[116] found HTL to be superior to other processes. Alhassan et al.[117] investigated HTL of dry biomass and obtained bio-oil with properties equivalent to pyrolysis oil. Costanzo et al.[118] obtained bio-oil derived from wet biomass and reported the composition of the bio-oil was equivalent to conventional gasoline. Lavanya

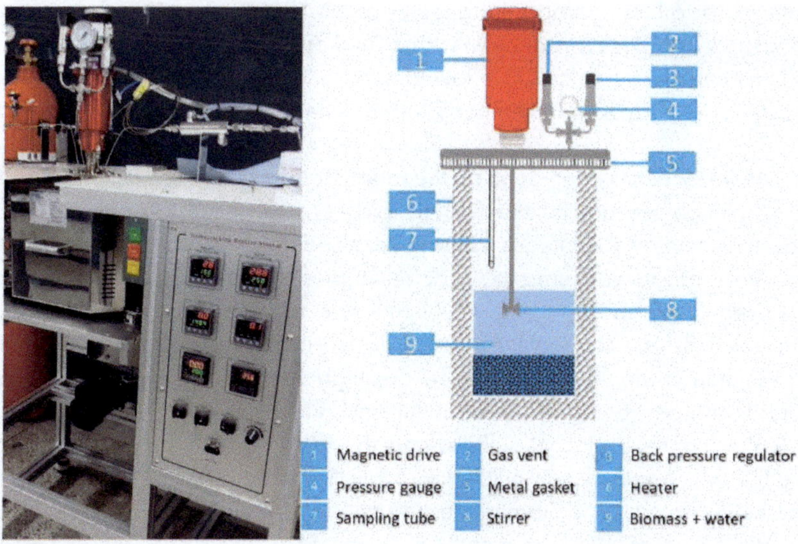

Figure 14 Hydrothermal liquefaction experimental setup and schematic.

et al.[119] have blended algal crude with Petro crude bio-oil. Pedersen et al.[120] have obtained aspen wood-derived bio-oil with a quality equivalent to gasoline. Elliot et al.[121] have stressed the importance of introducing continuous flow reactors rather than the conventional bath process to achieve higher yields of HTL products.

6 Chemical conversion routes

Transesterification involves interchanging an ester organic group with an alcohol organic group, catalyzed with an acid or base catalyst. A two-step transesterification process is preferable in which the acid catalyst helps remove fatty acids, and then a base catalyst can be used for the production of biodiesel.[94] An optimal temperature range for transesterification can be 50–60°C. The transesterification process requires a drying stage due to the high moisture content in the feed. Pretreatment steps include the removal of foreign objects, particle size reduction to increase reactivity, as well as dewatering.[94] Extraction can be achieved via mechanical methods (presses or expellers), solvent-based extraction or the use of pressurized liquid extraction.[94] Extraction of lipids is dependent on various factors such as the moisture content of the feed, the solvent employed for extraction, the ratio of the biomass and solvent used, the time allowed for extraction, as well as the temperature and agitation speed used for the extraction process.[122] Khan[123] reported a biodiesel yield of 88% by using soaking as an extraction process and transesterification in the presence of industrial dust as a catalyst for 6 h at a temperature range of 80–100°C using *Ulva fasciata* as a feedstock. Biodiesel production with a yield of 91% has been reported using *Scinaia hatei* as a feedstock, followed by alkaline esterification using KOH as a catalyst in anhydrous conditions with constant stirring at a temperature of 65°C.[124]

More work has been done on the production of biodiesel from microalgae than macroalgae due to the higher lipid content in the former. There are, however, several advantages of using seaweed as a feedstock. The quality of biodiesel produced from seaweeds resembles that of petrol-diesel and can be used in combustion engines, with fewer hazardous gasses emitted.[123] The residue from the extraction process can be further used for bioethanol production using AD.[122] However, the high metal content of macroalgae retards the conversion process. Sulfur and nitrogen in macroalgae result in a higher concentration of sulfides and the formation of nitrogen-containing oxides, which reduce the efficiency of biodiesel.[116] One option to optimize the use of seaweed is biorefineries, which can produce both fuels and other products such as feed ingredients.[94]

7 Large-scale production case studies and technoeconomical assessments

7.1 Case study 1: industrial-scale bioconversion platform

Ethanol can be added to gasoline up to a maximum ratio of 10% without affecting engine efficiency and emissions and can be added at a higher concentration in fuel flexible vehicles. There has not been much economic analysis of seaweed-to-ethanol production. Fasahati et al.[125] performed the first detailed design and cost-benefit analysis based on experimental data for an industrial-scale ethanol production from brown algae *L. japonica* at plant scales of 100 000 and 400 000 dry tons/year. The conversion process includes pretreatment, enzymatic hydrolysis, fermentation and ethanol recovery stages as shown in Fig. 1. Fasahati et al. designed a new pretreatment process that uses an HWW at 50°C prior to enzymatic hydrolysis and compared it with a classical ATH at 120°C. The use of acid as a catalyst disrupts cell wall structure and breaks down the crystalline structure of carbohydrates making them more easily digestible by downstream enzymatic hydrolysis.[126] However, acid is a corrosive chemical specifically at elevated temperatures. This leads to increased capital costs of pretreatment reactors as equipment in contact with acid should be cladded by a corrosion-resistant material such as Inconel, which is expensive. In addition, unlike lignocellulose biomass, seaweed species (apart from some species of red algae) lack lignin in their chemical structure. The conversions rate obtained by ATH is not as good for seaweed as it is for lignocellulose biomass. Addition of an acid treatment requires a downstream neutralization stage, which further increases capital and chemical costs.

In evaluating the relative benefits of HWW and ATH, Fasahati et al. found that using HWW decreased capital, energy and raw material costs, resulting in a lower minimum ethanol selling price (MESP). As shown in Fig. 1, an MESP value of $2.39 and $2.08/gal for HWW (and $2.85 and $2.33/gal for ATH) was calculated at plant scales of 100 000 and 400 000 dry ton/year, respectively, based on a 10% internal rate of return and 20 years of plant life. The operating costs for HWW and ATH were almost the same for the plants of identical scales, suggesting that plant economy is not strongly influenced by the difference in the pretreatment temperature between the two methods. The total capital investments of HWW and ATH were $30 million and $50 million, respectively, at 100 000 dry ton/year and $89 million and $142 million, respectively, at 400 000 dry ton/year, with the difference being caused primarily by pretreatment reactors used in ATH. They used TE models to calculate a maximum dry seaweed price (MDSP) that leads to a break-even point after 20 years considering an ethanol wholesale price of $2.56/gal for the year 2012. They obtained an MDSP value of $64.6 and $91.3/ton (dry) for HWW and $26.1 and $71.5/ton (dry) for ATH at plant scales of 100 000 and 400 000 dry ton/year, respectively. The values of MDSP obtained for different technologies could be used to understand the upper limits for seaweed purchase prices at biorefinery gates in order to produce ethanol at current market selling prices. The MDSP also could be used to compare different conversion technologies.

Soleymani and Rosentrater[127] used TE models to understand the minimum biorefinery scale required to produce ethanol at $0.93/L. They concluded that a minimum plant scale of 680 000 dry ton seaweed is required to meet the target ethanol price. They further identified that the main limiting parameter is high cultivation costs that must be reduced by increasing cultivation yield per unit area. In another study, Konda et al.[128] evaluated the economic potential of ethanol-only and ethanol-alginate coproduction with seaweed prices of $50 and $100/ton. Depending on conversion yield, solid loading and enzyme loading, MESP for an ethanol-only process could vary between $3.6 and $8.5/gal for $100/ton and between $2.9 and $7.5/gal for $50/ton. However, with the coproduction process, Konda et al. calculated a minimum alginate selling price of $3.1/gal for a fixed ethanol selling price of $2.5/gal, which is much lower than current alginate selling prices of $9–12/kg. Coproduction of alginate thus improves the economics of production. However, alginate market saturation is a potential risk as one coproduction biorefinery could produce 130 000–220 000 ton/year, which is much higher than current global demand at 26 500 ton/year.

Korzen et al.[129] evaluated the economic feasibility of bioethanol production using *Ulva rigida* seaweed co-cultured with fish in an intensive offshore aquaculture unit considering a maximum cultivation area limit of 600 ha. The process produces dried distillers grains with solubles (DDGS) as by-products to be sold as animal feed. Results of the analysis showed that a minimum production range of 77–240 ton/day is needed to make the process economically viable. Korzen et al. identified that DDGS price and biomass yield per unit cultivation area are the key factors that could significantly impact the economics of the biorefinery (Fig. 15).

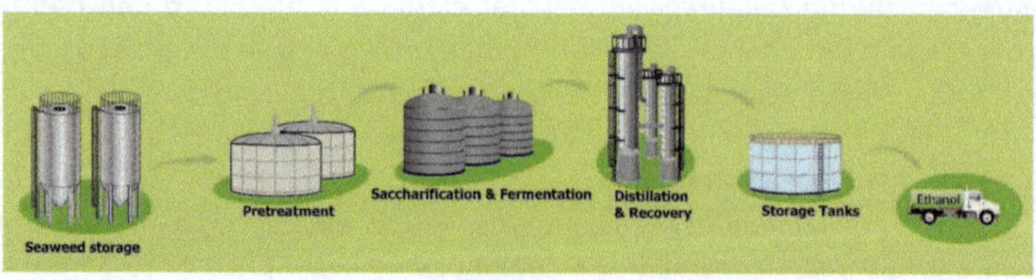

Figure 15 Process flow diagram for fermentation of seaweed biomass to ethanol.

7.2 Case study 2: industrial-scale thermochemical platform

The pyrolysis heat and power (PHP) process can potentially be used to produce biodiesel from macroalgae alongside heat and power generation. Process design, simulation and technoeconomical assessments have been made for industrial-scale production (400 000 dry tons per year) using the brown macroalgae *S. japonica* to assess a minimum product selling price (MPSP). Two designs were simulated: on-site hydrogen production and processing using purchased hydrogen. The design is based on experimental data produced by Choi et al. for fixed-bed pyrolysis of *S. japonica* in a fixed-bed reactor, with acid wash pretreatment and catalytic upgrading of the product.[26,108,130] The feedstock is seaweed with moisture content of 85 wt. % and ash content of 21 wt. % (dry basis).

In both cases, the process can be divided into seven stages:

- A100-1 Pretreatment;
- A100-2 PHP;
- A200 Phase separation;
- A300 Product upgrading;
- A400 Hydrogen production and/or hydrogen recycling;
- A500-Utility management; and
- A600 Product recovery and storage.

Pretreatment begins with size reduction in a hammer mill. Seaweed is then mixed with sulfuric acid with an acid-to-wet feed mass ratio of 1:24. This makes a diluted acid solution of no more than 2.5 wt.% H_2SO_4. The mixture is heated to 90°C and pumped into a stirred vessel for ash removal. After washing, most of the moisture is removed in a series of rotary vacuum filters, where the material is also washed with fresh water to remove residual acid and dissolved minerals. After this operation, the seaweed has 20–30 wt. % moisture and 2.5 wt.% ash. The seaweed is then conveyed to a hot air convective dryer where the moisture is reduced to under 5 wt.%, which finalizes the pretreatment section (Fig. 16).

The seaweed is then moved to a fixed-bed pyrolizer where it is converted into pyrolysis vapors, light gasses and char at a temperature of 400°C and at atmospheric pressure.[131] Gaseous products are chilled to 110°C and transported to the phase separation section. Bio char is pulverized and burnt in a furnace. This process design utilizes the Rankine steam cycle for heat and power production by combustion of biochar and combustible gasses in a mass ratio of 1:0.4. This design can utilize biochar since almost all the mineral content is removed, thus increasing the calorific value of the char to a sufficient level for the Rankine cycle.

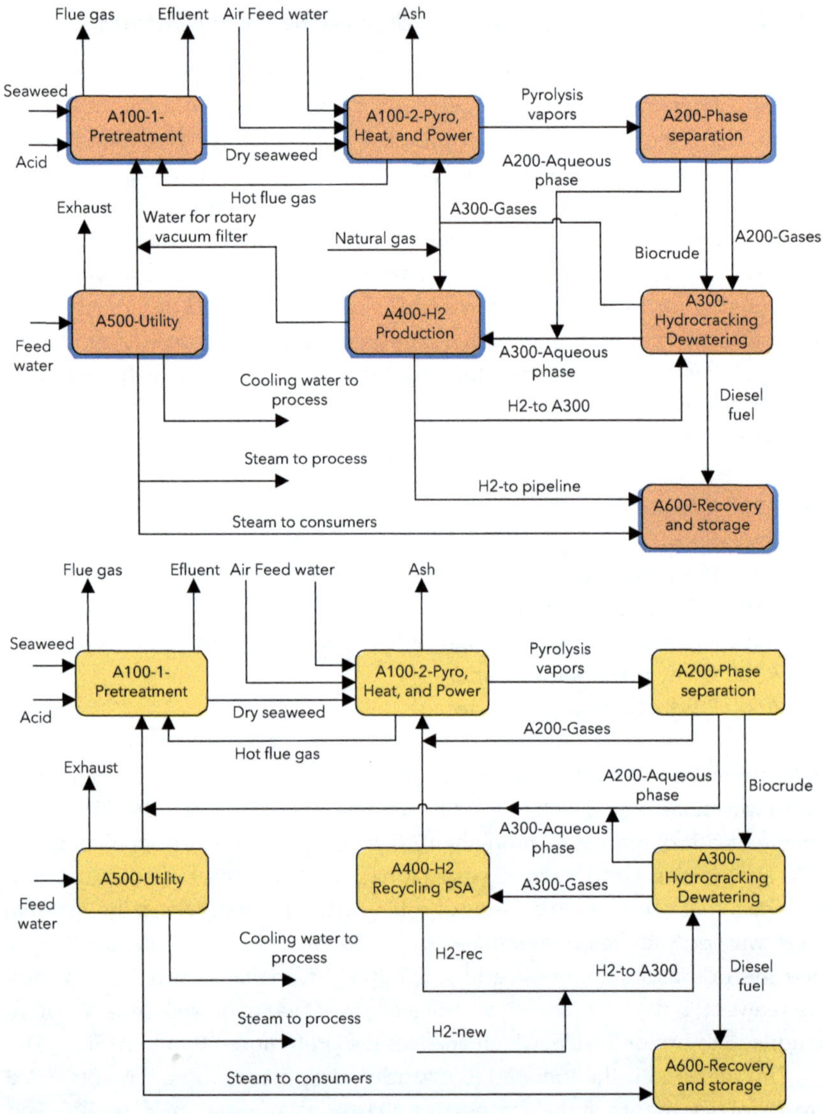

Figure 16 Block flow diagrams of two design cases for thermochemical platform.

In the phase separation section, vapors are cooled to 70°C, at which point most of the vapors are condensed. Non-condensable gasses such as syngas and light hydrocarbons are separated in a flash drum. At this point, condensed pyrolysis liquid product has a high water content of 65 wt. %, which must be removed before catalytic upgrading. A series of two distillation columns dewater the liquid product and separate the liquid into biocrude and aqueous phases.

Biocrude is pressurized to a reaction pressure of 100 bar for catalytic upgrading and heated up to 350°C. Hydrogen is introduced to the reaction system. The diesel fuel is cooled and degassed in the recovery section and stored. Gasses are mixed with the gas phase of the phase separation section. Of the gas stream, 35 wt. % is used in the hydrogen production section, and the rest is sent for combustion in the heat and power section. A mixture of pyrolysis gasses and natural gas is used as a feedstock for hydrogen production through steam reforming. The aqueous phase from the phase separation section is vaporized and mixed with the feed. Mass ratios of natural gas, pyrolysis gasses and steam are set to 1:1.8:6.3, respectively. Hydrogen is purified by pressure swing absorption (PSA) and the off-gasses are burnt in a reformer furnace. Of the hydrogen produced, 64 wt. % is utilized for catalytic upgrading, and the rest is sold as a product. Net power produced from this process is 9.8 MW and 7 tons/h of medium pressure steam (MPS) is generated as net useful heat. The process yields 50 000 L/h of diesel fuel or 5.7 tons/h at 1 atm and 25°C. The main waste streams of the process are brine water from the pretreatment section and flue gasses from combustion which are both treated as non-hazardous waste.

The second design case uses purchased hydrogen to reduce capital investment. A PSA system is still utilized for purification of residual hydrogen after catalytic upgrading. This process produces more net power (10.9 MW) but less useful heat (4 tons/h of MPS). It is also vulnerable to fluctuations in hydrogen prices (Table 4). MPSP has been calculated at $2.82/L for the first case and $2.77/L for the second case. These prices are significantly higher than the global average diesel-selling price of $1.06/L (as at August 2018). However, if the carbon fixation potential of seaweed is considered, these processes, assuming low CO_2 emissions during cultivation and transport, emit 7–45 times less per kg of seaweed processed compared to conventional fossil fuel production (on average 0.32 kg CO_2/kg of crude oil extracted and processed[133,134]).

7.3 Case study 3: industrial-scale integrated platform

This case study is based on a conceptual design and simulation of an integrated process combining biochemical and thermochemical pathways utilizing the brown seaweed *S. japonica* as a feedstock. The aim of the study is to assess the potential economic and environmental benefits of combining two pathways into a seaweed-based biorefinery.[67] The main products of the process were ethanol and mixed alcohols, produced via hydrogenation of VFAs. The major product process streams and pricing are summarized in Table 5. The process was designed and simulated using Aspen Plus software on an industrial scale of 500 000 tons of dried seaweed (18 wt. % moisture) per year. Combined mixed alcohol yield of the process compared to bone dry seaweed weight was 29 wt. %, of which 75 wt. %

Table 4 Technoeconimical breakdown of Cases 1 and 2

	Case 1	Case 2
	Costing; mil. USD (2018)	
Inside battery limit (ISBL)	119.86	87.36
Total installed cost (TIC)	120.64	88.15
Total direct costs (TDC)	141.76	103.57
Total indirect costs	85.05	62.14
Fixed capital investment (FCI)	226.81	165.71
Fixed operating costs	24.7	18.04
Total capital investment (TCI)	**245.39**	**179.29**
	Annual product (diesel) output	
Million liters	28.6	
Bbl	179 709	
	Water utilization; net heat and power output	
Utility water; ton/h	45.7	45.7
Process water; ton/h	1.5	2
MP steam output; ton/h	5.7	4
Net power; MW	9.8	10.6
	CO_2 emission; kg/kg of seaweed	
Processing	1.243	1.207
Harvesting and transport (assumption)	0.5	0.5
kg (CO_2) / kg seaweed (1 kg seaweed = 1.7 kg fixed CO_2) [132]	**0.043**	**0.007**

was ethanol after separation. The process is highly water intensive and consumes 532 T/h. The solid residue of AD or digestate is produced at a rate of 41.5 T/h (35 wt. % moisture) for use as a crop fertilizer. The total installation cost of the process was estimated to be \$85.9 million, and the calculated MPSPs of ethanol and mixed alcohols were \$0.589/L and \$1.143/L, respectively.

The thermochemical conversion process of seaweed digestate to diesel equivalent fuel (Fig. 1) was simulated in Aspen Plus at an industrial scale of 320 000 dry tons per year. The major product process streams and pricing are outlined in Table 5. Experimental data for the simulation model was based on Choi et al.[105,107] for fast pyrolysis of anaerobic digestate of brown algae *S. japonica* in a fluidized bed reactor. Feed with water content of 35 wt.% and mineral content of 12 wt.% is pressed and dried to a water content of 5 wt.%. The seaweed is pyrolyzed in a fluidized bed reactor, after which pyrolysis vapors are separated from char in a cyclone and sent to the phase separation section. Here, pyrolysis vapors are liquified, and organic pyrolysis liquid (biocrude) is separated from the aqueous phase and non-condensable gasses. Biocrude

Table 5 Summary of major streams and pricing of bio and thermal pathway

Bio	Flow kg/h	Price USD/kg	Thermal	Flow kg/h	Price USD/kg
Raw materials					
Seaweed (dry weight)	50 800	0.07	Digestate	27 050	0.0
Iodoform	10.2	5.68	H_2	250	1.8
MTBE	420	1.18	Nat. Gas	9227	0.2
Catalyst	0.12	55.82	Water	351 000	2.00E-04
H_2	680	1.80			
Water	532 400	2.00E-04			
Cooling chemicals	1.7	0.36			
Waste products					
Effluent	403 200		Flue gas	25 927	
Marketable products					
Digestate	27 050	/	Diesel eqv.	3600	MPSP
Mixed alcohols	6325	MPSP	Power	1.5 MW	
Ethanol	8420	MPSP			

is upgraded to a diesel equivalent fuel via catalytic hydrogenation. The hydrogen needed is partly purchased (45 wt.%), while the rest can be obtained by recycling the gasses from the hydrogenation section via PSA. The yield of diesel-upgraded fuel compared to dry seaweed (absent mineral content) is 14.5 wt.%. The thermochemical process is very heat intensive and consumes 9.2 T/h of natural gas and produces 29.1 T/h of aqueous phase. Net-produced power in the heat and power section was 1.5 MW. The estimated installation cost of the process was $56.8 million with an MPSP of $1.257/L.

The design concept for the integrated biochemical and thermochemical pathways is illustrated in Fig. 17. The two pathways are integrated at three key points. Firstly, anaerobic digestate is utilized as a feedstock for the fast pyrolysis process. This produces higher value-added product, as well as reducing the mineral content in the pyrolysis feedstock due to washing with water during fermentation. Secondly, both pathways require the same type of product upgrading process, as well as hydrogen recycling, which reduces costs. Thirdly, utilization of the aqueous phases in the pyrolysis section would reduce water consumption during biochemical conversion. Finally, the heat and power production section of the pyrolysis section can provide most of the heat and power required for the combined process.

In the AD section, seaweed with 85 wt.% moisture is digested to a point of maximum VFA concentration, at which point inhibitor is added (iodoform) and solid digestate is separated by a screen filter. VFA mixture with concentration

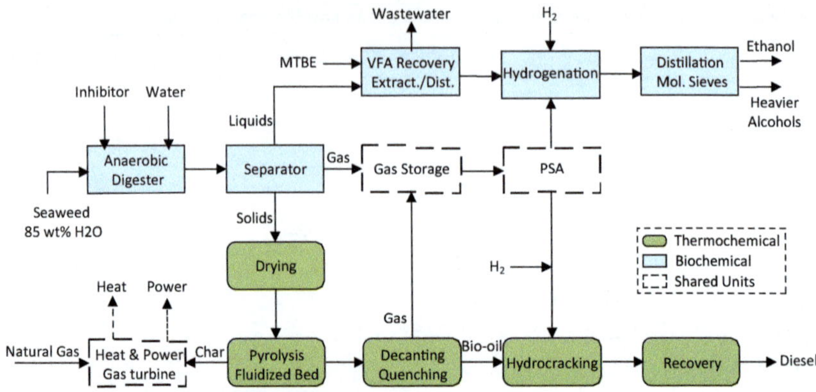

Figure 17 Block flow diagram of integrated process for macroalgal biofuels.

of 5 wt. % enters the section.. The main separation units are an extraction and rectification column, decanter and a stripping column. The extraction agent chosen is methyl tert-butyl ester (MTBE). Continuous hydrogenation of carboxylic acids to alcohols is performed in the gas phase (due to corrosive action of liquid acetic acid) at 290°C at a pressure of 60 bar with a cobalt-based catalyst. Molecular sieves are used for dehydration of alcohols (azeotrope removal), after which ethanol is separated from heavier alcohols in a distillation column. The solid digestate is dried with hot flue gasses from the gas turbine and pyrolyzed in a circulating fluidized bed reactor with olivine sand used as a heat carrier. In the quenching section, fluid pyrolysis vapors are condensed and light gasses are separated, after which water is removed in a series of distillation columns. Hydrocracking requires operating conditions of 350°C at 100 bar using a Cr-based catalyst. A Siemens SGT-A30 RB 33.2 MW turbine was chosen because it can utilize the multiple fuel sources cogenerated in this process.

Analysis of the integrated process is based on 400 000 tons of dry seaweed per year producing ethanol (76 mL), mixed alcohols (66 mL) and ethanol (70 mL) at MPSPs of \$0.86, \$0.51 and \$1.07/L, respectively. The process also produces 5.8 MW of net power, as well as 25 T/h MPS as useful heat. The total installation cost of the integrated process was estimated at \$173.64 million with a total capital investment of \$353.2 million per year. The viability of the process is most sensitive to market fluctuations in purchase prices of seaweed, natural gas and hydrogen. A Monte Carlo analysis was used to determine probable product selling prices (Fig. 18). This analysis shows that, even at higher NPVs, the price ranges are quite close to global price ranges. The process outputs 1.66 kg CO_2/kg of seaweed, which, combined with CO_2 fixation by the seaweed and cultivation emissions, comes to 0.01 kg CO_2/kg of seaweed. Compared to the average value for crude oil extraction and refining (0.32 kg CO_2/kg of crude oil), the process is more environmentally sustainable.

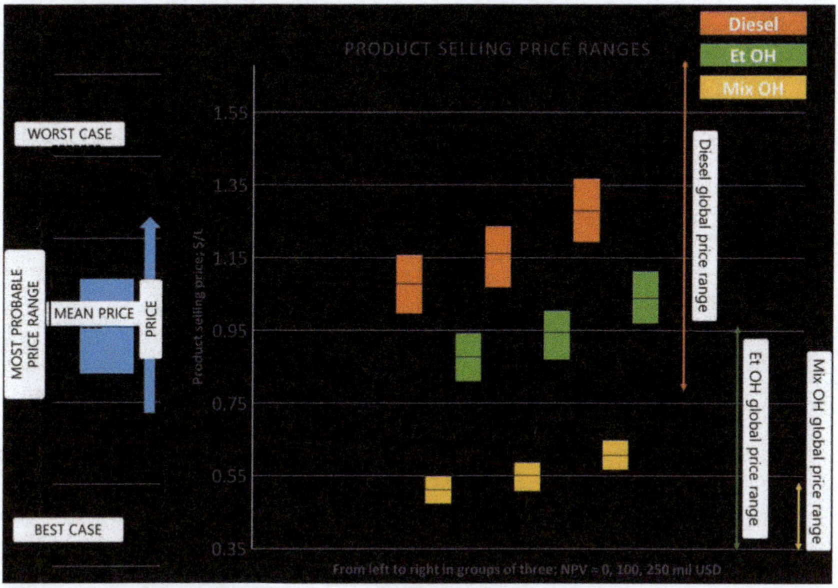

Figure 18 Product selling price ranges for various NPVs.

8 Conclusion and future trends

As a biofuel, seaweed has a number of advantages compared to first- and second-generation feedstocks. Seaweed does not require land for food production. Superior carbon fixation and growth rates could be the key to carbon neutral or even carbon-negative biofuel production processes. The biological structure of seaweeds does not contain rigid natural polymers (such as lignin), as most land-based plants do, which makes conversion processes easier and cheaper. On the other hand, high moisture and mineral content provides technological challenges for conversion processes due to the need for expensive pretreatment. This is especially the case for thermochemical conversion routes (except liquefaction) which generally require dry feedstocks. The answer to these challenges may very well lie in integrated processes which utilize the high moisture content of seaweeds for their advantage, such as AD, as well as the use of solid products as biofuel feedstocks (thermochemical conversion).

Since large-scale seaweed biofuel production is still relatively novel, there are still many challenges to overcome, particularly in the standardization of harvesting techniques, optimization of the supply chain and optimizing the feedstock conversion process. If these challenges are properly addressed, seaweed may very well become one of the dominant biofuel feedstock supplies of the 21st century.

9 References

1. Giampietro, M., and K. Mayumi, 2009, *The Biofuel Delusion: The Fallacy of Large Scale Agro-Biofuels Production*, Routledge.
2. Jang, B. W., R. Gläser, C. Liu, M. C. Capel-Sanchez, J. M. Campos-Martin, L. G. Fierro, E. E. Sci, A. Léonard, J. C. Rooke, C. F. Meunier, H. Sarmento, J. Descy, B. Su, B. G. Harvey, R. L. Quintana, J. Jae, G. A. Tompsett, Y. Lin, R. Torren, J. Shen, T. Zhang, B. Yang, and E. Charles, 2010, Fuels of the future, *Energy and Environmental Science*, 3, 253.
3. Cole, A., Y. Dinburg, B. S. Haynes, Y. He, M. Herskowitz, C. Jazrawi, M. Landau, X. Liang, M. Magnusson, T. Maschmeyer, A. F. Masters, N. Meiri, N. Neveux, R. de Nys, N. Paul, M. Rabaev, R. Vidruk-Nehemya, and A. K. L. Yuen, 2016, From macroalgae to liquid fuel via waste-water remediation, hydrothermal upgrading, carbon dioxide hydrogenation and hydrotreating, *Energy and Environmental Science*, 9, 1828–40.
4. Chen, W. T., Y. Zhang, J. Zhang, L. Schideman, G. Yu, P. Zhang, and M. Minarick, 2014, Co-liquefaction of swine manure and mixed- culture algal biomass from a wastewater treatment system to produce bio-crude oil, *Applied Energy*, 128, 209–16.
5. Gao, K., and K. R. McKinley, 1994, Use of macroalgae for marine biomass production and CO_2 remediation: a review, *Journal of Applied Phycology*, 6, 45–60.
6. Adams, J. M. M., A. B. Ross, K. Anastasakis, E. M. Hodgson, J. A. Gallagher, J. M. Jones, and I. S. Donnison, 2011, Seasonal variation in the chemical composition of the bioenergy feedstock Laminaria digitata for thermochemical conversion, *Bioresource Technology*, 102, 226–34.
7. Rowbotham, J. S., P. W. Dyer, H. C. Greenwell, and M. K. Theodorou, 2012, Thermochemical processing of macroalgae: a late bloomer in the development of third-generation biofuels? *Biofuels*, 3, 441–61.
8. Kim, S. K. 2011, *Handbook of Marine Macroalgae: Biotechnology and Applied Phycology*, Wiley.
9. Tiwari, B. K., and D. Troy, 2015, *Seaweed Sustainability: Food and Non-Food Applications*, Elsevier Science.
10. Suutari, M., E. Leskinen, K. Fagerstedt, J. Kuparinen, P. Kuuppo, and J. Blomster, 2015, Macroalgae in biofuel production, *Phycological Research*, 63, 1–18.
11. Roesijadi, G., S. B. Jones, L. J. Snowden-Swan, Y. Zhu, 2010, Macroalgae as a biomass feedstock: A preliminary analysis, Technical Report, Pacific Northwest National Lab. (PNNL), Richland, WA.
12. Pomin, V. H., 2011, *Seaweed: Ecology, Nutrient Composition and Medicinal Uses*, Nova Science.
13. Kraan, S., 2013, Mass-cultivation of carbohydrate rich macroalgae, a possible solution for sustainable biofuel production, *Mitigation and Adaptation Strategies for Global Change*, 18, 27–46.
14. Sudhakar, K., R. Mamat, M. Samykano, W. H. Azmi, W. F. W. Ishak, and T. Yusaf, 2018, An overview of marine macroalgae as bioresource, *Renewable and Sustainable Energy Reviews*, 91, 165–79.
15. Song, M., H. Duc Pham, J. Seon, and H. Chul Woo, 2015, Marine brown algae: A conundrum answer for sustainable biofuels production, *Renewable and Sustainable Energy Reviews*, 50, 782–92.
16. Jung, K. A., S. R. Lim, Y. Kim, and J. M. Park, 2013, Potentials of macroalgae as feedstocks for biorefinery, *Bioresource Technology*, 135, 182–90.

17. Chen, H., D. Zhou, G. Luo, S. Zhang, and J. Chen, 2015, Macroalgae for biofuels production: Progress and perspectives, *Renewable and Sustainable Energy Reviews*, 47, 427–37.
18. Hui, Z., Y. Huaxiao, Z. Mengmeng, and Q. Song, 2010, Pyrolysis characteristics and kinetics of macroalgae biomass using thermogravimetric analyzer, *World Academy of Science, Engineering and Technology*, 65, 1161–6.
19. Ceylan, S., and J. L. Goldfarb, 2015, Green tide to green fuels: TG-FTIR analysis and kinetic study of Ulva prolifera pyrolysis, *Energy Conversion and Management*, 101, 263–70.
20. Tabarsa, M., M. Rezaei, Z. Ramezanpour, and J. R. Waaland, 2012, Chemical compositions of the marine algae *Gracilaria salicornia* (Rhodophyta) and *Ulva lactuca* (Chlorophyta) as a potential food source, *Journal of the Science of Food and Agriculture*, 92, 2500–6.
21. Briand, X., and P. Morand, 1997, Anaerobic digestion of Ulva sp.1 relationship between Ulva composition amd methanisation, *Journal of Applied Phycology*, 9, 511–24.
22. Lorenzetti, C., R. Conti, D. Fabbri, and J. Yanik, 2016, A comparative study on the catalytic effect of H-ZSM5 on upgrading of pyrolysis vapors derived from lignocellulosic and proteinaceous biomass, *Fuel*, 166, 446–52.
23. Yaich, H., H. Garna, S. Besbes, M. Paquot, C. Blecker, and H. Attia, 2011, Chemical composition and functional properties of Ulva lactuca seaweed collected in Tunisia, *Food Chemistry*, 128, 895–901.
24. Fasahati, P. 2015, *Techno-Economic, Energy, and Environmental Assessment of Bio-alcohol Production from Brown Algae*, Pukyong National University.
25. Fasahati, P., C. M. Saffron, H. C. Woo, and J. J. Liu, 2017, Potential of brown algae for sustainable electricity production through anaerobic digestion, *Energy Conversion and Management*, 135, 297–307.
26. Choi, J. H., S. S. Kim, H. V. Ly, J. Kim, and H. C. Woo, 2017, Thermogravimetric characteristics and pyrolysis kinetics of high-density-aquacultured *Saccharina japonica*: Effects of water-washing, *Fuel*, 193, 159–67.
27. Ly, H. V., S. S. Kim, J. H. Choi, H. C. Woo, and J. Kim, 2016, Fast pyrolysis of *Saccharina japonica* alga in a fixed-bed reactor for bio-oil production, *Energy Conversion and Management*, 122, 526–34.
28. Fasahati, P., and J. J. Liu, 2016, Application of MixAlco® processes for mixed alcohol production from brown algae: Economic, energy, and carbon footprint assessments, *Fuel Processing Technology*, 144, 262–73.
29. Yu, L. J., S. Wang, X. M. Jiang, N. Wang, and C. Q. Zhang, 2008, Thermal analysis studies on combustion characteristics of seaweed, *Journal of Thermal Analysis and Calorimetry*, 93, 611–7.
30. Lee, S. Y., J. H. Chang, and S. B. Lee, 2014, Chemical composition, saccharification yield, and the potential of the green seaweed *Ulva pertusa*, *Biotechnology and Bioprocess Engineering*, 19, 1022–33.
31. Wang, S., X. M. Jiang, X. X. Han, and H. Wang, 2008, Fusion characteristic study on seaweed biomass ash, *Energy and Fuels*, 22, 2229–35.
32. Huesemann, M., G. Roesjadi, J. Benemann, and F. B. Metting, 2010, Biofuels from microalgae and seaweeds, *in* A. A. Vertès, N. Qureshi, H. P. Blaschek, and H. Yukawa (Eds), *Biomass to Biofuels: Strategies for Global Industries*, John Wiley and Sons, pp. 165–84.

33. FAO, 2018, *The State of World Fisheries and Aquaculture 2018 - Meeting the Sustainable Development Goals*, Rome.
34. Ghadiryanfar, M., K. A. Rosentrater, A. Keyhani, and M. Omid, 2016, A review of macroalgae production, with potential applications in biofuels and bioenergy, *Renewable and Sustainable Energy Reviews*, 54, 473–81.
35. Peteiro, C., N. Sánchez, C. Dueñas-Liaño, and B. Martínez, 2014, Open-sea cultivation by transplanting young fronds of the kelp *Saccharina latissima*, *Journal of Applied Phycology*, 26, 519–28.
36. Rodriguez, C., A. Alaswad, J. Mooney, T. Prescott, and A. G. Olabi, 2015, Pre-treatment techniques used for anaerobic digestion of algae, *Fuel Processing Technology*, 138, 765–79.
37. Miura, T., A. Kita, Y. Okamura, T. Aki, Y. Matsumura, T. Tajima, J. Kato, and Y. Nakashimada, 2014, Evaluation of marine sediments as microbial sources for methane production from brown algae under high salinity, *Bioresource Technology*, 169, 362–6.
38. Miura, T., A. Kita, Y. Okamura, T. Aki, Y. Matsumura, T. Tajima, J. Kato, and Y. Nakashimada, 2015, Improved methane production from brown algae under high salinity by fed-batch acclimation, *Bioresource Technology*, 187, 275–81.
39. Tedesco, S., K. Y. Benyounis, and A. G. Olabi, 2013, Mechanical pretreatment effects on macroalgae-derived biogas production in co-digestion with sludge in Ireland, *Energy*, 61, 27–33.
40. Sharma, N., and G. Pellizzi, 1991, Anaerobic biotechnology and developing countries - I. Technical status, *Energy Conversion and Management*, 32, 447–69.
41. Wu, B., 2012, Integration of mixing, heat transfer, and biochemical reaction kinetics in anaerobic methane fermentation, *Biotechnology and Bioengineering*, 109, 2864–74.
42. Gupta, S., N. Abu-Ghannam, and A. G. M. Scannell, 2011, Growth and kinetics of *Lactobacillus plantarum* in the fermentation of edible Irish brown seaweeds, *Food and Bioproducts Processing*, 89, 346–55.
43. Nallathambi Gunaseelan, V. 1997, Anaerobic digestion of biomass for methane production: A review, *Biomass and Bioenergy*, 13, 83–114.
44. Costa, J. C., P. R. Gonçalves, A. Nobre, and M. M. Alves, 2012, Biomethanation potential of macroalgae *Ulva* spp. and *Gracilaria* spp. and in co-digestion with waste activated sludge, *Bioresource Technology*, 114, 320–6.
45. Maneein, S., J. J. Milledge, B. V. Nielsen, and P. J. Harvey, 2018, A review of seaweed pre-treatment methods for enhanced biofuel production by anaerobic digestion or fermentation, *Fermentation*, 4, 100.
46. Anjaneyulu, K., S. J. Tarwadi, and D. J. Mehta, 2007, Anaerobic digestion of seaweed for biogas: A kinetic evaluation, *Journal of Chemical Technology and Biotechnology*, 45, 5–14.
47. Chynoweth, D. P., C. E. Turick, J. M. Owens, D. E. Jerger, and M. W. Peck, 1993, Biochemical methane potential of biomass and waste feedstocks, *Biomass and Bioenergy*, 5, 95–111.
48. Chynoweth, D. P., K. F. Fannin, and V. J. Srivastava, 1987, Biological gasification of marine algae, *in Seaweed Cultivation for Renewable Resources*, Elsevier, New York.
49. Troiano, R. A., D. L. Wise, D. C. Augenstein, R. G. Kispert, and C. L. Cooney, 1976, Fuel gas production by anaerobic digestion of kelp, *Resource Recovery and Conservation*, 2, 171–6.

50. Hanssen, J. F., M. Indergaard, K. Østgaard, O. A. Bævre, T. A. Pedersen, and A. Jensen, 1987, Anaerobic digestion of *Laminaria* spp. and *Ascophyllum nodosum* and application of end products, *Biomass*, 14, 1-13.
51. Chynoweth, D. P., 2002, *Review of Biomethane from Marine Biomass: History, Results and Conclusions of the "US Marine Biomass Energy Program (1968–1990)"*, San Francisco.
52. Chang, H. N., N.-J. Kim, J. Kang, and C. M. Jeong, 2010, Biomass-derived volatile fatty acid platform for fuels and chemicals, *Biotechnology and Bioprocess Engineering*, 15, 1–10.
53. Feng, L., Y. Chen, and X. Zheng, 2009, Enhancement of waste activated sludge protein conversion and volatile fatty acids accumulation during waste activated sludge anaerobic fermentation by carbohydrate substrate addition: The effect of pH, *Environmental Science and Technology*, 43, 4373–80.
54. Qi, B. C., G. M. Wolfaardt, C. Aldrich, and L. Lorenzen, 2003, Methanogenic digestion of lignocellulose residues under conditions of high-rate acidogenic fermentation, *Industrial and Engineering Chemistry Research*, 42, 1845–9.
55. Moen, E., B. Larsen, and Ø. Kjetill, 1997, Aerobic microbial degradation of alginate in *Laminaria hyperborea* stipes containing different levels of polyphenols, *Journal of Applied Phycology*, 45.
56. Moen, E., S. Horn, and K. Østgaard, 1997, Alginate degradation during anaerobic digestion of Laminaria hyperborea stipes, *Journal of Applied Phycology*, 9, 157–66.
57. Østgaard, K., M. Indergaard, S. Markussen, S. H. Knutsen, and A. Jensen, 1993, Carbohydrate degradation and methane production during fermentation of *Laminaria saccharina* (Laminariales, Phaeophyceae), *Journal of Applied Phycology*, 5, 333–42.
58. Dave, A., Y. Huang, S. Rezvani, D. McIlveen-Wright, M. Novaes, and N. Hewitt, 2013, Techno-economic assessment of biofuel development by anaerobic digestion of European marine cold-water seaweeds, *Bioresource Technology*, 135, 120–7.
59. Matsui, T., T. Amano, Y. Koike, A. Saiganji, and H. Saito, 2006, *Methane Fermentation of Seaweed Biomass*, AICHE.
60. Staniforth, J., and K. Kendall, 1998, Biogas powering a small tubular solid oxide fuel cell, *Journal of Power Sources*, 71, 275–7.
61. Persson, M., O. Jonsson, and A. Wellinger, 2006, Biogas upgrading to vehicle fuel standards and grid injection. *IEA Bioenergy Task* 37, 1–34.
62. Gunnerman, R. W., and P. W. Gunnerman, 2010, Continuous process and plant design for conversion of biogas to liquid fuel, US12212968.
63. Lee, S.-M., M.-H. Choi, W.-K. Hong, J. M. Park, A. Yu, J.-S. Lee, J.-W. Seo, and C. H. Kim, 2014, Production of mixed acids from non-pretreated red algae Gelidium amansii, *Renewable Energy*, 63, 366–9.
64. Chalima, A., L. Oliver, L. Fernández de Castro, A. Karnaouri, T. Dietrich, and E. Topakas, 2017, Utilization of volatile fatty acids from microalgae for the production of high added value compounds, *Fermentation*, 3, 54.
65. Fasahati, P., and J. Liu (2014). Techno-economic analysis of production and recovery of volatile fatty acids from brown algae using membrane distillation. *Computer Aided Chemical Engineering* 34, 303–8.
66. Fasahati, P., and J. J. Liu, 2015, Economic, energy, and environmental impacts of alcohol dehydration technology on biofuel production from brown algae, *Energy*, 93, 2321–36.

67. Fasahati, P., and J. J. Liu, 2015, Impact of volatile fatty acid recovery on economics of ethanol production from brown algae via mixed alcohol synthesis, *Chemical Engineering Research and Design*, 98, 107–22.
68. Holtzapple, M. T., R. R. Davison, M. K. Ross, S. Aldrett-Lee, M. Nagwani, C.-M. Lee, C. Lee, S. Adelson, W. Kaar, D. Gaskin, H. Shirage, N.-S. Chang, V. S. Chang, and M. E. Loescher, 1999, Biomass conversion to mixed alcohol fuels using the MixAlco process, *in Twent. Symp. Biotechnol. Fuels Chem.*, Humana Press, Totowa, NJ, pp. 609–31.
69. Taco Vasquez, S., J. Dunkleman, S. K. Chaudhuri, A. Bond, and M. T. Holtzapple, 2014, Biomass conversion to hydrocarbon fuels using the MixAlco™ process at a pilot-plant scale, *Biomass and Bioenergy*, 62, 138–48.
70. Pham, V., M. Holtzapple, and M. El-Halwagi, 2010, Techno-economic analysis of biomass to fuel conversion via the MixAlco process, *Journal of Industrial Microbiology and Biotechnology*, 37, 1157-68.
71. Landoll, M. P., and M. T. Holtzapple, 2011, Thermal decomposition of mixed calcium carboxylate salts: Effects of lime on ketone yield, *Biomass and Bioenergy*, 35, 3592–603.
72. Granda, C. B., M. T. Holtzapple, G. Luce, K. Searcy, and D. L. Mamrosh, 2009, Carboxylate platform: The MixAlco process part 2: Process economics, *Applied Biochemistry and Biotechnology*, 156, 107–24.
73. Holtzapple, M., R. Davison, C. Granda, F. Agbogbo, and Z. Fu, 2007, Production plant and method for converting biomass. WO2007009085A2, World Intellectual Property Organization, Application filing.
74. Jennings, E., A. Mohagheghi, D. Schell, and N. Dowe, 2008, Technical and economic performance of an ammonium hydroxide-based conditioning technology. E Milestone Report, Document #10379, National Renewable Energy Laboratory, Golden, CO, November 2008.
75. Lara Ruiz, J. H. J., 2006, *An Advanced Vapor-Compression Desalination System*, Texas AandM University.
76. Pham, T. N., T. Sooknoi, S. P. Crossley, and D. E. Resasco, 2013, Ketonization of carboxylic acids: Mechanisms, catalysts, and implications for biomass conversion, *ACS Catalysis*, 3, 2456–73.
77. Sankaranarayanapillai, S., S. Sreekumar, J. Gomes, A. Grippo, G. E. Arab, M. Head-Gordon, F. D. Toste, A. T. Bell, 2015, Catalytic upgrading of biomass-derived Methyl Ketones to liquid transportation fuel precursors by an organocatalytic approach, *Angewandte Chemie - International Edition*, 54, 4673–7.
78. Ju, I. B., H.-W. Lim, W. Jeon, D. J. Suh, M.-J. Park, and Y.-W. Suh, 2011, Kinetic study of catalytic esterification of butyric acid and N-butanol over Dowex 50Wx8-400, *Chemical Engineering Journal*, 168, 293-302.
79. Calvar, N., B. González, and A. Dominguez, 2007, Esterification of acetic acid with ethanol: Reaction kinetics and operation in a packed bed reactive distillation column, *Chemical Engineering and Processing: Process Intensification*, 46, 1317–23.
80. Filachione, E. M., and E. J. Costello, 1952, Lactic esters by reaction of ammonium lactate with alcohols, *Industrial and Engineering Chemistry*, 44, 2189–91.
81. Pouilloux, Y., F. Autin, and J. Barrault, 2000, Selective hydrogenation of methyl oleate into unsaturated alcohols. Relationships between catalytic properties and composition of cobalt-tin catalysts, *Catalysis Today*, 63, 87–100.
82. Zhang, J., G. Leitus, Y. Ben-David, and D. Milstein, 2006, Efficient homogeneous catalytic hydrogenation of esters to alcohols, *Angewandte Chemie*, 118, 1131–3.

83. Wargacki, A. J., E. Leonard, M. N. Win, D. D. Regitsky, C. N. S. Santos, P. B. Kim, S. R. Cooper, R. M. Raisner, A. Herman, A. B. Sivitz, A. Lakshmanaswamy, Y. Kashiyama, D. Baker, and Y. Yoshikuni, 2012, An engineered microbial platform for direct biofuel production from brown macroalgae, *Science*, 335, 308–13.
84. Horn, S. J., I. M. Aasen, and K. Østgaard, 2000, Ethanol production from seaweed extract, *Journal of Industrial Microbiology and Biotechnology*, 25, 249–54.
85. Horn, S. J., I. M. Aasen, and K. Østgaard, 2000, Production of ethanol from mannitol by *Zymobacter palmae*, *Journal of Industrial Microbiology and Biotechnology*, 24, 51-7.
86. Dijken, J. P., and W. A. Scheffers, 1986, Redox balances in the metabolism of sugars by yeasts, *FEMS Microbiology Letters*, 32, 199–224.
87. Ge, L., P. Wang, and H. Mou, 2011, Study on saccharification techniques of seaweed wastes for the transformation of ethanol, *Renewable Energy*, 36, 84–9.
88. Wang, X., X. Liu, and G. Wang, 2011, Two-stage Hydrolysis of Invasive Algal Feedstock for Ethanol Fermentation, *Journal of Integrative Plant Biology*, 53, 246–52.
89. Kim, N. J., H. Li, K. Jung, H. N. Chang, and P. C. Lee, 2011, Ethanol production from marine algal hydrolysates using *Escherichia coli* KO11, *Bioresource Technology*, 102, 7466–9.
90. Yanagisawa, M., K. Nakamura, O. Ariga, and K. Nakasaki, 2011, Production of high concentrations of bioethanol from seaweeds that contain easily hydrolyzable polysaccharides, *Process Biochemistry*, 46, 2111-6.
91. Lee, S. M., and J. H. Lee, 2011, The isolation and characterization of simultaneous saccharification and fermentation microorganisms for Laminaria japonica utilization, *Bioresource Technology*, 102, 5962-7.
92. Offei, F., M. Mensah, A. Thygesen, and F. Kemausuor, 2018, Seaweed bioethanol production: A process selection review on hydrolysis and fermentation, *Fermentation*, 4, 99.
93. Luque, R., and J. Clark, 2010, *Handbook of Biofuels Production: Processes and Technologies*, Elsevier.
94. Michalak, I. 2018, Experimental processing of seaweeds for biofuels, *Wiley Interdisciplinary Reviews: Energy and Environment*, 7, 1–25.
95. Brigljevic, B., and Liu, J. 2017, Process design and simulation of industrial scale biofuel production via pyrolysis of saccharina japonica, *in 2017 6th International Symposium on Advanced Control of Industrial Processes AdCONIP 2017*.
96. Bridgwater, T., 2006, Biomass for energy, *Journal of the Science of Food and Agriculture*, 86, 1755–68.
97. Karayıldırım, T., A. Sınağ, and A. Kruse, 2008, Char and Coke formation as unwanted side reaction of the hydrothermal biomass gasification, *Chemical Engineering and Technology*, 31, 1561-8.
98. Demirbas, M. F. 2011, Biofuels from algae for sustainable development, *Applied Energy*, 88, 3473–80.
99. Schumacher, M., J. Yanık, A. Sınağ, and A. Kruse, 2011, Hydrothermal conversion of seaweeds in a batch autoclave, *The Journal of Supercritical Fluids*, 58, 131–5.
100. Cherad, R., J. A. Onwudili, U. Ekpo, P. T. Williams, A. R. Lea-Langton, M. Carmargo-Valero, and A. B. Ross, 2013, Macroalgae supercritical water gasification combined with nutrient recycling for microalgae cultivation, *Environmental Progress and Sustainable Energy*, 32, 902–9.

101. Okoli, C. O., T. A. Adams, B. Brigljević, and J. J. Liu, 2016, Design and economic analysis of a macroalgae-to-butanol process via a thermochemical route, *Energy Conversion and Management*, 123, 410–22.
102. Jahirul, M. I., M. G. Rasul, A. A. Chowdhury, and N. Ashwath, 2012, Biofuels production through biomass pyrolysis – A technological review, *Energies*, 5, 4952–5001.
103. Bridgwater, A. V. 2012, Review of fast pyrolysis of biomass and product upgrading, *Biomass and Bioenergy*, 38, 68–94.
104. Ross, A. B., K. Anastasakis, M. Kubacki, and J. M. Jones, 2009, Investigation of the pyrolysis behaviour of brown algae before and after pre-treatment using PY-GC/MS and TGA, *Journal of Analytical and Applied Pyrolysis*, 85, 3–10.
105. Ly, H. V., S. S. Kim, H. C. Woo, J. H. Choi, D. J. Suh, and J. Kim, 2015, Fast pyrolysis of macroalga *Saccharina japonica* in a bubbling fluidized-bed reactor for bio-oil production, *Energy*, 93, 1436–46.
106. Li, D., L. Chen, X. Yi, X. Zhang, and N. Ye, 2010, Pyrolytic characteristics and kinetics of two brown algae and sodium alginate, *Bioresource Technology*, 101, 7131–6.
107. Choi, J. H., S. S. Kim, H. V. Ly, J. Kim, and H. C. Woo, 2017, Effects of water-washing *Saccharina japonica* on fast pyrolysis in a bubbling fluidized-bed reactor, *Biomass and Bioenergy*, 98, 112–23.
108. Choi, J. H., S. S. Kim, D. J. Suh, E. J. Jang, K. Il Min, and H. C. Woo, 2016, Characterization of the bio-oil and bio-char produced by fixed bed pyrolysis of the brown alga *Saccharina japonica*, *Korean Journal of Chemical Engineering*, 33, 2691–8.
109. Qureshi, K. M., A. N. Kay Lup, S. Khan, F. Abnisa, and W. M. A. Wan Daud, 2018, A technical review on semi-continuous and continuous pyrolysis process of biomass to bio-oil, *Journal of Analytical and Applied Pyrolysis*, 131, 52–75.
110. Isahak, W. N. R. W., M. W. M. Hisham, M. A. Yarmo, and T. Y. Yun Hin, 2012, A review on bio-oil production from biomass by using pyrolysis method, *Renewable and Sustainable Energy Reviews*, 16, 5910–23.
111. Zhang, Y., 2010, Hydrothermal liquefaction to convert biomass into crude oil, *in Biofuels from Agricultural Wastes and Byproducts*, Blackwell Publishing, pp. 201–32.
112. Toor, S. S., L. Rosendahl, and A. Rudolf, 2011, Hydrothermal liquefaction of biomass: A review of subcritical water technologies, *Energy*, 36, 2328–42.
113. Jones, S. B., Y. Zhu, D. B. Anderson, R. T. Hallen, D. C. Elliott, A. J. Schmidt, K. O. Albrecht, T. R. Hart, M. G. Butcher, C. Drennan, L. J. Snowden-Swan, R. Davis, and C. Kinchin, 2014, Process design and economics for the conversion of algal biomass to hydrocarbons: whole algae hydrothermal liquefaction and upgrading, Technical Report, Pacific Northwest National Lab. (PNNL), Richland, WA. doi:10.2172/1126336.
114. Zhou, D., L. Zhang, S. Zhang, H. Fu, and J. Chen, 2010, Hydrothermal liquefaction of macroalgae enteromorpha prolifera to bio-oil, *Energy and Fuels*, 24, 4054–61.
115. Anastasakis, K., and A. B. Ross, 2015, Hydrothermal liquefaction of four brown macro-algae commonly found on the UK coasts: An energetic analysis of the process and comparison with bio-chemical conversion methods, *Fuel*, 139, 546–53.
116. Pérez L., I. Blanco, J. L. Salgueiro, Á. Cancela, Á. Sánchez, 2017, Macroalgae for bio-crude production , *Journal of Environmental Science*, 6 (2), 1–3.
117. Alhassan, Y., N. Kumar, and I. M. Bugaje, 2016, Hydrothermal liquefaction of de-oiled *Jatropha curcas cake* using Deep Eutectic Solvents (DESs) as catalysts and co-solvents, *Bioresource Technology*, 199, 375–81.

118. Costanzo, W., R. Hilten, U. Jena, K. C. Das, and J. R. Kastner, 2016, Effect of low temperature hydrothermal liquefaction on catalytic hydrodenitrogenation of algae biocrude and model macromolecules, *Algal Research*, 13, 53-68.
119. Lavanya, M., A. Meenakshisundaram, S. Renganathan, S. Chinnasamy, D. M. Lewis, J. Nallasivam, and S. Bhaskar, 2016, Hydrothermal liquefaction of freshwater and marine algal biomass: A novel approach to produce distillate fuel fractions through blending and co-processing of biocrude with petrocrude, *Bioresource Technology*, 203, 228–35.
120. Pedersen, T. H., I. F. Grigoras, J. Hoffmann, S. S. Toor, I. M. Daraban, C. U. Jensen, S. B. Iversen, R. B. Madsen, M. Glasius, K. R. Arturi, R. P. Nielsen, E. G. Søgaard, and L. A. Rosendahl, 2016, Continuous hydrothermal co-liquefaction of aspen wood and glycerol with water phase recirculation, *Applied Energy*, 162, 1034–41.
121. Elliott, D. C., P. Biller, A. B. Ross, A. J. Schmidt, and S. B. Jones, 2015, Hydrothermal liquefaction of biomass: Developments from batch to continuous process, *Bioresource Technology*, 178, 147–56.
122. Ashokkumar, V., M. R. Salim, Z. Salam, P. Sivakumar, C. T. Chong, S. Elumalai, V. Suresh, and F. N. Ani, 2017, Production of liquid biofuels (biodiesel and bioethanol) from brown marine macroalgae *Padina tetrastromatica*, *Energy Conversion and Management*, 135, 351-61.
123. Khan, A. M., N. Fatima, M. S. Hussain, and K. Yasmeen, 2016, Biodiesel production from green seaweed Ulva fasciata catalyzed by novel waste catalysts from Pakistan Steel Industry, *Chinese Journal of Chemical Engineering*, 24, 1080–6.
124. Khan, A. M., M. Obaid, and R. Sultana, 2015, Production of biodiesel from marine algae to mitigate environmental pollution, *Journal of the Chemical Society of Pakistan*, 37.
125. Fasahati, P., H. C. Woo, and J. J. Liu, 2015, Industrial-scale bioethanol production from brown algae: Effects of pretreatment processes on plant economics, *Applied Energy*, 139, 175–87.
126. Humbird, D., R. Davis, L. Tao, C. Kinchin, D. Hsu, A. Aden, P. Schoen, J. Lukas, B. Olthof, M. Worley, and D. Sexton, 2011, Process design and economics for biochemical conversion of lignocellulosic biomass to ethanol: dilute-acid pretreatment and enzymatic hydrolysis of corn stover (No. NREL/TP-5100-47764), National Renewable Energy Lab.(NREL), Golden, CO.
127. Soleymani, M., and K. Rosentrater, 2017, Techno-economic analysis of biofuel production from Macroalgae (seaweed), *Bioengineering*, 4, 92.
128. Konda, N. V. S. N. M., S. Singh, B. A. Simmons, and D. Klein-Marcuschamer, 2015, An investigation on the economic feasibility of macroalgae as a potential feedstock for biorefineries, *BioEnergy Research*, 8, 1046–56.
129. Korzen, L., Y. Peled, S. Z. Shamir, M. Shechter, A. Gedanken, A. Abelson, and A. Israel, 2015, An economic analysis of bioethanol production from the marine macroalga *Ulva* (Chlorophyta), *Technology*, 3, 114–8.
130. Choi, J. H., 2015, *Pyrolysis and Catalytic Upgrading of Microalgal Biomass for Liquid Biofuel Production*, Pukyong National University.
131. Brigljević, B., P. Žuvela, J. J. Liu, H. C. Woo, and J. H. Choi, 2018, Development of an automated method for modelling of bio-crudes originating from biofuel production processes based on thermochemical conversion, *Applied Energy*, 215, 670–8.
132. Singh, S. P., and P. Singh, 2014, Effect of CO_2 concentration on algal growth: A review, *Renewable and Sustainable Energy Reviews*, 38, 172–9.

133. Kwasniewski, V., J. Blieszner, and R. Nelson, 2015, Petroleum refinery greenhouse gas emission variations related to higher ethanol blends at different gasoline octane rating and pool volume levels, *Biofuels, Bioproducts and Biorefining*, 10, 36–46.
134. Gavenas, E., K. E. Rosendahl, and T. Skjerpen, 2015, CO_2-emissions from Norwegian oil and gas extraction, *Energy*, 90, 1956–66.

Index